Multidimensional Analytical Techniques in Environmental Research

Multidimensional Analytical Techniques in Environmental Research

Edited by

Regina M.B.O. Duarte
Armando C. Duarte

ELSEVIER

Elsevier
Radarweg 29, PO Box 211, 1000 AE Amsterdam, Netherlands
The Boulevard, Langford Lane, Kidlington, Oxford OX5 1GB, United Kingdom
50 Hampshire Street, 5th Floor, Cambridge, MA 02139, United States

Library of Congress Cataloging-in-Publication Data
A catalog record for this book is available from the Library of Congress

British Library Cataloguing-in-Publication Data
A catalogue record for this book is available from the British Library

ISBN: 978-0-12-818896-5

For information on all Elsevier publications visit our website at https://www.elsevier.com/books-and-journals

Publisher: Susan Dennis
Acquisitions Editor: Kathryn Eriylmaz
Editorial Project Manager: Redding Morse
Production Project Manager: R. Vijay Bharath
Cover Designer: Christian J. Bilbow

Typeset by SPi Global, India

Contents

Contributors

Antoine S. Almeida Department of Chemistry & CESAM, University of Aveiro, Aveiro, Portugal

Pedro F. Brandão Department of Chemistry & CESAM, University of Aveiro, Aveiro, Portugal

Marie-Cecile Chalbot New York City College of Technology, Biological Sciences Department, Brooklyn, NY, United States

Xi Chen Anhui Province Key Laboratory of Farmland Ecological Conservation and Pollution Prevention, School of Resources and Environment, Anhui Agricultural University, Hefei, China

Wenying Chu Department of Chemistry and Biochemistry, Old Dominion University, Norfolk, VA, United States

Armando C. Duarte Department of Chemistry & CESAM, University of Aveiro, Aveiro, Portugal

Regina M.B.O. Duarte Department of Chemistry & CESAM, University of Aveiro, Aveiro, Portugal

Hongjian Gao Anhui Province Key Laboratory of Farmland Ecological Conservation and Pollution Prevention, School of Resources and Environment, Anhui Agricultural University, Hefei, China

Jeffrey A. Hawkes Uppsala University, Uppsala, Sweden

Dean Hesterberg Department of Crop and Soil Sciences, North Carolina State University, Raleigh, NC, United States

Ilias Kavouras CUNY Graduate School of Public Health & Health Policy, Department of Environmental, Occupational and Geospatial Health Sciences, New York, NY, United States

William Kew Environmental Molecular Sciences Laboratory, Pacific Northwest National Laboratory, Richland, WA, United States

Jingdong Mao Department of Chemistry and Biochemistry, Old Dominion University, Norfolk, VA, United States

Cátia Martins Department of Chemistry & QOPNA/LAQV-REQUIMTE, University of Aveiro, Aveiro, Portugal

Mauro Mecozzi Laboratory of Chemometrics and Environmental Applications, ISPRA, Rome, Italy

Jennifer Mejia Department of Chemistry and Biochemistry, Old Dominion University, Norfolk, VA, United States

Carina Pedrosa Costa Department of Chemistry & QOPNA/LAQV-REQUIMTE, University of Aveiro, Aveiro, Portugal

Sílvia M. Rocha Department of Chemistry & QOPNA/LAQV-REQUIMTE, University of Aveiro, Aveiro, Portugal

Yingxin Shang Northeast Institute of Geography and Agroecology, Chinese Academy of Sciences, Changchun; University of Chinese Academy of Sciences, Beijing, China

Aakriti Sharma Department of Crop and Soil Sciences, North Carolina State University, Raleigh, NC, United States

Kaishan Song Northeast Institute of Geography and Agroecology, Chinese Academy of Sciences, Changchun, China

Sophia Viar 2505 Tiswood Court, Chesapeake, VA, United States

Zhidan Wen Northeast Institute of Geography and Agroecology, Chinese Academy of Sciences, Changchun, China

Ying Zhao Northeast Institute of Geography and Agroecology, Chinese Academy of Sciences, Changchun, China

Multidimensional analytical techniques in environmental research: Evolution of concepts

1

Regina M.B.O. Duarte, Armando C. Duarte
Department of Chemistry & CESAM, University of Aveiro, Aveiro, Portugal

Challenges in environmental research

Nowadays, most of the environmental challenges are associated with the increased release of pollutants into the air, water, and soil, modifications on the global cycling of nutrients and contaminants, and climate change issues. The advancements made thus far in environmental research have originated either from the need to understand the abovementioned issues or to seek solutions and regulations. Either way, most studies focus on understanding the interactions within and among atmospheric, terrestrial, aquatic, and living compartments of ecosystems. This is, however, an extremely challenging task, mainly due to the variability of those ecosystems and the high degree of heterogeneity, both in terms of composition and concentration, of the samples and analytes of interest taken from the different environmental compartments. This complexity represents a true analytical challenge. It is, therefore, not surprising that the development of new analytical strategies to unravel such complex matrices has occupied a central role in the effort of researchers.

The dramatic development during the past decade in a diverse suite of analytical tools, using a second or third dimension or multiscale hyphenated methods (i.e., separative and detection methods), have contributed to advances in environmental research. These advances include significant improvements in (i) analytical sensitivity and accuracy for the targeted, semitargeted, and untargeted screening of complex organic matrices (e.g., high-resolution mass spectrometry, HR-MS [1–6]); (ii) the use and/or combination of spectroscopic [e.g., one- and two-dimensional (2D) liquid- and solid-state nuclear magnetic resonance (NMR) spectroscopy [7–12], and excitation-emission matrix (EEM) fluorescence spectroscopy [13–16]], HR-MS [17–20], and chromatographic separation (e.g., Ref. [21]) to acquire compositional, geographic, and time evolution information on complex organic structures and interactions; (iii) development of powerful comprehensive multidimensional chromatographic tools for the resolution of complex organic matrices (e.g., Refs. [22–30]); (iv) use of synchrotron radiation-based methods to elucidate the speciation and spatial arrangement of toxic elements and nutrients in complex environmental matrices (e.g., Refs. [31–36]); and (v) development of more "user-friendly" data processing and treatment software to deal with the complexity of multidimensional data gathered

Multidimensional Analytical Techniques in Environmental Research. https://doi.org/10.1016/B978-0-12-818896-5.00001-6

from the environmental samples in order to glean the desired information (e.g., Refs. [37, 38]), to name a few of the many. The complementarity and technological advances of these multidimensional analytical tools have been key to allow a wider range of complex environmental matrices to be analyzed, enabling the acquisition of innovative data and transformative advances in environmental research.

This chapter aims to introduce the reader to the underlying concepts that have driven the development and use of sophisticated multidimensional and multiscale hyphenated methods for unraveling complex organic mixtures from different environmental matrices. The focus is on second- and third-dimensional spectroscopic, spectrometry, and chromatographic methods, and how these state-of-the-art multidimensional analytical strategies are being used for the targeted and untargeted profiling of such complex organic mixtures. This is not a comprehensive review on the use of these analytical methodologies but instead a broad overview and an introduction to the subsequent chapters, where the most popular multidimensional analytical techniques used in environmental biogeochemistry research are carefully addressed.

Coping with environmental organic matrices complexity

In recent years, there has been an increasing concern for environmental monitoring and development of new analytical procedures for dealing with the huge number of analytes and tackling the great complexity of environmental samples. These complex organic mixtures exhibit a diversity of constituents with different molecular sizes, structures, and chemical properties, which makes their analysis one of the enduring challenges in analytical chemistry. For example, while solving the chemical structure of high molecular size analytes, such as proteins or other natural polymers, requires exploring the relatively well-organized composition of their smaller molecular subunits (i.e., monomers), the analysis of smaller molecules in a mixture is rather more difficult. In the latter situation, the analyst faces a broad chemical and structural diversity, which requires different types of analyses if aiming at the full structural identification of each organic compound [i.e., elemental composition, spatial structure (i.e., its isomers), and/or spatial configuration]. Nevertheless, not all environmental problems require the full identification of all organic compounds present in a sample. Fig. 1.1 illustrates how different levels of compositional information can be distinguished, depending on the purpose of investigation: (i) functional group analysis, which copes with the highest level of molecular diversity (number of organic compounds, $n \geq 1000$) at the expenses of chemical resolution, is typically employed when interested in understanding specific properties of complex organic assemblies (e.g., structural average information [7, 8, 10, 12], chemical processes [9, 39], optical properties [13, 40], and fine-scale spatial arrangement of organic carbon forms [31, 34, 35]); (ii) resolve the chemical composition of complex organic mixtures into different organic components or molecular structures ($10 \geq n \geq 100$) is usually chosen to unraveling the molecular codes [11, 21, 41–45], the organic precursors [20, 46, 47], and reactivity [1, 48, 49] of these highly complex mixtures; (iii) target analysis of molecular organic

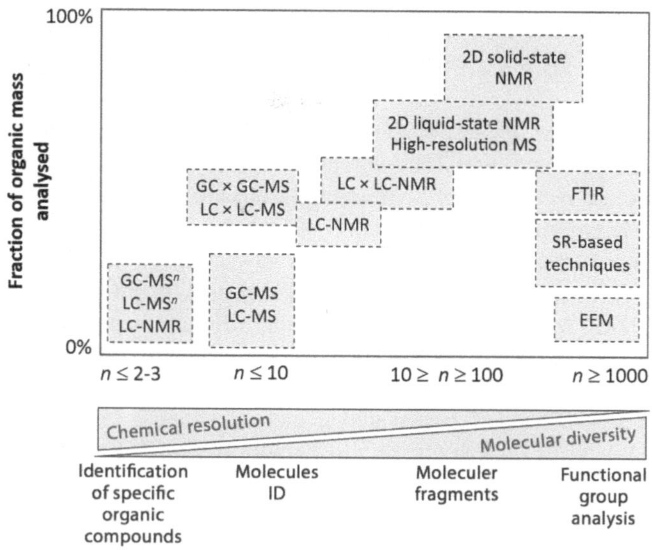

Fig. 1.1 Levels of organic compositional identification in the analysis of complex mixtures from diverse environmental matrices, highlighting the quantification attained by different advanced analytical techniques (*n*: number of organic compounds identified and/or measured).

markers ($n \leq 10$) is typically used to accurately quantitate and/or monitoring known formation processes or sources of the target compounds in different environmental matrices [23, 24, 28, 29, 50–54]; and (iv) corresponding to the highest level of chemical resolution, the identification of up to three specific organic compounds ($n \leq 2$–3) when studying, for example, unknown formation processes or sources of organic particles in the atmosphere (Ref. [55] and references therein) or identifying emerging organic pollutants in industrial wastewater [56] or freshwater [57].

In environmental research, the aim of the analysis and the choice of a fit for purpose analytical methodology are strongly interconnected and should be thoroughly assessed beforehand. Identifying specific organic compounds (known or unknown) in a complex environmental sample (i.e., $n \leq 10$ in Fig. 1.1, such as the identification and quantification of organic pollutants in a water sample) is different from a global characterization of the whole environmental sample [i.e., $n \geq 1000$ in Fig. 1.1, such as the characterization of natural organic matter (NOM)]. Regardless of using a targeted or untargeted analytical approach, a key objective of analytical chemistry has been the continuous improvement and development of analytical methodologies capable of reducing a complex problem into manageable data sets. As the environmental problems continue to grow ever more challenging, the level of compositional identification has evolved toward the integration of different analytical dimensions to reach the resolution necessary for the detection and identification of a broader range of molecular structures. The following sections intend to highlight those multidimensional analytical approaches that can entice the researchers to cope with the complexity of environmental samples and, thus, discover yet unknown new molecules.

Multidimensional nuclear magnetic resonance (NMR) spectroscopy in environmental research

NMR spectroscopy has unquestionable merits in the structure elucidation of organic structures in complex mixtures. The high reproducibility, as well as the nondestructive and noninvasive characteristics of NMR spectroscopy are key advantages for employing this technique in environmental research. NMR can be applied for in-depth studies of most environmental matrices, including liquid, gels, and solid samples, or even for the elucidation of organic structures present in all phases in unaltered environmental samples [58]. This feature of NMR relies on the different techniques available, thus making NMR spectroscopy a pivotal analytical tool to unravel the complexity of the countless molecular structures typically found in various environmental samples (Fig. 1.2). The NMR techniques available include solution-state NMR, solid-state NMR, gel-phase NMR, and comprehensive multiphase (CMP) NMR spectroscopy. For a more in-depth discussion of all these techniques, experimental protocols, and applications in the analysis of environmental complex matrices, the reader is encouraged to refer to the review works of Simpson et al. [7, 59], Mao et al. [8], and Duarte and Duarte [10], as well as to Chapters 2 and 3. Here, it is intended to highlight the advantage of using these NMR techniques, particularly 2D NMR, to acquire a wealth of information on the molecular bonds, structures, and interactions within the complex organic fraction present in water, soils, sediments, and air particles.

Solution-state NMR spectroscopy in environmental research

Solution-state NMR is ideally suited to acquire comprehensive molecular information of complex organic matrices that are naturally soluble, such as the dissolved organic matter (DOM) from ice [60] and water [20, 61], but also the organic matter isolated from soils [7] and air particles [11, 62–64]. Undoubtedly, solution-state one-dimensional (1D) ^{1}H NMR technique has a prime position as a tool for rapid screening and determination of the general structural properties of such complex organic mixtures. Although providing a relatively broad 1D profile of DOM, one can still withdraw excellent compositional information on the sample, including near-quantitative data on the different ^{1}H functional groups with C-H bonds, as long as the spectra are carefully acquired, processed, and interpreted. These semiquantitative approaches have been used, for example, to assess the molecular divergence within DOM from different wetlands [61] or to shed light on the dominant sources of atmospheric organic aerosols at different locations (i.e., source apportionment) [65]. The well-known downside of solution-state 1D ^{1}H NMR of complex organic matrices is that accurate qualitative and quantitative structural assessment is hampered by the high degree of overlap characterizing these spectra. Three main reasons can explain this spectral overlap: (1) the resonances are dispersed over a limited ^{1}H chemical shift range (δ_H 0–10 ppm), (2) the presence of organic compounds with resembling structural features, for which the corresponding ^{1}H NMR spectra are very similar, and (3) the presence of a high number of compounds resonating in the same limited spectral region.

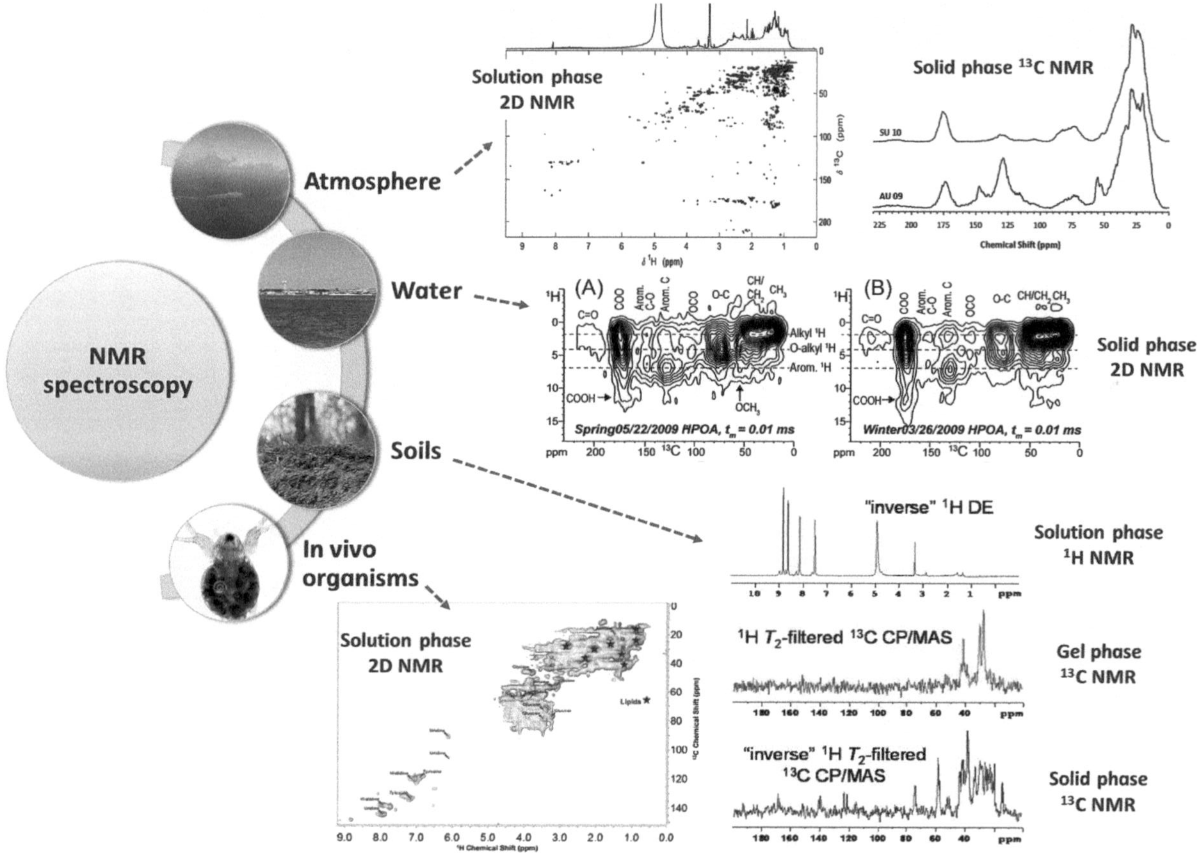

Fig 1.2 See figure legends on next page

One appealing solution to overcome the spectral overlapping issue is to rely on solution-state multidimensional NMR spectroscopy. The multidimensional approach has the high advantage of offering a much better discrimination of resonances than 1D NMR as the peaks are spread along a second or third dimension (^{1}H or ^{13}C frequencies), thus enhancing the reliability of NMR assignments and allow the identification of molecular fragments, via homonuclear (^{1}H-^{1}H) and heteronuclear (^{1}H-^{13}C) connectivity information [7, 66]. Undoubtedly, the most important multidimensional solution-state NMR experiments applied into environmental research are the 2D NMR techniques, including (a) ^{1}H-^{1}H homonuclear COSY and TOCSY, which provide connectivity information between protons that are directly attached to adjacent carbons (COSY), or regarding a given proton that is interacting with other protons of the same structure which are within the spin system (unbroken chain of couplings) of the atom (TOCSY); (b) ^{1}H-^{13}C HSQC, which detects H-C couplings over one bond and provides chemical shift data for both atoms in a C-H unit; and (c) ^{1}H-^{13}C HMBC, which provides direct evidence about the bonding of H-C fragments over two- and three-bond range (i.e., H-C-$\underline{C}$ or H-C-C-$\underline{C}$) [67]. The combination of ^{1}H-^{1}H homonuclear (COSY and/or TOCSY) with ^{1}H-^{13}C heteronuclear (HSQC and HMBC) connectivity information is a powerful approach for assignment of signals, allowing a higher spectral resolution and, therefore, greater detail on the C-H backbone of the substructures present in complex organic matrices such as those of NOM [7, 11, 20, 61–63, 65, 68]. Recently, an isotope-filtered *n*D NMR methodology—a combination of isotopic tagging and *n*D NMR—was developed to characterize phenolic moieties of humic molecules [69]. The principle was illustrated

Fig. 1.2 Solution-state, solid-state, and comprehensive multiphase NMR spectroscopy employed in the structural characterization of different environmental matrices.
Reprinted (adapted) with permission from J.T.V. Matos, R.M.B.O. Duarte, S.P. Lopes, A.M.S. Silva, A.C. Duarte, Persistence of urban organic aerosols composition: decoding their structural complexity and seasonal variability, Environ. Pollut. 231 (2017) 281–90, https://doi.org/10.1016/j.envpol.2017.08.022 (Copyright (2017), with permission from Elsevier), R.M.B.O. Duarte, S.M.S.C. Freire, A.C. Duarte, Investigating the water-soluble organic functionality of urban aerosols using two-dimensional correlation of solid-state 13C NMR and FTIR spectral data, Atmos. Environ. 116 (2015) 245–52, https://doi.org/10.1016/j.atmosenv.2015.06.043 (Copyright (2015), with permission from Elsevier), X. Cao, G.R. Aiken, R.G.M. Spencer, K. Butler, J. Mao, K. Schmidt-Rohr, Novel insights from NMR spectroscopy into seasonal changes in the composition of dissolved organic matter exported to the Bering Sea by the Yukon River, Geochim. Cosmochim. Acta 181 (2016) 72–88. https://doi.org/10.1016/j.gca.2016.02.029 (Copyright (2016), with permission from Elsevier), D. Courtier-Murias, H. Farooq, H. Masoom, A. Botana, R. Soong, J.G. Longstaffe, et al., Comprehensive multiphase NMR spectroscopy: basic experimental approaches to differentiate phases in heterogeneous samples, J. Magn. Reson. 217 (2012) 61–76, https://doi.org/10.1016/j.jmr.2012.02.009 (Copyright (2012), with permission from Elsevier), and M. Tabatabaei Anaraki, R. Dutta Majumdar, N. Wagner, R. Soong, V. Kovacevic, E.J. Reiner, et al., Development and application of a low-volume flow system for solution-state in Vivo NMR, Anal. Chem. 90 (2018) 7912–21, https://doi.org/10.1021/acs.analchem.8b00370 (Copyright (2018) American Chemical Society).

using a 4D $^{13}CH_3O$-filtered NMR experiment, which correlates chemical shifts of four nuclei—the aromatic CH atoms *ortho* to methoxy groups and those of $^{13}CH_3O$ atoms. The information gathered on the multiple chemical shifts and coupling constants have led to the identification of the major substitution patterns of nine phenolic aromatic moieties of a peat soil fulvic acid [69], and the prospect of applying other tags containing NMR-active nuclei (e.g., such as ^{15}N and ^{31}P).

Regardless of the solution-state multidimensional NMR experiment employed in the structural characterization of complex organic mixtures, it is advantageous to extract/isolate/preconcentrate the organic component from the original environmental matrix, particularly when dealing with water, soil, or sediment samples. The outcome of the preprocessing sample procedure is twofold: (1) it decreases the heterogeneity of the sample, by enriching the isolated fractions in those organic species that are targeted by the physicochemical mechanisms governing their extraction, and (2) removes the paramagnetic species that interfere with NMR signal acquisition, thus enhancing both the sensitivity and resolution of the spectra. An alternative way of improving NMR detection of unique molecular structures within complex organic mixtures, such as those of natural organic matter, is through the chromatographic separation of these matrices into simplified fractions prior to offline NMR detection. As shown by Woods et al. [21, 45], improved discrete structural assignments within DOM are readily attainable using multidimensional [1D, 2D, and three-dimensional (3D)] NMR for the characterization of simplified chromatographic DOM fractions. Multidimensional NMR data provided a range of connectivity and chemical shift information that is not apparent from the unfractionated DOM material [21, 45].

It has been also shown that solution-state multidimensional NMR can be used to characterize complex environmental samples with limited [e.g., water-soluble organic matter (WSOM) from atmospheric aerosols [62]] or even with no preconcentration procedure (e.g., DOM from ice [60], as well as rivers, lakes, and the ocean [70–72]). The application of improved water suppression techniques has allowed the acquisition of meaningful NMR spectra and the subsequent characterization of the organic matter at its natural abundance in almost unaltered environmental samples. Although providing compositional information on the organic constituents without pretreatment, these structural data are acquired at the expenses of long times of analysis, which usually prevents the application of this procedure on a routine basis.

Solid-state NMR spectroscopy in environmental research

Solid-state NMR is traditionally performed on dried samples (100–500 mg of sample mass is required) and also widely employed to investigate the structure of NOM from diverse environmental matrices. In this regard, the reader is encouraged to refer to the review works of Mao et al. [8], Cook [73], and Duarte et al. [10] on the application of solid-state NMR spectroscopy to NOM studies from water, soils, and atmospheric particulate organic matter. In a similar way to solution-state NMR, high-quality solid-state NMR data of environmental samples can be obtained if concentrating the organic matter by removing the paramagnetic species from the complex matrix, typically by using a solid-phase extraction procedure.

^{13}C is the most commonly detected nucleus in solid-state NMR of environmental samples. Due to the low natural abundance and, therefore, low sensitivity of ^{13}C detection, cross-polarization (CP) in combination with magic angle spinning (MAS) is often used to enhance the ^{13}C signal. During CP, the magnetization is passed from proton to carbon for enhancing the signal; however, this feature is also the main drawback of CP-MAS, since it does not detect nonprotonated carbons (e.g., carbon atoms of carboxylic groups, or carbon from fused aromatic rings) or mobile segments with weak H–C dipolar couplings [74]. To achieve a quantitative assessment of all carbon functional groups present in a sample, direct polarization (DP) combined with MAS should be performed [75]. However, the acquisition of a solid-state ^{13}C DP-MAS NMR spectrum is more time-consuming than that of a CP-MAS spectrum. Recently, a new method has been developed that yields quantitative solid-state MAS ^{13}C NMR spectra of organic materials with good signal-to-noise ratios. The multiple cross-polarization (multiCP) technique developed by Johnson and Schmidt-Rohr [74] provides quantitative information about all carbon atoms, typically reducing the measuring time by more than a factor of 50 compared to quantitative ^{13}C DP/MAS [74]. The solid-state multiCP ^{13}C NMR technique aid by the application of suitably designed radio frequency pulse sequences allows targeting subspectra of specific types of functional groups, such as sp^3-hybridized only, nonprotonated carbons (e.g., aromatic C-C, and anomeric O-C-O and anomeric O-C(R,R′)-O groups), mobile CH_3 groups, OCH_3, immobile CH_n-only (i.e., CH_2 and CH), CH_2-only, and CH-only carbons in NOM from various origins [48, 75, 76]. The combination of different spectral-editing techniques, which have been described in detail by Mao et al. [8], could allow the identification of at least 27 different functional groups in ^{13}C NMR spectra of complex NOM, in contrast to less than 10 typically distinguished in the literature based on simple, routine ^{13}C CP-MAS NMR spectroscopy. Additional advantages of the solid-state ^{13}C NMR approach has been recently reviewed by Duarte et al. [10], and include (1) the distinctive feature of being a nondestructive technique, leaving the sample available for other complementary chemical analyses; (2) it facilitates a much higher sample concentration than solution-state NMR, enhancing signals and saving instrument time; (3) the technique does not have some of the problems reported for solution-state NMR analyses of NOM, including solvent effects on the chemical shifts of the sample, potential masking of certain sample chemical shifts due to solvent signals, and limited solubility of the organic material in the selected solvent; (4) the detection of nonprotonated carbons using solid-state ^{13}C NMR is straightforward; and (5) the macromolecular structures and/or colloids within NOM slow the tumbling of these molecules, leading to T_2 values that are too short to allow many of the pulse sequences of solution NMR to be successfully used [8, 73].

NOM applications of nuclei other than ^{13}C have been also reported for solid-state NMR, including both ^{15}N and ^{31}P nuclei. As recently reviewed by Mao and coworkers [8], ^{15}N CP-MAS has been the primary solid-state NMR technique used for studying organic nitrogen forms in NOM from soil, water, sediments, coal, and kerogen. However, acquiring a meaningful ^{15}N CP-MAS NMR spectra of such complex NOM matrices is rather difficult. An alternative solid-state NMR technique, $^{13}C\{^{14}N\}$

saturation pulse-induced dipolar exchange with recoupling (SPIDER), has been successfully employed to investigate the chemical nature of nitrogen in NOM by detecting ^{13}C bonded to nitrogen [8, 77, 78]. ^{31}P solids NMR have been reported for marine DOM [79] and bulk soils [80], being successfully used for the identification of different phosphorus forms and for the evaluation of their dynamics in the studied samples.

Solid-state 2D ^{1}H-^{13}C heteronuclear correlation (HETCOR) NMR is possible and has proved extremely useful for assessing through-space ^{1}H-^{13}C correlations and, therefore, acquire valuable information on the structure of the surroundings of carbon functional groups [8, 48, 75, 78]. This 2D solid-state NMR technique allows the identification of connectivities or proximities of different functional groups (e.g., aromatic and alkyl), being particularly useful for the identification of the nearest protons for nonprotonated carbons, such as COO/NC=O [75] or quaternary carbons [48].

Comprehensive multiphase and in vivo NMR for analysis of natural samples

Comprehensive multiphase (CMP) NMR, which integrates the capabilities of solution-state, solid-state, and gel-state NMR into a single approach, allows to detect and differentiate all liquids, solutions, and gels in unaltered samples in their natural state. Gel-phase NMR, also referred to as high-resolution magic angle spinning (HR-MAS) NMR, involves the study of samples that are "swellable" and/or in the gel phase [7]. The samples constituents are analyzed after being swollen in a penetrating solvent (e.g., DMSO-d_6) or they can be analyzed in their undried natural state with water acting as the natural "solvent" [7]. For example, the HR-MAS has huge potential for the analysis of soil, plant materials, atmospheric particles, and small organisms in their swollen sate [59]. In soil, for example, HR-MAS provides information on the structures and associations of organic components at the solid-water interface [81]. Combined with solution- and solid-state NMR techniques, as well as editing-based experiments as in CMP-NMR, it can provide a multidimensional detailed insight into the organization of soil components and how the domains and associations change with pH and solvent [12], using samples that are in their unaltered state. This approach has been also applied to examine oil-contaminated soil [82], to study the molecular interactions and fate during contaminant sequestration in urban soil [9], as well as for in vivo 2D ^{1}H-^{13}C HSQC identification of metabolites in ^{13}C enriched living organisms [59, 83–86], and examination of plants structure and function in their native state [87].

High-resolution mass spectrometry in environmental research

High-resolution mass spectrometry (HR-MS) (addressed in Chapter 4) is another significant analytical advance and holds great promise in studies of complex materials, such as NOM from aqueous [1, 2, 20, 61, 77, 88–90], soils [6, 91, 92], extraterrestrial organic matter [19], and organic matter in atmospheric aerosols [18, 47, 93] and rainwater [94] samples. The most significant advantage of HR-MS techniques in the analysis of complex mixtures is their ability to provide high peak capacity and high measurement throughput necessary to assign accurate molecular weights and, thus, molecular formulas to the individual components, without the need for prior separation. As discussed by May and McLean [44], multidimensional separations based on HR-MS techniques exhibit peak capacities approaching 100,000 or greater and are capable of very high peak production rates ranging from 100,000 peaks per second for Orbitrap MS [Fourier transform MS (FTMS)] to over 100 million peaks per second for time-of-flight (TOF)MS. Electrospray ionization (ESI) combined with Fourier transform ion cyclotron resonance (FT-ICR)-MS has become a prevailing method to assign molecular formulas to thousands of molecules in a single complex organic matrix [89]. ESI is a "soft" ionization technique that transfers ions from solution to the gas phase with minimal fragmentation before they are subjected to MS analysis [89]. While the application of this approach is fairly straightforward for water-soluble NOM fractions, it is of limited utility for poorly soluble materials, such as soil organic matter unless one can extract this organic component with little or no chemical alteration [6].

A typical FT-ICR mass spectrum of complex organic mixtures, such as those of NOM, contains thousands of individual peaks, each representing a unique molecular mass, signal magnitude, and a specific molecular formula [89]. Once such HR mass spectra are obtained, two important issues need to be addressed: (1) separate noise from analyte peaks in order to avoid assigning false molecular formulas [90] and (2) find adequate ways to visualize and reduce the acquired complex multidimensional data sets [89]. Integrating additional separation dimensions with FT-ICR-MS provides additional compositional information, but it adds to the complexity of analyzing large data sets produced by the hyphenated HR-MS method [95].

In order to address the first issue, Riedel and Dittmar [90] have recently proposed a new detection limit method for the analysis of NOM via FT-ICR-MS, allowing to identify peaks that can reliably be distinguished from noise. As explained by the authors, this method requires the analysis of replicate blanks, a procedure usually implemented to check for impurities or contaminations. The noise peaks found in the blanks are then used to define the signal uncertainty of the noise, and peaks that are indistinguishable from this noise can readily be removed from real samples, with software help [90]. The second issue in FT-ICR-MS studies of complex samples is data presentation and exploitation. As pointed out by Reemtsma [89, 96], FT-ICR-MS data sets of complex organic matrices, such as those of NOM, are not only large but also multidimensional, where for one molecule the number of several elements is known (C, H, O as a minimum, but also of N, S, and/or P), together with its molecular mass, signal intensity, and retention time in case that chromatographic separation is employed. The 2D van Krevelen diagram (Fig. 1.3), which plots the H/C ratios of the molecules against the respective O/C ratios [89, 96, 97], is the most widely used graphical representation of FTICR-MS data, producing an illustration of different compound classes based on the molecular formula data of the molecules within the complex matrix. The 2D van Krevelen diagram can be further expanded to a 3D representation, by adding ion abundance or another molar ratio (N/C, S/C) as the z-axis [97]. As explained by Reemtsma [89, 96], the van Krevelen diagram has its own disadvantages because it normalizes to the carbon number, thus discarding a large set of information: different molecules that exhibit similar O/C and H/C ratios plot at the

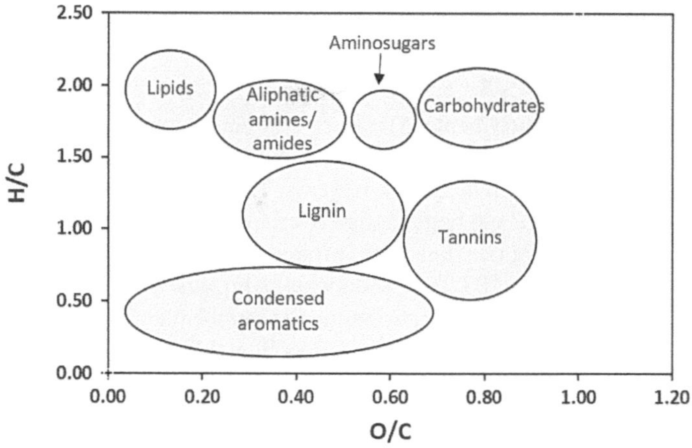

Fig. 1.3 Schematic representation of the Van Krevelen diagram of major compound classes identified in DOM samples.
Adapted from the works of R.L. Sleighter, P.G. Hatcher, The application of electrospray ionization coupled to ultrahigh resolution mass spectroscopy for the molecular characterisation of NOM, J. Mass Spectrom. 43 (2008) 854–64, https://doi.org/10.1002/jms, and A. Nebbioso, A. Piccolo, Molecular characterization of dissolved organic matter (DOM): a critical review, Anal. Bioanal. Chem. 405 (2013) 109–24, https://doi.org/10.1007/s00216-012-6363-2.

same point in the diagram, thus losing any mass-dependent information. Reemtsma [96] suggested an alternative graphical representation, by plotting the number of carbons in each formula vs its nominal mass (C vs M), where the molecules are classified into different categories based on their sum of carbon and oxygen atoms. However, this C vs M approach has not been employed as much as the van Krevelen diagram in the FTICR-MS analysis of complex organic matrices.

Although FT-ICR-MS has been proved to be highly suitable for resolving thousands of molecular formulas within a complex organic mixture, its use in combination with other techniques, for example, 2D NMR spectroscopy (e.g., ^{1}H-^{1}H COSY and TOCSY and ^{1}H-^{13}C HSQC) offers unsurpassed molecular resolution allowing an in-depth description of molecular skeleton and functional groups of the studied samples [19, 20, 61, 92, 98]. For example, Hertkorn et al. [61] used this multidimensional FT-ICR-MS and 2D NMR approach to describe and compare the compositional features of DOM in subtropical wetlands from different regions around the globe [Everglades (USA), Pantanal (Brazil), and the Okavango Delta (Botswana)], which are under the influence of different organic matter sources and flooding events. With this new multidimensional analytical window, the authors concluded that wetland DOM samples share various molecular features; however, each DOM sample was unique in its composition, reflecting specific environmental drivers and/or specific biogeochemical processes [61]. As the need to get a deeper insight into the molecular diversity of unknown complex environmental matrices increases, the combined use of multidimensional HR-MS and NMR tools represents a major step toward an improved understanding of the environmental importance of such complex mixtures.

Two-dimensional correlation spectroscopy in environmental research

2D correlation spectroscopy (Chapter 5) is a versatile chemometric technique introduced by Isao Noda (e.g., Refs. [99–102] references therein), which has started to become popular in environmental research in the last 10 years. The 2D correlation spectroscopy has been mostly used to resolve and better assign overlapped peaks typically shown in conventional 1D NMR and Fourier transform infrared (FTIR) spectra of complex organic mixtures from water [17, 42, 103, 104], sediments [105], atmospheric particles [43], and biofilms [106]. In 2D correlation spectroscopy, the sample under spectroscopic study is subjected to an external perturbation (e.g., temperature, pH, or salinity), which induces systematic variations in the spectral signal intensity. The obtained set of spectra observed as a function of the perturbation variable is then transformed into a set of 2D correlation spectra by a form of cross correlations, which define structural relationships [99] (an example of such 2D correlation spectra is shown Fig. 1.4; in this case for WSOM in atmospheric particles [43]). For example, Abdulla et al. [42, 103] applied the

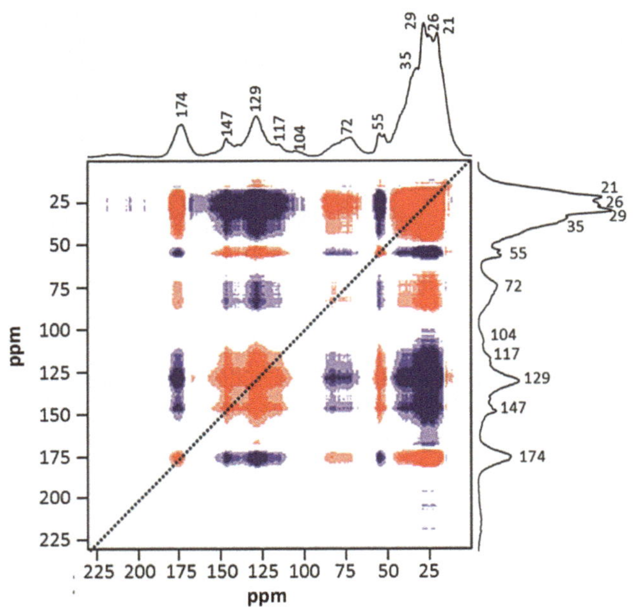

Fig. 1.4 Example of a synchronous map generated from CP-MAS ^{13}C NMR spectra of atmospheric aerosol WSOM samples collected during different seasons, where the top and the right side are the average ^{13}C NMR spectra. Red and blue represent positive and negative correlations, respectively.
Reprinted (adapted) with permission from R.M.B.O. Duarte, S.M.S.C. Freire, A.C. Duarte, Investigating the water-soluble organic functionality of urban aerosols using two-dimensional correlation of solid-state 13C NMR and FTIR spectral data, Atmos. Environ. 116 (2015) 245–52, https://doi.org/10.1016/j.atmosenv.2015.06.043 (Copyright (2015), with permission from Elsevier).

2D correlation technique on a set of ^{13}C NMR, ^{1}H NMR, and Fourier transform infrared (FTIR) spectra of high molecular-weight DOM samples isolated along a salinity transect. By combining insights from these spectral probes, either correlating the same or different spectroscopic probes (as in hetero-spectral 2D correlations) along the same perturbation (salinity), the authors concluded that the DOM samples consists of three major components [i.e., heteropolysaccharides (HPS), carboxyl-rich alicyclic molecules (CRAM), and amide/amino sugars] that have different biogeochemical reactivities [42]. The 2D correlation maps involving ^{1}H NMR spectra further revealed the major compound classes within each component—for example, it was suggested that HPS encompass three major compound classes (*N*-acetyl amino sugars, 6-deoxy sugars, and sulfated polysaccharide compounds), whereas CRAM consists of at least two compound classes (lignin-like and carboxylic functional groups of aliphatic nature) [103]. Another example of coupling between ^{13}C NMR and FTIR probes through 2D correlation analysis has been used to glean new structural information on WSOM from fine urban air particles collected during different seasons, with the median of air temperature within each season as the perturbation variable that prompts the observed spectral features [43] (Fig. 1.4). It was concluded that the WSOM samples consists of at least two classes of compounds: one is rich in both carboxylic and hydroxyl functional groups and it has an aliphatic character, and the other entails lignin-derived structures [43].

The 2D correlation analysis has been also performed between FT-ICR-MS and ^{13}C NMR spectra of DOM samples from a salinity transect [88]. The generated 2D correlation map distributed the mass spectral peaks based on their correlation with specific carbon functional groups (namely, HPS and lignin/CRAM-like components) detected in the ^{13}C NMR spectra. As stated by Abdulla et al. [88], this type of hetero-spectral 2D correlation analysis has the potential to expand our analytical window toward a deeper understanding of complex organic mixtures containing thousands of components, thus allowing to glean in-depth knowledge on the molecular structural features and dynamics of such complex mixtures. Regardless of the selected spectroscopic probe, it is clear that the structural information obtained by means of 2D correlation analysis can be hardly retrieved by using a single spectroscopic technique alone.

Fluorescence spectroscopy in the characterization of environmental samples

Excitation-emission matrix (EEM) fluorescence spectroscopy is a high sensitivity and nondestructive 3D technique, widely used to compare and discern the dynamics and transformations of chromophoric dissolved or water-extracted organic matter in diverse environmental matrices (freshwater [1, 16, 107], estuaries [108, 109], wetlands [61], soil [110], and atmospheric particles [13, 14, 111]). One additional example of the use of EEM fluorescence spectroscopy for the characterization of DOM in an aquatic ecosystem can be found in Chapter 6.

The acquisition of an EEM fluorescence spectrum involves the collection of sequential fluorescence emission (Em) spectra at successively increasing excitation

(Ex) wavelengths. The Em spectra obtained are concatenated to produce a plot in which the fluorescence intensity is displayed as a function of Ex and Em wavelengths. The main fluorescent groups in DOM studies that have been identified correspond to humic-like, protein-like (tyrosine- and tryptophan-like), and pigment-like substances [112, 113]. EEM fluorescence spectroscopy combined with parallel factor analysis (PARAFAC) modeling has made it possible to further resolve the complex 3D EEM spectra into its dominant fluorescent components and quantify each component's contribution to the total fluorescence. In this regard, the reader is encouraged to refer to the tutorial of Murphy et al. [114] in the practical application of PARAFAC to fluorescence data sets, using a DOM fluorescence data set. For example, Singh et al. [109] used EEM-PARAFAC to examine the compositional distribution and chromophoric DOM variability in an estuarine system. Four components were identified by the PARAFAC model, with the PARAFAC sample scores being used to examine probable linkages to wetlands, agricultural sources, and other water bodies. Hertkorn et al. [61] also used EEM-PARAFAC to discriminate chromophoric DOM samples from three subtropical wetlands [Everglades (USA), Pantanal (Brazil), and Okavango Delta (Botswana)]. In this case study, the authors reported analogies in the DOM fluorescence properties for the three wetlands in such a way that the generated EEM-PARAFAC model was perfectly applicable to the three wetlands. When applied to investigate the fluorescence features of chromophoric WSOM in atmospheric aerosols from different environments (urban, forest, marine, and pristine), EEM-PARAFAC aided in the classification and source identification of chromophores in atmospheric organic aerosols [13, 14, 111]. The acquired EEM-PARAFAC data is of utmost importance to shed light on the pivotal role played by the WSOM in the optical properties and photochemical reactivity of atmospheric organic aerosols.

An additional interesting application of EEM-PARAFAC analysis was shown by Woods et al. [21], combining chromatographic separation (by polarity) with offline solution-state multidimensional NMR into the characterization of Suwannee River DOM fractions. Findings suggested that both the structural features from NMR and EEM-PARAFAC components vary with polarity [21]. The majority of the fluorescence signals were dominant in the most hydrophobic fractions which were found to be enriched in structures derived from both cyclic and linear terpenoids [21]. The hydrophilic material, on the other hand, was highly correlated with carbohydrate-type structures as well as high contributions from amino acid fluorescence [21]. This combined application of multidimensional analytical techniques set the basis for new analytic strategies aiming at the molecular-level identification and further understanding of the origin, structure, fate, and chemical reactivity of complex environmental matrices.

Comprehensive two-dimensional chromatography in environmental analysis

One-dimensional gas and liquid chromatography (1D-GC and 1D-LC, respectively) are undoubtedly successful analytical separation tools in environmental analysis. When 1D chromatographic separation, even after an optimization process, still is insufficient to achieve a good resolution between co-eluted compounds, particularly when dealing with complex environmental samples, it becomes necessary to upgrade the analytical process by adding two or more different separation mechanisms in order to take full advantage of coupling advanced detection systems (e.g., HR-MS). In this context, comprehensive 2D gas and liquid chromatography (GC × GC and LC × LC, respectively) have become attractive analytical approaches, offering increased peak capacity and selectivity relative to conventional 1D-GC and 1D-LC separations, respectively. Both GC × GC and LC × LC involve the use of two independent separation mechanisms (i.e., orthogonal), separated by an interface called modulator, often referred to as the "heart" of GC × GC or LC × LC [27, 115]. This modulation interface has the function of transferring fractions of the first-dimension (^{1}D) effluent to the second-dimension (^{2}D) column, while preserving the integrity of ^{1}D separation. In both GC × GC and LC × LC, the whole sample is subjected to multidimensional and independent separation mechanisms, ensuring that the separation achieved in any previous dimension is maintained in the following one. In this methodology, the obtained chromatogram must be representative of the entire sample and, for this reason, it is necessary that the whole sample passes through the detector or at least in a percentage that guarantees its representativeness [116]. Multiple reviews have discussed the theoretical and practical aspects of GC × GC and LC × LC, including the fundamental principles, instrumental innovations, parameter optimization, and data processing approaches [26–28, 30, 37]. In this regard, interested readers should refer to these review works for additional information on the aspects of LC × LC and/or GC × GC method development. This section solely aims to provide a brief flavor of the diversity of studies using LC × LC and GC × GC in environmental analysis as these are addressed in Chapters 7 and 8, respectively.

LC × LC has been applied in a diversity of areas, including biochemical analysis, pharmaceutical analysis, analysis of Traditional Chinese Medicines, and polymer analysis. Readers interested in one of these LC × LC applications should refer to the recent review works of Stoll and Carr [26] and Pirok et al. [27]. Although there is a great potential to apply LC × LC in environmental research, this area is still in its early stages. The huge complexity of environmental matrices, in fact, places a great demand in terms of resolution power, challenging analysts to choose the most appropriate columns with orthogonal selectivities for each separation dimension, as well as an information-rich detector. Nonetheless, one of the attractive features for applying LC × LC in the analysis of complex environmental samples is that it adds additional information on these samples (e.g., polarity, size, and electrophoretic mobility), which enhances the interpretation of their physicochemical composition, especially when hyphenated with MS detection. Most environmental studies using LC × LC coupled to high-resolution detectors have focused either on the identification and quantification of a small group of polar compounds (i.e., targeted analysis) [53, 117–119] or in the nonselective search (i.e., untargeted analysis) and characterization of unknown components in a sample [23–25, 51]. Both analytical approaches require LC × LC method optimization (e.g., coupling of highly complementary (i.e., orthogonal) separation modes, mobile phases composition in both dimensions and their compatibility, flow rates, and time of analysis), as well as separation of background interferences from the analytes and accuracy in the generated LC × LC data.

Given its universal applicability, the importance of LC × LC is expected to grow rapidly in environmental research. In fact, untargeted analyses using LC × LC coupled to MS detection is a powerful tool with the capability of revealing new compositional and yet hidden structural details of complex environmental samples, thus availing new pathways of investigation—and this research field is only at its beginning.

The astounding separation power offered by GC × GC hyphenated to either universal or selective detectors, entices the researchers to use this approach for the targeted nonpolar compound analysis in environmental matrices [28, 30]. Unlike LC × LC, one can find in the literature a huge number of environmental studies more focused on the chemical information provided by GC × GC coupled to an H-MS detector rather than on the performance and optimization of the whole multidimensional analytical technique. Recently, Muscalu and Górecki [28] presented a systematic review of the most recent applications of GC × GC coupled with HR-MS detectors in the analysis of persistent organic pollutants in water, wastewater, leachates, soil, sediments, sludge, and biota. The targeted analytes include hydrocarbons, polycyclic aromatic hydrocarbons and its derivatives, polychlorinated biphenyls and pesticides, benzothiazoles, benzotriazoles and benzosulfonamides, nonylphenols and their derivatives (e.g., from production of plastics and surfactants), steroids, synthetic musks, personal care products, and pharmaceuticals [28]. Complex atmospheric organic aerosols containing volatile and semivolatile compounds have been also successfully determined with GC × GC-based methods [55]. As highlighted by Muscalu and Górecki [28], GC × GC also allows the separation of many constituents of previously unresolved complex mixtures of contaminants. When hyphenated with a HR-MS or other MS detectors, this GC × GC approach offers unsurpassed resources for the nonselective search of a diversity of organic pollutant groups that might be environmentally relevant but are not routinely analyzed. For example, Chapter 8 highlights how GC × GC coupled with HR-MS detection enables a better understanding of the impact of environmental exposures on human health. This chapter discusses how GC × GC-HR-MS methods are being used as key analytical resources for the identification and quantification of a wide range of analytes (e.g., persistent organic pollutants, dibenzo-*p*-dioxins, and aromatic amines), sometimes found only in trace amounts, in body fluids (e.g. urine, blood, and breath).

Synchrotron-based techniques as multidimensional analytical tools

The literature shows that two important synchrotron (Sr)-based techniques—Sr-FTIR and near-edge X-ray absorption fine structure (NEXAFS)—can be used for elemental speciation as well as analysis of the bulk properties and spatial distribution of carbon forms in environmental samples, particularly in soils [31, 35, 120, 121]. An important strength of the Sr-based techniques is the high spatial resolution at the fine scale, which allows to draw a multidimensional map of the chemical environments of organic carbon, minerals, metals, and microbial habitats in different environmental matrices [31, 35, 36, 104, 120, 122, 123]. The ability to potentially discern organic carbon functional groups and their pixel-scale associations with other elements in the form of multidimensional maps and on scales of nano- and micrometers within any given sample, may significantly enhance the current understanding of the mechanisms responsible for nutrient and contaminant mobility, reactivity, bioavailability, and fate in the environment [35, 120]. There is, however, a downside of such grain-scale multidimensional distribution of organic carbon and associated elements. According to Lehmann and Solomon [31], significant constraints are encountered when information on single micro- or nanoscale locations within a sample needs to be scaled to processes observed at the macroscale (e.g., in soil or landscapes). Additional concerns that need to be considered and further optimized when using synchrotron-based techniques include sample preparation protocols that preserve the spatial assemblage, control of radiation damage, sectioning artifacts, and spectral quantification [31]. A major obstacle to the widespread use of synchrotron-based techniques in environmental studies is related to the limited number of analytical facilities available and associated expertise to conduct such research [121]. Therefore, researchers must be cautious about the environmental problems they want to address by means of these sophisticated analytical techniques.

Conclusions

Exploring the chemical composition of different environmental matrices (air, soil, sediments, water, and living organisms) and the biogeochemical processes taking place in those matrices, involves many unknowns and as nearly many challenges. The different multidimensional techniques introduced in this chapter and further explored in this book, constitute a great promise in environmental research. Integrating multidimensional separation and spectroscopic analyses at multiple scales, from small soluble molecules to macromolecules, nanoparticles or even larger sized samples is useful for decoding highly heterogeneous environmental media (air, soil, sediments, water, living organisms, and body fluids) with a high degree of specificity.

Notwithstanding the sophisticated multidimensional analytical techniques that are being used in environmental research, two additional major challenges remain: (1) lack of knowledge and analytical expertise to deal with such advanced multidimensional approaches, including processing and interpretation of the voluminous and complex data sets, and (2) the development of expertise for sampling and monitoring highly heterogeneous environmental matrices (e.g., air particles, waters, soils, and sediments) in an extensive and regular manner. Without improvements in these important areas, the use of multidimensional analytical techniques will not be noteworthy in environmental research.

Acknowledgments

Thanks are due to FCT/MCTES for the financial support to CESAM (UID/AMB/50017/2019) and project AMBIEnCE (PTDC/CTA-AMB/28582/2017), through national funds (OE). FCT/MCTES is also acknowledged for an Investigator FCT Contract (IF/00798/2015).

References

[1] A.M. Kellerman, F. Guillemette, D.C. Podgorski, G.R. Aiken, K.D. Butler, R.G. M. Spencer, Unifying concepts linking dissolved organic matter composition to persistence in aquatic ecosystems, Environ. Sci. Technol. 52 (2018) 2538–2548, https://doi.org/10.1021/acs.est.7b05513.

[2] M. Zark, T. Dittmar, Universal molecular structures in natural dissolved organic matter, Nat. Commun. 9 (2018) 1–8, https://doi.org/10.1038/s41467-018-05665-9.

[3] A.T. Lebedev, O.V. Polyakova, D.M. Mazur, V.B. Artaev, The benefits of high resolution mass spectrometry in environmental analysis, Analyst 138 (2013) 6946–6953, https://doi.org/10.1039/c3an01237a.

[4] J. Hollender, E.L. Schymanski, H.P. Singer, P.L. Ferguson, Nontarget screening with high resolution mass spectrometry in the environment: Ready to Go? Environ. Sci. Technol. 51 (2017) 11505–11512, https://doi.org/10.1021/acs.est.7b02184.

[5] M. Krauss, H. Singer, J. Hollender, LC-high resolution MS in environmental analysis: from target screening to the identification of unknowns, Anal. Bioanal. Chem. 397 (2010) 943–951, https://doi.org/10.1007/s00216-010-3608-9.

[6] M.M. Tfaily, R.K. Chu, N. Tolić, K.M. Roscioli, C.R. Anderton, L. Paša-Tolić, et al., Advanced solvent based methods for molecular characterization of soil organic matter by high-resolution mass spectrometry, Anal. Chem. 87 (2015) 5206–5215, https://doi.org/10.1021/acs.analchem.5b00116.

[7] A.J. Simpson, D.J. McNally, M.J. Simpson, NMR spectroscopy in environmental research: from molecular interactions to global processes, Prog. Nucl. Magn. Reson. Spectrosc. 58 (2011) 97–175, https://doi.org/10.1016/j.pnmrs.2010.09.001.

[8] J. Mao, X. Cao, D.C. Olk, W. Chu, K. Schmidt-Rohr, Advanced solid-state NMR spectroscopy of natural organic matter, Prog. Nucl. Magn. Reson. Spectrosc. 100 (2017) 17–51, https://doi.org/10.1016/j.pnmrs.2016.11.003.

[9] H. Masoom, D. Courtier-Murias, R. Soong, W.E. Maas, M. Fey, R. Kumar, et al., From spill to sequestration: the molecular journey of contamination via comprehensive multiphase NMR, Environ. Sci. Technol. 49 (2015) 13983–13991, https://doi.org/10.1021/acs.est.5b03251.

[10] R.M.B.O. Duarte, A.C. Duarte, NMR studies of organic aerosols, in: G.A. Webb (Ed.), Annual Reports on NMR Spectroscopy, 92, Academic Press, Oxford, 2017, pp. 83–135, https://doi.org/10.1016/bs.arnmr.2017.04.003.

[11] J.T.V. Matos, R.M.B.O. Duarte, S.P. Lopes, A.M.S. Silva, A.C. Duarte, Persistence of urban organic aerosols composition: decoding their structural complexity and seasonal variability, Environ. Pollut. 231 (2017) 281–290, https://doi.org/10.1016/j.envpol.2017.08.022.

[12] H. Masoom, D. Courtier-Murias, H. Farooq, R. Soong, B.P. Kelleher, C. Zhang, et al., Soil organic matter in its native state: unravelling the most complex biomaterial on Earth, Environ. Sci. Technol. 50 (2016) 1670–1680, https://doi.org/10.1021/acs.est.5b03410.

[13] J.T.V. Matos, S.M.S.C. Freire, R.M.B.O. Duarte, A.C. Duarte, Natural organic matter in urban aerosols: comparison between water and alkaline soluble components using excitation–emission matrix fluorescence spectroscopy and multiway data analysis, Atmos. Environ. 102 (2015) 1–10, https://doi.org/10.1016/j.atmosenv.2014.11.042.

[14] Q. Chen, Y. Miyazaki, K. Kawamura, K. Matsumoto, S. Coburn, R. Volkamer, et al., Characterization of chromophoric water-soluble organic matter in urban, forest, and marine aerosols by HR-ToF-AMS analysis and excitation–emission matrix spectroscopy, Environ. Sci. Technol. 50 (2016) 10351–10360, https://doi.org/10.1021/acs.est.6b01643.

[15] Á. Andrade-Eiroa, M. Canle, V. Cerdá, Environmental applications of excitation-emission spectrofluorimetry: an in-depth review II, Appl. Spectrosc. Rev. 48 (2013) 77–141, https://doi.org/10.1080/05704928.2012.692105.
[16] S.A. Baghoth, S.K. Sharma, G.L. Amy, Tracking natural organic matter (NOM) in a drinking water treatment plant using fluorescence excitation-emission matrices and PARAFAC, Water Res. 45 (2011) 797–809, https://doi.org/10.1016/j.watres.2010.09.005.
[17] H.A.N. Abdulla, P.G. Hatcher, Dynamics of dissolved organic matter: a view from two dimensional correlation spectroscopy techniques, J. Mol. Struct. 1069 (2014) 313–317, https://doi.org/10.1016/j.molstruc.2014.03.038.
[18] A.S. Willoughby, A.S. Wozniak, P.G. Hatcher, Detailed source-specific molecular composition of ambient aerosol organic matter using ultrahigh resolution mass spectrometry and 1H NMR, Atmosphere 7 (2016), https://doi.org/10.3390/atmos7060079.
[19] N. Hertkorn, M. Harir, P. Schmitt-Kopplin, Nontarget analysis of Murchison soluble organic matter by high-field NMR spectroscopy and FTICR mass spectrometry, Magn. Reson. Chem. 53 (2015) 754–768, https://doi.org/10.1002/mrc.4249.
[20] N. Hertkorn, M. Harir, B.P. Koch, B. Michalke, P. Schmitt-Kopplin, High-field NMR spectroscopy and FTICR mass spectrometry: powerful discovery tools for the molecular level characterization of marine dissolved organic matter, Biogeosciences 10 (2013) 1583–1624, https://doi.org/10.5194/bg-10-1583-2013.
[21] G.C. Woods, M.J. Simpson, P.J. Koerner, A. Napoli, A.J. Simpson, HILIC-NMR: toward the identification of individual molecular components in dissolved organic matter, Environ. Sci. Technol. 45 (2011) 3880–3886, https://doi.org/10.1021/es103425s.
[22] P.F. Brandão, A.C. Duarte, R.M.B.O. Duarte, Comprehensive multidimensional liquid chromatography for advancing environmental and natural products research, TrAC Trends Anal. Chem. 116 (2019) 186–197, https://doi.org/10.1016/j.trac.2019.05.016.
[23] A.S. Paula, J.T.V. Matos, R.M.B.O. Duarte, A.C. Duarte, Two chemically distinct light-absorbing pools of urban organic aerosols: a comprehensive multidimensional analysis of trends, Chemosphere 145 (2016), https://doi.org/10.1016/j.chemosphere.2015.11.093.
[24] J.T.V. Matos, S.M.S.C. Freire, R.M.B.O. Duarte, A.C. Duarte, Profiling water-soluble organic matter from urban aerosols using comprehensive two-dimensional liquid chromatography, Aerosol Sci. Technol. 49 (2015), https://doi.org/10.1080/02786826.2015.1036394.
[25] T.A. Brown, B.A. Jackson, B.J. Bythell, A.C. Stenson, Benefits of multidimensional fractionation for the study and characterization of natural organic matter, J. Chromatogr. A 1470 (2016) 84–96, https://doi.org/10.1016/j.chroma.2016.10.005.
[26] D.R. Stoll, P.W. Carr, Two-dimensional liquid chromatography: a state of the art tutorial, Anal. Chem. 89 (2017) 519–531, https://doi.org/10.1021/acs.analchem.6b03506.
[27] B.W.J. Pirok, D.R. Stoll, P.J. Schoenmakers, Recent developments in two-dimensional liquid chromatography: fundamental improvements for practical applications, Anal. Chem. 91 (2019) 240–263, https://doi.org/10.1021/acs.analchem.8b04841.
[28] A.M. Muscalu, T. Górecki, Comprehensive two-dimensional gas chromatography in environmental analysis, Trends in Anal. Chem. 106 (2018) 225–245, https://doi.org/10.1016/j.trac.2018.07.001.
[29] D.C. Hilton, R.S. Jones, A. Sjödin, A method for rapid, non-targeted screening for environmental contaminants in household dust, J. Chromatogr. A 1217 (2010) 6851–6856, https://doi.org/10.1016/j.chroma.2010.08.039.
[30] J.V. Seeley, S.K. Seeley, Multidimensional gas chromatography: fundamental advances and new applications, Anal. Chem. 85 (2013) 557–578, https://doi.org/10.1021/ac303195u.

[31] J. Lehmann, D. Solomon, Organic carbon chemistry in soils observed by synchrotron-based spectroscopy, in: B. Singh, M. Gräfe (Eds.), Developments in Soil Science, vol. 34, Elsevier Masson SAS, 2010, pp. 289–312, https://doi.org/10.1016/S0166-2481(10)34010-4.
[32] D. Hesterberg, I. McNulty, J. Thieme, Speciation of soil phosphorus assessed by XANES spectroscopy at different spatial scales, J. Environ. Qual. 46 (2017) 1190–1197, https://doi.org/10.2134/jeq2016.11.0431.
[33] S. Legros, P. Chaurand, J. Rose, A. Masion, V. Briois, J.H. Ferrasse, et al., Investigation of copper speciation in pig slurry by a multitechnique approach, Environ. Sci. Technol. 44 (2010) 6926–6932, https://doi.org/10.1021/es101651w.
[34] S. Peth, C. Chenu, N. Leblond, A. Mordhorst, P. Garnier, N. Nunan, et al., Localization of soil organic matter in soil aggregates using synchrotron-based X-ray microtomography, Soil Biol. Biochem. 78 (2014) 189–194, https://doi.org/10.1016/j.soilbio.2014.07.024.
[35] J. Lehmann, D. Solomon, J. Kinyangi, L. Dathe, S. Wirick, C. Jacobsen, Spatial complexity of soil organic matter forms at nanometre scales, Nat. Geosci. 1 (2008) 238–242, https://doi.org/10.1038/ngeo155.
[36] P.M. Kopittke, P. Wang, E. Lombi, E. Donner, Synchrotron-based X-ray approaches for examining toxic trace metal(loid)s in soil-plant systems, J. Environ. Qual. 46 (2017) 1175–1189, https://doi.org/10.2134/jeq2016.09.0361.
[37] J.T.V. Matos, R.M.B.O. Duarte, A.C. Duarte, Trends in data processing of comprehensive two-dimensional chromatography: state of the art, J. Chromatogr. B Anal. Technol. Biomed. Life Sci. 910 (2012), https://doi.org/10.1016/j.jchromb.2012.06.039.
[38] B.K. Lavine, J. Workman, Chemometrics, Anal. Chem. 85 (2013) 705–714, https://doi.org/10.1021/ac303193j.
[39] M. Brege, M. Paglione, S. Gilardoni, S. Decesari, M. Cristina Facchini, L.R. Mazzoleni, Molecular insights on aging and aqueous-phase processing from ambient biomass burning emissions-influenced Po Valley fog and aerosol, Atmos. Chem. Phys. 18 (2018) 13197–13214, https://doi.org/10.5194/acp-18-13197-2018.
[40] A. Vergnoux, R. Di Rocco, M. Domeizel, M. Guiliano, P. Doumenq, F. Théraulaz, Effects of forest fires on water extractable organic matter and humic substances from Mediterranean soils: UV–vis and fluorescence spectroscopy approaches, Geoderma 160 (2011) 434–443, https://doi.org/10.1016/j.geoderma.2010.10.014.
[41] Y. Li, M. Harir, M. Lucio, M. Gonsior, B.P. Koch, P. Schmitt-Kopplin, et al., Comprehensive structure-selective characterization of dissolved organic matter by reducing molecular complexity and increasing analytical dimensions, Water Res. 106 (2016) 477–487, https://doi.org/10.1016/j.watres.2016.10.034.
[42] H.A.N. Abdulla, E.C. Minor, P.G. Hatcher, Using two-dimensional correlations of 13C NMR and FTIR to investigate changes in the chemical composition of dissolved organic matter along an estuarine transect, Environ. Sci. Technol. 44 (2010) 8044–8049, https://doi.org/10.1021/es100898x.
[43] R.M.B.O. Duarte, S.M.S.C. Freire, A.C. Duarte, Investigating the water-soluble organic functionality of urban aerosols using two-dimensional correlation of solid-state 13C NMR and FTIR spectral data, Atmos. Environ. 116 (2015) 245–252, https://doi.org/10.1016/j.atmosenv.2015.06.043.
[44] J.C. May, J.A. McLean, Advanced multidimensional separations in mass spectrometry: navigating the big data deluge, Annu. Rev. Anal. Chem. 9 (2016) 387–409, https://doi.org/10.1146/annurev-anchem-071015-041734.
[45] G.C. Woods, M.J. Simpson, A.J. Simpson, Oxidized sterols as a significant component of dissolved organic matter: evidence from 2D HPLC in combination with 2D and 3D

NMR spectroscopy, Water Res. 46 (2012) 3398–3408, https://doi.org/10.1016/j.watres.2012.03.040.
[46] P. Schmitt-Kopplin, A. Gelencsér, E. Dabek-Zlotorzynska, G. Kiss, N. Hertkorn, M. Harir, et al., Analysis of the unresolved organic fraction in atmospheric aerosols with ultrahigh-resolution mass spectrometry and nuclear magnetic resonance spectroscopy: organosulfates as photochemical smog constituents, Anal. Chem. 82 (2010) 8017–8026, https://doi.org/10.1021/ac101444r.
[47] P. Schmitt-Kopplin, G. Liger-Belair, B.P. Koch, R. Flerus, G. Kattner, M. Harir, et al., Dissolved organic matter in sea spray: a transfer study from marine surface water to aerosols, Biogeosciences 9 (2012) 1571–1582, https://doi.org/10.5194/bg-9-1571-2012.
[48] X. Cao, G.R. Aiken, R.G.M. Spencer, K. Butler, J. Mao, K. Schmidt-Rohr, Novel insights from NMR spectroscopy into seasonal changes in the composition of dissolved organic matter exported to the Bering Sea by the Yukon River, Geochim. Cosmochim. Acta 181 (2016) 72–88, https://doi.org/10.1016/j.gca.2016.02.029.
[49] L. Bliumkin, R. Dutta Majumdar, R. Soong, A. Adamo, J.P.D. Abbatt, R. Zhao, et al., Development of an in situ NMR photoreactor to study environmental photochemistry, Environ. Sci. Technol. 50 (2016) 5506–5516, https://doi.org/10.1021/acs.est.6b00361.
[50] M. Caldeira, R. Perestrelo, A.S. Barros, M.J. Bilelo, A. Morête, J.S. Câmara, et al., Allergic asthma exhaled breath metabolome: a challenge for comprehensive two-dimensional gas chromatography, J. Chromatogr. A 1254 (2012) 87–97, https://doi.org/10.1016/j.chroma.2012.07.023.
[51] T. Spranger, D. Van Pinxteren, H. Herrmann, Two-dimensional offline chromatographic fractionation for the characterization of humic-like substances in atmospheric aerosol particles, Environ. Sci. Technol. 51 (2017) 5061–5070, https://doi.org/10.1021/acs.est.7b00077.
[52] L.C. Brent, J.L. Reiner, R.R. Dickerson, L.C. Sander, Method for characterization of low molecular weight organic acids in atmospheric aerosols using ion chromatography mass spectrometry, Anal. Chem. 86 (2014) 7328–7336, https://doi.org/10.1021/ac403937e.
[53] J. Pól, B. Hohnová, M. Jussila, T. Hyötyläinen, Comprehensive two-dimensional liquid chromatography-time-of-flight mass spectrometry in the analysis of acidic compounds in atmospheric aerosols, J. Chromatogr. A 1130 (2006) 64–71, https://doi.org/10.1016/j.chroma.2006.04.050.
[54] H. Zhang, L.D. Yee, B.H. Lee, M.P. Curtis, D.R. Worton, G. Isaacman-VanWertz, et al., Monoterpenes are the largest source of summertime organic aerosol in the southeastern United States, Proc. Natl. Acad. Sci. U. S. A. 115 (2018) 2038–2043, https://doi.org/10.1073/pnas.1717513115.
[55] B. Nozière, M. Kalberer, M. Claeys, J. Allan, B. D'Anna, S. Decesari, et al., The molecular identification of organic compounds in the atmosphere: state of the art and challenges, Chem. Rev. 115 (2015) 3919–3983, https://doi.org/10.1021/cr5003485.
[56] A. Preiss, U. Sänger, N. Karfich, K. Levsen, C. Mügge, Characterization of dyes and other pollutants in the effluent of a textile company by LC/NMR and LC/MS, Anal. Chem. 72 (2000) 992–998, https://doi.org/10.1021/ac9903609.
[57] R. Loos, S. Tavazzi, G. Mariani, G. Suurkuusk, B. Paracchini, G. Umlauf, Analysis of emerging organic contaminants in water, fish and suspended particulate matter (SPM) in the Joint Danube Survey using solid-phase extraction followed by UHPLC-MS-MS and GC–MS analysis, Sci. Total Environ. 607–608 (2017) 1201–1212, https://doi.org/10.1016/j.scitotenv.2017.07.039.
[58] A.J. Simpson, M.J. Simpson, R. Soong, Nuclear magnetic resonance spectroscopy and its key role in environmental research, Environ. Sci. Technol. 46 (2012) 11488–11496, https://doi.org/10.1021/es302154w.

[59] A.J. Simpson, M.J. Simpson, R. Soong, Environmental nuclear magnetic resonance spectroscopy: an overview and a primer, Anal. Chem. 90 (2018) 628–639, https://doi.org/10.1021/acs.analchem.7b03241.
[60] B.G. Pautler, G.C. Woods, A. Dubnick, A.J. Simpson, M.J. Sharp, S.J. Fitzsimons, et al., Molecular characterization of dissolved organic matter in glacial ice: coupling natural abundance 1H NMR and fluorescence spectroscopy, Environ. Sci. Technol. 46 (2012) 3753–3761, https://doi.org/10.1021/es203942y.
[61] N. Hertkorn, M. Harir, K.M. Cawley, P. Schmitt-Kopplin, R. Jaffé, Molecular characterization of dissolved organic matter from subtropical wetlands: a comparative study through the analysis of optical properties, NMR and FTICR/MS, Biogeosciences 13 (2016) 2257–2277, https://doi.org/10.5194/bg-13-2257-2016.
[62] R.M.B.O. Duarte, M. Piñeiro-Iglesias, P. López-Mahía, S. Muniategui-Lorenzo, J. Moreda-Piñeiro, A.M.S. Silva, et al., Comparative study of atmospheric water-soluble organic aerosols composition in contrasting suburban environments in the Iberian Peninsula Coast, Sci. Total Environ. 648 (2019) 430–441, https://doi.org/10.1016/j.scitotenv.2018.08.171.
[63] R.M.B.O. Duarte, A.M.S. Silva, A.C. Duarte, Two-dimensional NMR studies of water-soluble organic matter in atmospheric aerosols, Environ. Sci. Technol. 42 (2008) 8224–8230, https://doi.org/10.1021/Es801298s.
[64] A.S. Almeida, R.M.P. Ferreira, A.M.S. Silva, A.C. Duarte, B.M. Neves, R.M.B. O. Duarte, Structural features and pro-inflammatory effects of water-soluble organic matter in inhalable fine urban air particles, Environ. Sci. Technol. (2019), https://doi.org/10.1021/acs.est.9b04596.
[65] R.M.B.O. Duarte, J.T.V. Matos, A.S. Paula, S.P. Lopes, G. Pereira, P. Vasconcellos, et al., Structural signatures of water-soluble organic aerosols in contrasting environments in South America and Western Europe, Environ. Pollut. 227 (2017) 513–525, https://doi.org/10.1016/j.envpol.2017.05.011.
[66] A.J. Simpson, W.L. Kingery, P.G. Hatcher, The identification of plant derived structures in humic materials using three-dimensional NMR spectroscopy, Environ. Sci. Technol. 37 (2003) 337–342, https://doi.org/10.1021/es025956j.
[67] A. Simpson, Multidimensional solution state NMR of humic substances: a practical guide and review, Soil Sci. 166 (2001) 795–809, https://doi.org/10.1097/00010694-200111000-00006.
[68] A.J. Simpson, J. Burdon, C.L. Graham, M.H.B. Hayes, N. Spencer, W.L. Kingery, Interpretation of heteronuclear and multidimensional NMR spectroscopy of humic substances, Eur. J. Soil Sci. 52 (2001) 495–509, https://doi.org/10.1046/j.1365-2389.2001.00402.x.
[69] N.G.A. Bell, A.A.L. Michalchuk, J.W.T. Blackburn, M.C. Graham, D. Uhrín, Isotope-filtered 4D NMR spectroscopy for structure determination of humic substances, Angew. Chem. Int. Ed. 54 (2015) 8382–8385, https://doi.org/10.1002/anie.201503321.
[70] B. Lam, A.J. Simpson, Direct 1H NMR spectroscopy of dissolved organic matter in natural waters, Analyst 133 (2008) 263–269, https://doi.org/10.1039/b713457f.
[71] G. Zheng, W.S. Price, Direct hydrodynamic radius measurement on dissolved organic matter in natural waters using diffusion NMR, Environ. Sci. Technol. 46 (2012) 1675–1680, https://doi.org/10.1021/es202809e.
[72] S.D. Whitty, D.C. Waggoner, R.M. Cory, L.A. Kaplan, P.G. Hatcher, Direct noninvasive 1 H NMR analysis of stream water DOM: insights into the effects of lyophilization compared with whole water, Magn. Reson. Chem. (2019) 1–14, https://doi.org/10.1002/mrc.4935.

[73] R.L. Cook, Coupling NMR to NOM, Anal. Bioanal. Chem. 378 (2004) 1484–1503, https://doi.org/10.1007/s00216-003-2422-z.
[74] R.L. Johnson, K. Schmidt-Rohr, Quantitative solid-state 13C NMR with signal enhancement by multiple cross polarization, J. Magn. Reson. 239 (2014) 44–49, https://doi.org/10.1016/j.jmr.2013.11.009.
[75] J. Mao, X. Kong, K. Schmidt-Rohr, J.J. Pignatello, E.M. Perdue, Advanced solid-state NMR characterization of marine dissolved organic matter isolated using the coupled reverse osmosis/electrodialysis method, Environ. Sci. Technol. 46 (2012) 5806–5814, https://doi.org/10.1021/es300521e.
[76] X. Cao, K. Schmidt-Rohr, Abundant nonprotonated aromatic and oxygen-bonded carbons make humic substances distinct from biopolymers, Environ. Sci. Technol. Lett. 5 (2018) 476–480, https://doi.org/10.1021/acs.estlett.8b00107.
[77] K. Mopper, A. Stubbins, J.D. Ritchie, H.M. Bialk, P.G. Hatcher, Advanced instrumental approaches for characterization of marine dissolved organic matter: extraction techniques, mass spectrometry, and nuclear magnetic resonance spectroscopy, Chem. Rev. 107 (2007) 419–442, https://doi.org/10.1021/cr050359b.
[78] J.D. Mao, L. Tremblay, J.P. Gagné, S. Kohl, J. Rice, K. Schmidt-Rohr, Humic acids from particulate organic matter in the Saguenay Fjord and the St. Lawrence Estuary investigated by advanced solid-state NMR, Geochim. Cosmochim. Acta 71 (2007) 5483–5499, https://doi.org/10.1016/j.gca.2007.09.022.
[79] S.T. Dyhrman, C.R. Benitez-Nelson, E.D. Orchard, S.T. Haley, P.J. Pellechia, A microbial source of phosphonates in oligotrophic marine systems, Nat. Geosci. 2 (2009) 696–699, https://doi.org/10.1038/ngeo639.
[80] P. Conte, D. Šmejkalová, A. Piccolo, R. Spaccini, Evaluation of the factors affecting direct polarization solid state 31P-NMR spectroscopy of bulk soils, Eur. J. Soil Sci. 59 (2008) 584–591, https://doi.org/10.1111/j.1365-2389.2008.01043.x.
[81] A.J. Simpson, W.L. Kingery, D.R. Shaw, M. Spraul, E. Humpfer, P. Dvortsak, The application of 1H HR-MAS NMR spectroscopy for the study of structures and associations of organic components at the solid—aqueous interface of a whole soil, Environ. Sci. Technol. 35 (2001) 3321–3325, https://doi.org/10.1021/es010607v.
[82] H. Farooq, D. Courtier-Murias, M.J. Simspon, W.E. Maas, M. Fey, B. Andrew, et al., Characterisation of oil contaminated soils by comprehensive multiphase NMR spectroscopy, Environ. Chem. 12 (2015) 227–235, https://doi.org/10.1071/EN14129.
[83] M. Tabatabaei Anaraki, R. Dutta Majumdar, N. Wagner, R. Soong, V. Kovacevic, E. J. Reiner, et al., Development and application of a low-volume flow system for solution-state in Vivo NMR, Anal. Chem. 90 (2018) 7912–7921, https://doi.org/10.1021/acs.analchem.8b00370.
[84] Y.L. Mobarhan, B. Fortier-McGill, R. Soong, W.E. Maas, M. Fey, M. Monette, et al., Comprehensive multiphase NMR applied to a living organism, Chem. Sci. 7 (2016) 4856–4866, https://doi.org/10.1039/c6sc00329j.
[85] A.J. Simpson, Y. Liaghati, B. Fortier-Mcgill, R. Soong, M. Akhter, Perspective: in vivo NMR-A potentially powerful tool for environmental research, Magn. Reson. Chem. 53 (2015) 686–690, https://doi.org/10.1002/mrc.4142.
[86] E.G. Nagato, B.P. Lankadurai, R. Soong, A.J. Simpson, M.J. Simpson, Development of an NMR microprobe procedure for high-throughput environmental metabolomics of Daphnia magna, Magn. Reson. Chem. 53 (2015) 745–753, https://doi.org/10.1002/mrc.4236.

[87] H.L. Wheeler, R. Soong, D. Courtier-Murias, A. Botana, B. Fortier-Mcgill, W.E. Maas, et al., Comprehensive multiphase NMR: a promising technology to study plants in their native state, Magn. Reson. Chem. 53 (2015) 735–744, https://doi.org/10.1002/mrc.4230.
[88] H.A.N. Abdulla, R.L. Sleighter, P.G. Hatcher, Two dimensional correlation analysis of fourier transform ion cyclotron resonance mass spectra of dissolved organic matter: a new graphical analysis of trends, Anal. Chem. 85 (2013) 3895–3902, https://doi.org/10.1021/ac303221j.
[89] T. Reemtsma, Determination of molecular formulas of natural organic matter molecules by (ultra-) high-resolution mass spectrometry. Status and needs, J. Chromatogr. A 1216 (2009) 3687–3701, https://doi.org/10.1016/j.chroma.2009.02.033.
[90] T. Riedel, T. Dittmar, A method detection limit for the analysis of natural organic matter via fourier transform ion cyclotron resonance mass spectrometry, Anal. Chem. 86 (2014) 8376–8382, https://doi.org/10.1021/ac501946m.
[91] T. Ohno, T.B. Parr, M.I. Gruselle, I.J. Fernandez, R.L. Sleighter, P.G. Hatcher, Molecular composition and biodegradability of soil organic matter: a case study comparing two new england forest types, Environ. Sci. Technol. 48 (2014) 7229–7236, https://doi.org/10.1021/es405570c.
[92] J. Zhong, R.L. Sleighter, E. Salmon, G.A. McKee, P.G. Hatcher, Combining advanced NMR techniques with ultrahigh resolution mass spectrometry: a new strategy for molecular scale characterization of macromolecular components of soil and sedimentary organic matter, Org. Geochem. 42 (2011) 903–916, https://doi.org/10.1016/j.orggeochem.2011.04.007.
[93] A.S. Wozniak, A.S. Willoughby, S.C. Gurganus, P.G. Hatcher, Distinguishing molecular characteristics of aerosol water soluble organic matter from the 2011 trans-North Atlantic US GEOTRACES cruise, Atmos. Chem. Phys. 14 (2014) 8419–8434, https://doi.org/10.5194/acp-14-8419-2014.
[94] B.A. Cottrell, M. Gonsior, L.M. Isabelle, W. Luo, V. Perraud, T.M. McIntire, et al., A regional study of the seasonal variation in the molecular composition of rainwater, Atmos. Environ. 77 (2013) 588–597, https://doi.org/10.1016/j.atmosenv.2013.05.027.
[95] S. Kim, D. Kim, S. Kim, S. Son, M.J. Jung, Application of online liquid chromatography 7 T FT-ICR mass spectrometer equipped with quadrupolar detection for analysis of natural organic matter, Anal. Chem. 91 (2019) 7690–7697, https://doi.org/10.1021/acs.analchem.9b00689.
[96] T. Reemtsma, The carbon versus mass diagram to visualize and exploit FTICR-MS data of natural organic matter, J. Mass Spectrom. 45 (2010) 382–390, https://doi.org/10.1002/jms.1722.
[97] R.L. Sleighter, P.G. Hatcher, The application of Electrospray ionization coupled to ultrahigh resolution mass spectroscopy for the molecular characterisation of NOM, J. Mass Spectrom. 43 (2008) 854–864, https://doi.org/10.1002/jms.
[98] N. Cortés-Francisco, M. Harir, M. Lucio, G. Ribera, X. Martínez-Lladó, M. Rovira, et al., High-field FT-ICR mass spectrometry and NMR spectroscopy to characterize DOM removal through a nanofiltration pilot plant, Water Res. 67 (2014) 154–165, https://doi.org/10.1016/j.watres.2014.08.046.
[99] I. Noda, Y. Ozaki, Two-Dimensional Correlation Spectroscopy: Applications in Vibrational and Optical Spectroscopy, John Wiley & Sons Ltd, 2004.
[100] I. Noda, Frontiers of two-dimensional correlation spectroscopy. Part 1. New concepts and noteworthy developments, J. Mol. Struct. 1069 (2014) 3–22, https://doi.org/10.1016/j.molstruc.2014.01.025.

[101] I. Noda, Frontiers of two-dimensional correlation spectroscopy. Part 2. Perturbation methods, fields of applications, and types of analytical probes, J. Mol. Struct. 1069 (2014) 23–49, https://doi.org/10.1016/j.molstruc.2014.01.016.

[102] I. Noda, Recent developments in two-dimensional (2D) correlation spectroscopy, Chin. Chem. Lett. 26 (2015) 167–172, https://doi.org/10.1016/j.cclet.2014.10.006.

[103] H.A.N. Abdulla, E.C. Minor, R.F. Dias, P.G. Hatcher, Transformations of the chemical compositions of high molecular weight DOM along a salinity transect: using two dimensional correlation spectroscopy and principal component analysis approaches, Geochim. Cosmochim. Acta 118 (2013) 231–246, https://doi.org/10.1016/j.gca.2013.03.036.

[104] F. Sun, Y. Li, X. Wang, Z. Chi, G. Yu, Using new hetero-spectral two-dimensional correlation analyses and synchrotron-radiation-based spectromicroscopy to characterize binding of Cu to soil dissolved organic matter, Environ. Pollut. 223 (2017) 457–465, https://doi.org/10.1016/j.envpol.2017.01.046.

[105] M. Mecozzi, E. Pietrantonio, M. Pietroletti, The roles of carbohydrates, proteins and lipids in the process of aggregation of natural marine organic matter investigated by means of 2D correlation spectroscopy applied to infrared spectra, Spectrochim. Acta A Mol. Biomol. Spectrosc. 71 (2009) 1877–1884, https://doi.org/10.1016/j.saa.2008.07.015.

[106] G.H. Yu, Z. Tang, Y.C. Xu, Q.R. Shen, Multiple fluorescence labeling and two dimensional FTIR-13C NMR heterospectral correlation spectroscopy to characterize extracellular polymeric substances in biofilms produced during composting, Environ. Sci. Technol. 45 (2011) 9224–9231, https://doi.org/10.1021/es201483f.

[107] N. Mladenov, R. Sommaruga, R. Morales-Baquero, I. Laurion, L. Camarero, M. C. Diéguez, et al., Dust inputs and bacteria influence dissolved organic matter in clear alpine lakes, Nat. Commun. 2 (2011) 405, https://doi.org/10.1038/ncomms1411.

[108] A. Nebbioso, A. Piccolo, Molecular characterization of dissolved organic matter (DOM): a critical review, Anal. Bioanal. Chem. 405 (2013) 109–124, https://doi.org/10.1007/s00216-012-6363-2.

[109] S. Singh, E.J. D'Sa, E.M. Swenson, Chromophoric dissolved organic matter (CDOM) variability in Barataria Basin using excitation-emission matrix (EEM) fluorescence and parallel factor analysis (PARAFAC), Sci. Total Environ. 408 (2010) 3211–3222, https://doi.org/10.1016/j.scitotenv.2010.03.044.

[110] J. Wu, M. Wu, C. Li, G. Yu, Long-term fertilization modifies the structures of soil fulvic acids and their binding capability with Al, PLoS ONE 9 (2014) e105567, https://doi.org/10.1371/journal.pone.0105567.

[111] P. Fu, K. Kawamura, J. Chen, M. Qin, L. Ren, Y. Sun, et al., Fluorescent water-soluble organic aerosols in the High Arctic atmosphere, Sci. Rep. 5 (2015) 1–8, https://doi.org/10.1038/srep09845.

[112] M. Tedetti, P. Cuet, C. Guigue, M. Goutx, Characterization of dissolved organic matter in a coral reef ecosystem subjected to anthropogenic pressures (La Réunion Island, Indian Ocean) using multi-dimensional fluorescence spectroscopy, Sci. Total Environ. 409 (2011) 2198–2210, https://doi.org/10.1016/j.scitotenv.2011.01.058.

[113] P.G. Coble, Marine optical biogeochemistry: the chemistry of ocean color, Chem. Rev. 107 (2007) 402–418, https://doi.org/10.1021/cr050350+.

[114] K.R. Murphy, C.A. Stedmon, D. Graeber, R. Bro, Fluorescence spectroscopy and multi-way techniques. PARAFAC, Anal. Methods 5 (2013) 6557–6566, https://doi.org/10.1039/c3ay41160e.

[115] M. Edwards, A. Mostafa, T. Górecki, Modulation in comprehensive two-dimensional gas chromatography: 20 years of innovation, Anal. Bioanal. Chem. 401 (2011) 2335–2349, https://doi.org/10.1007/s00216-011-5100-6.

[116] P. Schoenmakers, P. Marriott, J. Beens, Nomenclature and conventions in comprehensive multidimensional chromatography, LC GC Europe 16 (2003) 335–339.

[117] J. Leonhardt, T. Teutenberg, J. Tuerk, M.P. Schlüsener, T.A. Ternes, T.C. Schmidt, A comparison of one-dimensional and microscale two-dimensional liquid chromatographic approaches coupled to high resolution mass spectrometry for the analysis of complex samples, Anal. Methods 7 (2015) 7697–7706, https://doi.org/10.1039/c5ay01143d.

[118] J. Haun, J. Leonhardt, C. Portner, T. Hetzel, J. Tuerk, T. Teutenberg, et al., Online and splitless Nano LC × Capillary LC with quadrupole/time-of- flight mass spectrometric detection for comprehensive screening analysis of complex samples, Anal. Chem. 85 (2013) 10083–10090, https://doi.org/10.1021/ac402002m.

[119] X. Ouyang, P.E.G. Leonards, Z. Tousova, J. Slobodnik, J. De Boer, M.H. Lamoree, Rapid screening of acetylcholinesterase inhibitors by effect-directed analysis using LC × LC fractionation, a high throughput in vitro assay, and parallel identification by time of flight mass spectrometry, Anal. Chem. 88 (2016) 2353–2360, https://doi.org/10.1021/acs.analchem.5b04311.

[120] J. Wan, T. Tyliszczak, T.K. Tokunaga, Organic carbon distribution, speciation, and elemental correlations within soil microaggregates: applications of STXM and NEXAFS spectroscopy, Geochim. Cosmochim. Acta 71 (2007) 5439–5449, https://doi.org/10.1016/j.gca.2007.07.030.

[121] J. Lehmann, D. Solomon, J. Brandes, H. Fleckenstein, C. Jacobson, J. Thieme, Synchrotron-based near-edge X-ray spectroscopy of natural organic matter in soils and sediments. in: Biophysico-Chemical Processes Involving Natural Nonliving Organic Matter in Environmental Systems, 2009, pp. 729–781, https://doi.org/10.1002/9780470494950.ch17.

[122] G.M. Hettiarachchi, E. Donner, E. Doelsch, Application of synchrotron radiation-based methods for environmental biogeochemistry: introduction to the special section, J. Environ. Qual. 46 (2017) 1139–1145, https://doi.org/10.2134/jeq2017.09.0349.

[123] B.E. Etschmann, E. Donner, J. Brugger, D.L. Howard, M.D. De Jonge, D. Paterson, et al., Speciation mapping of environmental samples using XANES imaging, Environ. Chem. 11 (2014) 341–350, https://doi.org/10.1071/EN13189.

Environmental solution-state NMR spectroscopy: Recent advances, potential, and impacts

2

Marie-Cecile Chalbot[a], *Ilias Kavouras*[b]
[a]New York City College of Technology, Biological Sciences Department, Brooklyn, NY, United States, [b]CUNY Graduate School of Public Health & Health Policy, Department of Environmental, Occupational and Geospatial Health Sciences, New York, NY, United States

Introduction

Nuclear magnetic resonance (NMR) spectroscopy identifies nuclei with magnetic moments in a molecule including naturally abundant hydrogen (^{1}H), phosphorous (^{31}P), fluorine (^{19}F), and low abundance nuclei (carbon (^{13}C), nitrogen (^{15}N), oxygen (^{17}O), and sulfur (^{33}S) because of nuclei's ability to transition between two spin states (for example, $-1/2$ and $+1/2$ for ^{1}H, ^{13}C, ^{15}N, and ^{31}P) when placed in an external magnetic field (B_0). The energy needed for the transition between the two spin states of nuclei is provided by a radio frequency (rf). Upon relaxation, the energy is released at the same wavelength (i.e., resonance frequency) and is captured to generate the NMR spectra following appropriate signal processing. The resonance frequencies of nuclei in a molecule depend on location and the electronic environment around them. The chemical shift (i.e., the intensity of resonance frequencies of target nuclei in the molecule relative to the internal standard, reported in ppm (parts per million) and spin-spin couplings (i.e., the magnetic interactions between neighboring, non-equivalent target nuclei) patterns allow for the qualitative identification of the molecular structure and stereochemistry of the molecule as well as quantitative studies.

In contrast to the traditional use of NMR in chemical sciences aiming to determine the structure or properties of a single molecule, environmental solution-state NMR spectroscopy aims to determine the molecular content and dynamics of complex mixtures of organic and diamagnetic inorganic matter chemical species in environmental matrices [1]. In soil and aqueous matrices, it is referred to as natural organic matter (NOM) and it is composed of humic-like macromolecules. Organic carbon in the atmospheric aerosol is composed of a mixture of smaller molecules and macromolecules from the natural and man-made origin. The intrinsic property of NMR to uniquely identify all relevant nuclei in a molecule is also a crucial limitation in environmental solution-state NMR spectroscopy. Because of the concurrent presence of many organic species with different and multiple functional groups at variable concentrations (ranging from ppt to ppm), resultant NMR spectra may adequately

Multidimensional Analytical Techniques in Environmental Research. https://doi.org/10.1016/B978-0-12-818896-5.00002-8

resolve some resonances while others may be convoluted with resonances of other compounds. Moreover, in the case of ^{1}H NMR, resonances of exchangeable proton may not be detectable (e.g., in carboxylic and hydroxyl groups). Advanced methods including multidimensional homo- and heteronuclear NMR protocols were developed to address these limitations. Homonuclear $^{1}H—^{1}H$ or heteronuclear $^{1}H—^{13}C$ 2D-NMR spectroscopy, as well as 3D-NMR spectroscopy, enhanced greatly the recognition of molecular signatures in complex environmental media. There is a wealth of information on environmental solution-state ^{1}H NMR due to its natural abundance. The relatively low ^{13}C natural abundance (i.e., 1.1%) restrains the sensitivity and reliability of ^{13}C NMR spectroscopy; however, heteronuclear $^{1}H—^{13}C$ correlation spectroscopy is more sensitive and delineates information on carbon content without directly examining ^{13}C signals.

The analysis of environmental matrices by NMR spectrometry elucidated key features in understanding the mechanisms, interactions, multiphase processes, and the soil-water, water-air, and air-soil partitioning of organic compounds. These processes typically occur in the interface and are critical for the survival of living organisms and the fate and transformation of NOM [2]. Soil has the highest carbon content and is moist by water infiltration that continuously influences the dynamics across the soil-water interface and facilitates structural changes in the dissolved organic matter [3]. The spectral profiles of sediment and core samples map the evolution and changes in soil organic carbon content over long time periods [4, 5]. For example, spectral signatures typical of continental soil organic matter were observed in ocean sediments indicating the resuspension of soil particles by wind [6]. Better knowledge of organic composition in environmental samples would allow us to understand the global carbon cycle and variability of organic composition among environmental matrices [7]. Deforestation has decreased organic carbon in soils, consistent with the NMR spectral profiles of natural and agricultural soils [7a, 8]. In this chapter, the application of solution-state NMR spectroscopy in various environmental matrices to better understand the compositional features and properties of organic matter is presented. Furthermore, recent advances in multidimensional environmental NMR such as 3D-NMR, multiphase CMP-NMR to analyze liquid-gel-solid with the same probe as well as in vivo and in situ NMR applications will be discussed [9, 10].

NMR methodology

Operating principle

NMR spectroscopy utilizes the magnetic properties of nuclei including their charge and spinning that generate a magnetic dipole across its axis. The precessional frequency, ω_L (i.e., the number of times per second that the nuclei precesses in a complete circle) is expressed by the spin and its values are 0, ½, 2/2, 3/2, 4/2, 5/2, and so on depending on the nuclei. Nuclei with even atomic number and mass (e.g., $^{12}_{6}C$, $^{16}_{8}O$) do not spin and therefore are inactive in NMR spectroscopy. Nuclei with odd atomic mass and odd/even atomic number (e.g., $^{1}_{1}H$, $^{13}_{6}C$) do spin with values $n/2$, where $n = 1, 3, 5, 7$ and 9. Nuclei with even atomic mass and odd atomic number (e.g., ${}^{2}_{1}H$, ${}^{14}_{7}N$) have integer spin values. The two latter categories are active in NMR spectroscopy with ${}^{1}H$ and ${}^{13}C$ NMR spectroscopy accounting for the majority of NMR investigations.

NMR-active nuclei align parallel with the direction of the magnetic field which represents the low-energy ground spin state (for ${}^{1}H$, it has a spin of +1/2) (Fig. 2.1A).

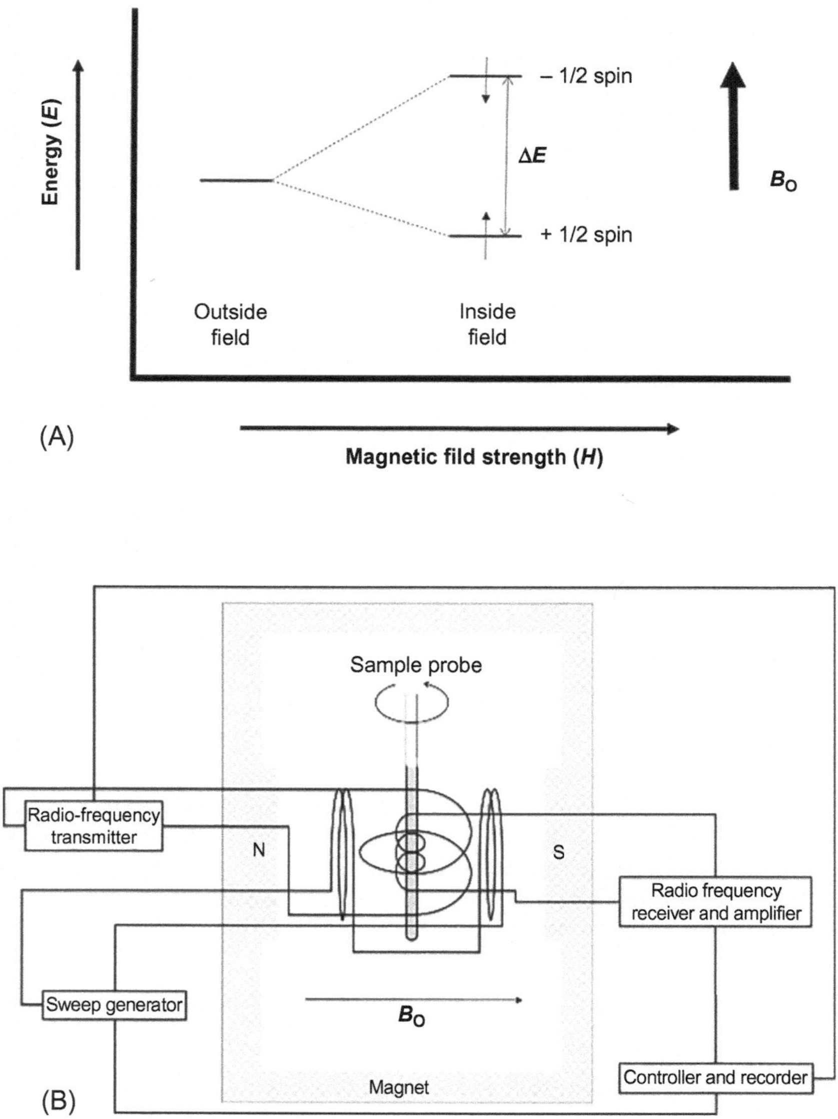

Fig. 2.1 (A) The ground and excited spin states of ${}^{1}H$ outside and inside a magnetic field; (B) schematic representation of NMR spectrometer.

The antiparallel orientation of the spin represents the excited spin state (for ^{1}H, it has a spin of $-1/2$). The difference in the energy between the ground and excited states (ΔE) is related to the precessional frequency, in the order of 10^{-2} cal/mol. NMR spectra are generated by the transition of nuclei between the two spin states by a sequence of radio-frequency pulses (i.e., pulse sequence) in a magnetic field that can cause resonance. To achieve this, there is a need for an excess of nuclei in the ground state, otherwise, the absorption of energy from nuclei in the ground state would be counterbalanced by the energy release of excited nuclei, resulting in no observable signal. The excess of ground state nuclei is typically achieved by nuclei relaxation mechanisms, which are described by the relaxation time (T in milliseconds) (i.e., the halftime needed to achieve a balance between ground and excited spin states). The spin-lattice relaxation (T_1) proceeds through the energy transfer of excited nuclei to increased vibration and rotation within the lattice. The spin-spin relaxation (T_2) the energy from the excited nuclei is transferred to nearby nuclei in the ground spin state. Sharp resonances are typically related to high T_2 values, while broader ones are indicative of low T_2 values.

NMR spectrometer

Fig. 2.1B shows a schematic drawing of an NMR spectrometer. It is composed of five essential components:

(a) a magnet for the generation of a powerful, homogenous, and uniform magnetic field across the sample with magnetic field strength being directly proportional to the resonance signal strength;

(b) sweep generator that is used to fine-tune the strength of the magnetic field by electromagnetic coils with sweeping velocity also directly affecting resonance signal strength (no resonance signals for slow sweep velocity) and wiggles (i.e., exponential decreasing signal waves) after the signal for high sweep velocities;

(c) sample tube, a magnetically isotropic borosilicate glass tube, during the NMR experiment, spins at a rate to 20–40 Hz to achieve a homogeneous distribution of the magnetic field across the entire sample; the sample tube is inserted into the sample probe that integrates both the radio-frequency generator and receiver coils;

(d) radio-frequency pulse generator to create the perpendicular to B_o magnetic field and provide the required energy to the sample;

(e) radio-frequency receiver that is connected to a coil surrounding the sample to detect the signal (i.e., free induction decay due to signal decay) and amplifier to strengthen and process the resonance signals through Fourier transformation. Repetitive signals obtained by pulsed radio frequencies can be summed to substantially improve the signal-to-noise (S/N) ratio, the resolution, and sensitivity of the resultant NMR spectra.

NMR methods

Due to phase differences and relatively low concentrations of organic matter in environmental media (i.e., water, soil, and air/particles), environmental samples need to undergo concentration and/or purification prior to NMR spectroscopy. For liquid/aquatic samples, lyophilization is typically used to remove water. Solid-phase extraction (SPE) also results in the concentration of organic analytes in water-based environmental samples and purification based on the extraction protocol. For solid samples (i.e., soils, atmospheric particles), the extraction of organic carbon is initially needed to be followed by concentration depending on the extraction solvent. Additional steps such as flash chromatography and SPE may be done for target compounds. In all cases, the dried extract is redissolved in mixed solvents containing at least one deuterated solvent to minimize/eliminate the interference of protons associated with the solvent and "lock" the NMR signal. Typical NMR solvents include H_2O/D_2O 90/10 (v/v) (water-deuterated water), D_2O/NaOD (deuterated water and sodium hydroxide) mixtures, DMSO-d_6 (deuterated dimethyl sulfoxide), $CDCl_3$ (deuterated chloroform), CD_2Cl_2-d_2 (deuterated methylene chloride), and CD_3OD-d_4 (deuterated methanol).

Water-based solvents are predominantly used in solution-state environmental NMR due to the environmental, biological, and toxicological relevance of water-soluble organic carbon (i.e., WSOC). Water is also ubiquitous and trace quantities (albeit at higher absolute levels than an organic proton) may be available in NMR samples. As a result, the resonances of water protons (δ_H 4.5–4.9 ppm) are very strong and interfere with the detection of resonances appearing in the same range. Although the overlap of water resonances with other resonances cannot be reconstructed, water suppression pulse sequences aim to reduce the intensity of water resonances relative to the NMR spectra. Newer water suppression pulse sequences also enable for the use of minimum D_2O amount for locking and shimming of the instrument.

Table 2.1 shows the types and pulse sequences commonly used in environmental solution-state NMR spectroscopy. A thorough review of solution-state NMR methods and optimization of important variables for the analysis of environmental samples have been published elsewhere [1].

1D ^1H and ^{13}C NMR

1D ^1H and ^{13}C NMR spectroscopy have been routinely used for the characterization of environmental samples. ^1H NMR aims to identify the major types of nonexchangeable protons based on the chemical shift of proton resonances relative to the internal standard. Terminal methyl (**CH_3**), methylene (**CH_2**), and methyne (C_H) protons are observed between δ_H 0.8–1.0 ppm, δ_H 1.2–1.4 ppm, and δ_H 1.4–1.8 ppm, respectively. Allylic protons (CH_2=CH—**CH_3**) (protons next to a double bond carbon) are detected in δ_H 1.8–3.0 ppm. This region may also include terminal vinylic protons (RCH=**CH_2**) and benzylic protons (phenyl-**CH_3**). Proton in α-position to a hydroxyl (R**CH_2**—OH) or methoxy (R**CH_2**—OCH_3) is detected in the δ_H 3.3–4.0 ppm range. Proton attached to hydroxyl (R—O**H**; δ_H 0.5–5.0 ppm), phenol (Ar—O**H**; δ_H 4.0–7.0 ppm), carboxylic acids (RC(=O)O**H**, δ_H 10–13 ppm), and nitrogen in amines (R—N**H_2**; δ_H 0.5–5.0 ppm) may only be observed in samples within DMSO and other organic solvents, due to rapid exchange of H/D in H_2O/D_2O solvents. Vinylic (RC**H**=C**H**R) and acetalic (O—C**H**—O) protons are identified in the δ_H 4.5–5.5 ppm range. Protons on phenyl rings (Ar—**H**) are typically observed in the δ_H 6.5–8.5 ppm. Amides protons (R(C=O)N**H**R) also resonate in the same range. Carbonyl protons (RC**H**O) may be observed at δ_H >9.0 ppm. It is noteworthy that this is an indicative

Table 2.1 Typical 1D and 2D NMR experiments and pulse sequences in solution-state environmental NMR spectroscopy.

Nuclei	Type	Sequence	Comments
1D ^{1}H and ^{13}C NMR experiments			
^{1}H-	Without water suppression	zg30	Spectrum acquired with a 30° pulse (instead of 90°)
^{1}H-	With water suppression	purge	Presaturation utilizing relaxation gradients and echoes (PURGE) [11]
^{1}H-	With water suppression	zggpwg	Watergate uses a single pulsed field gradient (PFG) spin echo by a applying two G1 gradients in the same sense
^{1}H-	With water suppression	zgesgp (ES element)	Gradient-based 1D excitation sculpting using 180° water selective pulses. Two PFG spin-echo are applied (G1 and G2)
^{1}H-	1D NOESY	noesypr1d	Water suppression achieved by presaturation, Rd. 5 s [12]
DE-^{1}H	Diffusion edited ^{1}H NMR	ledbpgp2s1d	Longitudinal encode decode or longitudinal eddy-current delay (LED) experiment using bipolar gradient. Small molecules signals that diffuse rapidly are attenuated [13]
^{13}C-	^{1}H-decoupled ^{13}C with NOE effect	zgdc30	Composite pulse decoupling (CPD). Improved sensitivity using NOE effect
^{13}C-	^{1}H-decoupled ^{13}C without NOE effect	zgig30	Inverse gated decoupling when quantification is required
^{13}C-	^{1}H-decoupled ^{13}C		Inverse gated WALTZ-16 decoupling [14]
2D ^{1}H—^{1}H NMR experiments			
^{1}H—^{1}H	Phase-sensitive 2D DQF-COSY using ES element	cosydfesgpph	Identifies J scalar couplings but also suppresses signals in the diagonal [15]
^{1}H—^{1}H	Magnitude-mode gradient-enhanced ge-2D COSY	cosygpppqf	This sequence using purge pulses before d_1. This method is used for concentrated samples, in D_2O to minimize water signal [16, 17]
^{1}H—^{1}H long-range	Phase-sensitive 2D TOCSY	dipsi2esgpph	with ES element using DIPSI-2

Table 2.1 Continued

Nuclei	Type	Sequence	Comments
$^1H—^1H$ long-range	Clean adiabatic TOCSY		with WURST-2 pulses; $d_1 = 0.5$ s (lowest $5 \times T_1$), mixing time from 50 to 100 ms [10, 18]
$^1H—^1H$ through space	Phase-sensitive 2D NOESY	noesyesgpph	with ES element; mixing time 300 ms [16]
$^1H—^1H$	Diffusion Ordered Spectroscopy DOSY	ledbpgp2s or stebpgp1s19	Bipolar pulse longitudinal eddy current delay [19] With or without WATERGATE
2D $^1H—^{13}C$ NMR experiments			
$^1H—^{13}C$	Phase sensitive ge-HSQC	hsqcetgp	Use echo-antiecho TPPI gradient selection. To remove the H_2O peaks in HSQC, the spectrum has to be centered at 4.7 ppm. GARP element is used during acquisition
$^1H—^{13}C$	Phase sensitive ge-HSQC sensitivity improved	hsqcetgpsisp2.2	Use PEP and adiabatic pulses for inversion and refocusing with gradient in back INEPT-^{13}C decoupling GARP sequence. WURST 180° ^{13}C inversion pulse [14, 20]
$^1H—^{13}C$	Phase sensitive ge-HSQC sensitivity improved	hsqcetgpsp	Use echo-antiecho and adiabatic pulses for inversion. $d_1 = 1$ s, anaerobic digester, adiabatic pulses have a wider band and improves sensitivity [21]
$^1H—^{13}C$	Gradient-based magnitude mode HMBC	hmbcgplpndqf	HMBC with low-pass J-filter for better water suppression
$^1H—^{13}C$	ge-2D 2J,3J HMBC, STAR-HMBC	hmbcacbigpl2nd	Differentiates two-bond from three-bonds long range correlations, using a STAR operator
3D heteronuclear $^1H—^1H—^{13}C$ NMR experiments			
$^1H—^1H—^{13}C$	HMQC-TOCSY	hmqcbimlqf	Spectra acquired using a BIRD pulse train during the relaxation delay [22] which selectively inverts protons bond to ^{12}C
$^1H—^1H—^{13}C$	HSQC-TOCSY	hsqcdietgpsisp2 with dipsi-2	PEP and adiabatic inversion and refocusing pulses are used. A DIPSI mixing time of 70 ms was applied [14]

Sequence name relies on Bruker monograms [23].

list of the most abundant proton types detected in environmental samples as compared to comprehensive resonances for different types of protons associated with synthetic chemistry.

The low natural abundance of ^{13}C implies long NMR experiments since many of the carbon signals are split into multiplets. However, signal resolution and sensitivity may increase by applying broadband ^{1}H-decoupling that results in the increased population of ^{13}C from close in space saturated ^{1}H nuclei because of the nuclear Overhauser effect (NOE). The *cpd* (i.e., carbon-proton decoupling) scheme is applied during decoupling time d_1 and acquisition in order to achieve $^{1}H—^{13}C$ transfer through the NOE effect during acquisition. The existing programs are designed to collect the information needed as well as removal (or cancellation) of unwanted data. The chemical shifts of ^{13}C range from 0 to 250 ppm. Unsubstituted saturated aliphatic carbon ($R\mathbf{C}H_3$, $\mathbf{C}H_2$, and $\mathbf{C}H$) is observed in δ_{13C} 10–45 ppm. Aliphatic carbon bonded to a nitrogen or oxygen atom with a single bond (e.g., N-alkyl: $R_2N\mathbf{C}H_3$ or O-alkyl: $\mathbf{C}—O$) are observed in δ_{13C} 45–60 ppm and δ_{13C} 60–95 ppm, respectively. Carbon linked to two oxygen atoms with single bonds (e.g., O—**C**—O) resonates at δ_{13C} 95–110 ppm. Carbon in sp^2-hydridized form (**C**=**C** including aromatic, C=**C**—(C,H) and C=**C**—(O,N)) is detected in δ_{13C} 110–160 ppm. Finally, the carbon in esters, carboxyl, amide (O=**C**—(OH/NH_2)) and carbonyl (O=**C** and O=**C**—H) functionalities resonates at lower fields with δ_{13C} 160–190 ppm and δ_{13C} 190–220 ppm, respectively.

High-resolution-magic angle spinning (HR-MAS) spectroscopy is applied to materials with anisotropic or reduced (i.e., viscous) mobility and solubility yielding high-resolution NMR spectra. It typically requires a mixture of solvents such as D_2O or DMSO-d_6. NMR analysis is achieved by spinning the sample at an angle equal to 54.7° (termed as the magic angle) along the field direction, to narrow the line broadening in NMR spectra. This allows for the analysis of matrices that are not fully soluble, and complement solid-state NMR, including lipids, tissue samples, resins, plant, and food extracts [24].

2D $^{1}H—^{1}H$ and $^{1}H—^{13}C$ NMR

One-dimensional ^{1}H and ^{13}C NMR frequently are sufficient to determine the structure of individual chemical species, but more complex chemicals or mixtures of chemicals with convoluted resonances complicate NMR analysis relying simply on the detection of specific proton (or carbon) types. To address this limitation of 1D NMR spectroscopy, 2D NMR techniques have been developed. In principle, this is achieved by varying the length of time and the system is allowed to evolve following the first pulse. The generated signal from this process is recorded and processed allowing for a second dimension to the NMR spectrum, providing data on the coupling of different types of nuclei in the molecule. In environmental applications, 2D NMR experiments include homonuclear ($^{1}H—^{1}H$) and heteronuclear ($^{1}H—^{13}C$) configurations [1, 8, 25–27].

Homonuclear $^{1}H—^{1}H$ correlation spectroscopy (COSY) aims to delineate the interactions between protons attached to adjacent carbons (**HC—CH**) through scalar couplings (via chemical bonds), while homonuclear $^{1}H—^{1}H$ total correlation spectroscopy (TOCSY) reveals information on the coupling between protons attached to distant carbon atoms (i.e., two to three C—C bonds). TOCSY experiments add a composite pulse sequence or isotropic mixing sequence to the COSY experiment to induce a magnetization transfer between coupled spins distant by several bonds. COSY experiments can be carried out in magnitude mode where the phase cannot be modified, or in the phase-sensitive mode where the phase needs to be corrected. COSY double-quantum filtered (COSY-DQF) experiments are frequently used for complex mixtures to reduce the noise around the diagonal line [12]. TOCSY mixing pulse sequences used to detect long-range proton-proton correlation include WURST-2, MLEV, or DIPSI, where the mixing time must be adjusted depending on sample characteristics.

The nuclear Overhauser effect spectroscopy (NOESY) is used to detect $^{1}H—^{1}H$ couplings through space for adjacent and distant protons. It is generated from direct dipolar couplings as opposed to scalar couplings in other homo and heteronuclear 2D spectra. As a result, the NOESY spectra show cross peaks that would not be detected with other 2D experiments. Mixing time setting is important for NOESY experiments to be most effective. Mixing time differ for environmental samples that contain small organic molecules, from environmental samples containing peptides, proteins, and other large molecules. For this reason, it is sometimes useful to calculate the recycle delay before the run, as environmental samples differ from purified organic compounds. Recycle delay (Rd) can be calculated for environmental samples with larger molecules and is usually set between 0.5 and 2 s. Rd. values range from 1 s in ^{1}H to 1.5 s in 2D experiments for atmospheric samples, from as low as 0.5 s for Suwannee river samples up to 5 s for dissolved organic matter [1, 16, 28]. Further processing of spectra included apodization with sine-squared bell function: ($\pi/2$) for HSQC and ($\pi/9$) for COSY.

The diffusion-ordered spectroscopy (DOSY) is a 2D experiment that shows in F1 dimension the 1D proton and the diffusion coefficient of the compound (as calculated by the Stokes-Einstein equation) in the F2 dimension. This method is employed in environmental samples to investigate the molecular size, binding, and aggregation in humic-like macromolecules. In diffusion experiments, the pulse program contains a spin-echo diffusion sequence and molecules are marked depending on their position in the NMR tube along the axis of the pulsed-field gradient applied. The marking occurs during the diffusion delay Δ and during this time if molecules move, their position is decoded by a second gradient. Then the dephasing and rephasing of the spins introduced during the sequence are not matching and this leads to a decrease of the signal amplitude. Rapidly diffused molecules are typically associated with faster signal decay [29]. The homonuclear J-resolved (J_{res}) experiment shows the proton-proton coupling in the F1 dimension and provides information on multiplicity and frequency of J coupling in Hz, while the projection spectrum in the F2 dimension is interpreted as the decoupled spectrum. This method indicates multiplicity of one single signal from/and intricate of several signals [30].

Heteronuclear correlation spectroscopy, usually $^{1}H—^{13}C$, are inverse chemical shift correlation experiments based on the detection of the nuclei of the highest gyromagnetic ratio, i.e., ^{1}H, to yield high-sensitivity spectra. Heteronuclear single-quantum correlation spectroscopy ($^{1}H—^{13}C$ HSQC experiments) detects the correlation between directly bonded proton and carbon in a molecule. The pulse sequence requires an insensitive nuclei-enhanced pulse train (INEPT) block that produce polarization transfer from ^{1}H that is highly sensitive to ^{13}C, increasing the sensitivity of the carbon, followed by a ^{1}H-decoupled ^{13}C evolution $t1$ period, a retro-INEPT block to transfer magnetization back to in-phase ^{1}H magnetization, and proton acquisition with ^{13}C decoupling. ^{13}C decoupling during acquisition is done with a globally optimized alternative-phase rectangular-pulses (GARP) composite pulse decoupling sequence. The $^{1}J_{C—H}$ coupling constant of 145 Hz is generally set to optimize the evolution delay d_4 to $1/(4 \times {}^{1}J_{C—H})$ of the INEPT block. This parameter allows us to observe an appropriate cross peak signal for compounds containing alkanes and alkenes where the coupling constant varies from 125 to 170 Hz for aliphatic (C—H) and alkene (C=C—H), respectively. Alkynes (≡C – H) have a coupling constant ranging from 240 to 250 Hz resulting in a lower cross peak intensity. A sequence without water suppression can be used; the residual water signal would be minimized by setting the frequency offset O1P at 4.7 ppm. Projections of the correlations in the F1 axis allow us to extract a ^{13}C spectrum; however, it does not contain information on the quaternary carbon chemical shift. Gradient-enhanced-HSQC (ge-HSQC) experiment incorporates a dephasing gradient resulting in the record of a magnitude-mode spectrum. The echo-antiecho (or P-/N-type selection) version of ge-HSQC experiment such as hsqcetgp, results in the acquisition of a phase-sensitive spectra but with a loss of sensitivity. The decreased sensitivity may be addressed by adding adiabatic pulses with a wider band (e.g., hsqcetgpsp pulse) and second retro-INEPT block in the sensitivity improved HSQC (e.g., hsqcetgpsisp2.2 pulse). The heteronuclear multiple bond correlation (HMBC) detects multiple bond coupling over two or three bonds and the experiment is designed to suppress direct C—H correlations by means of a low-pass J filter, as well as proton signals not correlated to carbon (such as water protons). This experiment also provides information about the chemical shift of carbonyl groups including carboxylic acids, ester, and aldehydes, and quaternary carbons on aromatic rings. The delay has to be set based on the long-range heteronuclear coupling constants (${}^{n}J_{C—H}$) with values ranging from 5 to 20 Hz. The HMBC experiments are recorded in the magnitude mode. The relative intensities of cross peaks from long-range C—H interactions in the same spin system (such as in an aliphatic chain, or in aromatic compounds) depend directly on the magnitude of the coupling constant. The HMBC experiment uses an arbitral long-range delay to detect the long-range C—H interactions. Recently designed accordion-optimized experiment incorporating a STAR operator to generate a variable delay was capable of differentiating between $^{2}J_{C—H}$ and $^{3}J_{C—H}$ correlations [31].

3D $^{1}H—^{1}H—^{13}C$ NMR

Three-dimensional (3D) NMR experiment can be generated by NMR experiments consisting of two sequential 2D experiments or through the implementation of triple resonance experiments. Typical examples of sequential 2D experiments include

NOESY (or TOCSY)-HSQC and where the NOESY (or TOCSY) experiment is extended by an HSQC pulse. Triple resonance experiments aim to resolve the correlations between three different nuclei (^{1}H, ^{13}C, ^{15}N). They are typically used to determine protein structure because of the reduced overlap of resonances in space and high sensitivity. 3D NMR requires a high concentration of organic compounds, which is rarely the case in environmental media. It has been occasionally used to characterize the plant-derived structures in soil organic matter [22].

Soil organic matter analysis

Characteristics of soil organic matter and humic substances

Soil organic matter (SOM) is the predominant component of soil, composed of a complex mixture of compounds produced by plant and living microorganisms through their metabolism and activity, as well as organic compounds formed by humification, a term that refers to the decomposition of plant and dead animal remains [32]. The magnitude of organic compound humification in soils influences soil's properties including its hydrophilicity. Recalcitrant organic matter formed in soils during slow organic matter decomposition is hydrophobic, whereas labile soils are hydrophilic [21, 33]. Recalcitrant organic matter includes proteins, lipids, lignin residues, cellulose in clay, silk, and sand aggregates. The abundance of chemical species for different types of soils (e.g., agricultural soils, forest soils, grasslands, peats, road dust) and depth has been studied by NMR spectroscopy [8, 20, 25, 34]. Solution-state NMR profiles were obtained for litter and decomposing soils that contain a greater amount of water soluble, giving valuable information of the fate of organic matter in soils and reconstructing the fresh organic matter composition [20]. In the case of agricultural soil, chemical, and biological composition strongly depends on its treatment, such as organic farming and the occurrence of spills [35, 36].

SOM samples preparation

Target macromolecules isolated from SOM and studied by solution-state NMR are the two main components of humic substances: the humic acids (HA) that are the soil fraction extracted in alkaline aqueous solution, and the fulvic acids that extracted in both alkaline and acidic aqueous solution. SOM samples are extracted with diluted (0.1 M) NaOH aqueous solution to increase the solubility of soil macromolecules. The surfactant is collected after centrifugation to remove any insoluble material and passed through a 0.2-µm hydrophilic filter to avoid growth, and the extract undergoes cation exchange to remove Na^{+} ions and is freeze dried [37]. Humic acids may also be isolated by precipitation upon the addition of an acidic solution such as the final pH of the solution is lower than 2. Fulvic acids are extracted from soil by exhaustive extraction with NaOH at pH 12 and separation from humic acids by acidic precipitation at pH 2. Further SPE with XAD-8 and XAD-4 resins in tandem complete the fulvic acid isolation procedure [22, 25].

The insoluble fraction that remains after the extraction process is the humin (HU) fraction [38]. Part of HU that is soluble in DMSO-H_2SO_4 (94:6 v/v) has been characterized by solution-state NMR and HR-MAS in DMSO-d_6 [10, 39]. DMSO-d_6 dissolves rapidly large molecules and limits the proton exchanges between solvent and organic compounds, making amide, hydroxyl, and carboxylic acid functional groups visible in the NMR spectrum. The disadvantage of using DMSO-d_6 is its nonvolatility, such as it is difficult to get rid of by solvent exchange, and its hygroscopic character. The sample and solvent must be thoroughly dried before NMR analysis because the water residual peak that resonates at 2.5 ppm in DMSO-d_6 might interfere with important peaks.

NMR characterization

The NMR characterization of SOM has revealed strong signatures attributed to humic substances that are composed of a heterogeneous mixture of relatively high-molecular-weight molecules and also the presence of nonhumic substances corresponding to low-molecular-weight molecules such as amino acids, carbohydrates, organic acids, and lipids components (fatty acids, resin, and waxes) of plant origin [32, 39]. Humic substances in soil extracts affect the relaxation times T_1 and T_2 of the organic constituents of the sample resulting in broad signal envelopes. This is due to the aggregation of the compounds in the soil, their molecular conformation, their charge, and the intermolecular interactions in the mixture [25].

Fig. 2.2 shows the characteristic 2D $^1H—^{13}C$ NMR spectra of SOM acquired in DMSO-d_6. 2D NMR spectra from soils samples show well distinct C—H correlations from functional group that can be categorized in regions as follows: alkyl (δ_H 0.5–3.0/δ_C 5–45 ppm), nonanomeric carbohydrate (δ_H 3.0–6.0/δ_C 50–90 ppm), anomeric carbohydrate (δ_H 3.8–6.0/δ_C 90–110 ppm), and aromatics (δ_H 6.0–8.6/δ_C 100–150 ppm) in soils rich in lipids, carbohydrate, and lignin compounds. Alkyl functional groups include terminal methyl (CH_3) in aliphatic or branched aliphatic of lipids and amino acids (or peptides), CH_3 in acetyl groups, methylene groups (CH_2) in α, β, and γ position of carboxylic acids and ester R—O—(C=O)—CH_2—CH_2—CH_2—R or in α-position of double bond in alkene or aromatic compound. Peaks from the α-carbon of amino acids and methoxy groups present in lignins resonate in the nonanomeric carbohydrate region. Peaks in the anomeric carbohydrate region indicate the presence of mono-, di-, and polysaccharides. In the aromatic region from 6 to 9 ppm, the aromatic cycle of lignin or side chains of amino acid residues and amide functional group are found [40].

Signals arise for a large part from plant residues formed from either microbial decomposition that have enzymes able to break down strong molecular linkages such as those present in lipids, cellulose or hemicellulose and lignin fragments, or photooxidation processes [20, 33]. The presence of lipids is also shown by methyl and methylene groups of long-chain fatty acid residues and epicuticular waxes that resonate in the aliphatic region, and other methylene groups in aliphatic-containing heteroatoms regions (methylene group adjacent to an ester

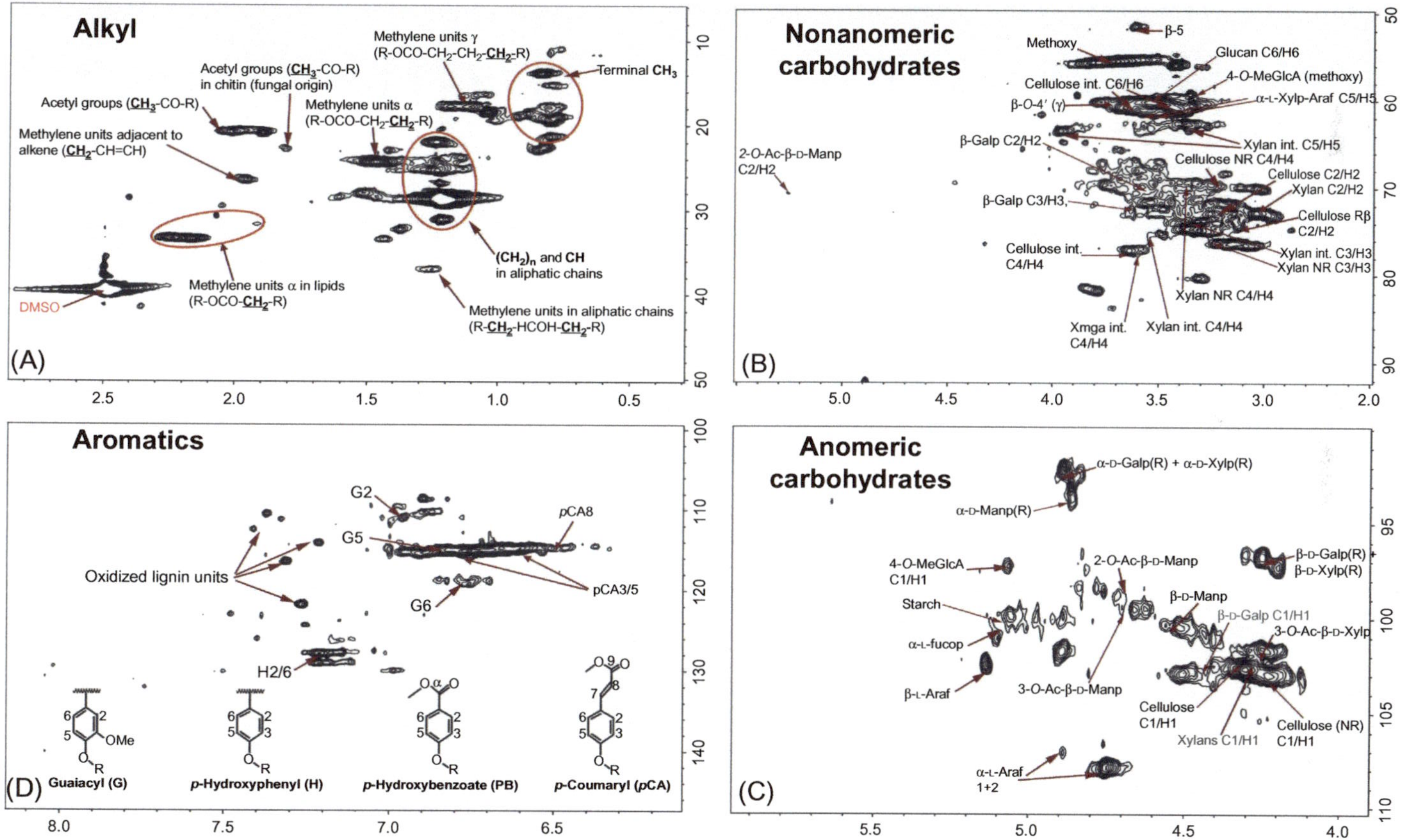

Fig. 2.2 Signal assignments for the four spectral regions of the HSQC spectrum of the superficial spruce soil DMSO-extract. Assignments labeled in gray are tentative.

With permission from L.N. Soucémarianadin, B. Erhagen, M.B. Nilsson, M.G. Öquist, et al., Two dimensional NMR spectroscopy for molecular characterization of soil organic matter: application to boreal soils and litter, Org. Geochem. 113 (2017) 184–195.

or hydroxyl group). Steroids also contribute to signals in the aromatic region. Cellulose is a polymer of glucose linked together through β(1–4)-glycosidic bonds and the main constituent of the plant cell wall, being very resistant to hydrolysis. Cellulose degradation products give resonance in the polyols region. Lignins are constituents of the cell wall of a vascular plant that is present in wood and bark [41]. Lignin residues are found in the aromatic region, being associated with $O—CH_3$ signal as well. Lignins mainly consist of *p*-hydroxyphenyl (H), guaiacyl (G), and syringyl (S) units that are linked through different side-chain linkages with several types of carbon-carbon and ether bonds. ^{13}C-labeled lignocellulose combustion products were characterized by 2D-NMR spectroscopy [42] showing characteristic fingerprints of linkage structures types in the aromatic and carbohydrate regions. The combustion products of lignin fragments, cellulose, and hemicellulose are found and also described in analytical studies of biomass burning aerosols. Some of the combustion products of plant residues emitted in the atmosphere redeposit and can be found in sediments at the bottom of close-bye water bodies [4]. Biomass burning markers in smoke particles accumulate in soils through sedimentation, with time forming layers of soil that differ in chemical composition and give an indication of the frequency of fires in the past.

Solution-state NMR spectroscopy of dissolved organic matter in water bodies and sediments

Characteristics of dissolved organic matter

Dissolved organic matter (DOM) constitutes the major form of reduced carbon in the ocean and other water bodies [43]. DOM transports organic matter from land to aquatic systems through rivers [44]. In terrestrial ecosystems where infiltration of water exists, it constitutes a small fraction of SOM but plays an important role in biological and geochemical processes while their content strongly depends on land use, management practices, climatic, and soil factors [33, 45–47]. It is a dynamic component in biogeochemical cycling and can sorb reversibly or irreversibly on minerals and contaminants depending on its functional group composition and mineral content [3, 48]. DOM also influences the formation of reactive oxygen species that can be used for the degradation of contaminants or remediation on wetlands [49]. It can also interact with contaminants such as NEPs (nano-enabled nanoparticles), hydrophobic organic contaminants, metals, thereby changing their bioavailability and toxicity. DOM organic composition has been characterized by UV-Vis, FT-IR, or fluorescence spectroscopy and NMR spectroscopy [50] but the structural information of DOM is still limited. The fluorescent fraction of DOM is composed of tryptophan-like (an aromatic amino-acid that absorbs light efficiently), humic-like, and protein-like components. Humic and fulvic-like substances have been isolated from DOM in rivers accounting for a large fraction of DOM, contributing from 1/3 to 2/3 of dissolved organic carbon (DOC) in water bodies [51].

DOM sample preparation

Analysis of DOM by solution-state NMR spectroscopy is a relatively straightforward process that requires steps to concentrate DOM (or upscaling process) to obtain sufficient quantities for NMR experiments. Water samples are filtered through 0.22 μm hydrophilic filters to remove biological contamination and microorganism. DOM can be isolated by passing through ion-exchange resin such as DEAD-cellulose, recovered using 0.1 M NaOH, ion-exchanged, adjusting of pH, removal of excess of salts and freeze dried [52]. Most recent DOM studies use ultrafiltration or solid-phase extraction (SPE) to isolate organic matter from water bodies (ocean, lakes, rivers) [12, 28]. SPE is used for organic component upscaling, using different cartridges with physicochemical properties and sorption mechanisms that vary as a function of the nature of DOM constituents (N-containing, for example). The choice of SPE cartridge depends on recovery and selectivity, and the possibility of purifying high-molecular-weight molecules [28]. Attention has been paid for eventual leaching of SPE cartridges into DOM by comparing sample NMR spectra with structural blanks and measuring real-recovery values. Polymer-based sorbents, such as functionalized styrene-divinylbenzene, are used as the solid matrix of SPE in order to collect hydrophobic and certain polar compounds (XAD-2, XAD-8). Functionalized PPL (styrene-divinylbenzene) is now generally used over XAD-2 resins. Purification of DOM into many fractions through hydrophilic interaction chromatography has also been used to elute and concentrate target analytes including carboxylic acids, alpha-hydroxy acids [53].

NMR characterization

The 1D ^{1}H NMR spectrum of DOM is characterized by the presence of resonance in four regions assigned to aromatic and phenolic compounds, δ_H 6.5–8.4 ppm; carbohydrates, δ_H 3.2–4.6 ppm; carboxylic rich alicyclic material (CRAM), δ_H 1.6–3.2 ppm; and material derived from linear terpenoids (MDLT), δ_H 0.6–1.6 ppm [50]. CRAM is a refractory component of DOM that is resistant to heat, high pressure, and other chemicals that exist in aquatic environments. It has been defined as "a complex mixture of carboxylic and fused alicyclic structures" [54]. The carboxylic acid and amide functional groups resonance were detected in DMSO-d_6 under the total exclusion of moisture in order to have no proton exchanges. The composition of the DOM characteristic functional group varies in freshwater as compared to marine waters, as well as considering deep seawater or surface seawater. Surface marine DOM is constituted of a large fraction of oxygenated aliphatics and acetate derivatives, while deep marine DOM was mostly constituted with purely aliphatic compounds and CRAM abundance increased as compared to surface marine DOM [14]. Other NMR studies of DOM with conventional ^{1}H-NMR and DE-NMR found similarities with NMR spectra of SOM extracts, with the presence of broad unresolved signals and a few well-resolved peaks from the most concentrated small organic compounds [47, 55]. The resolved peaks are not visible in DE-NMR spectra, confirming the rapid diffusion of those molecules, while the large envelopes in DE-NMR spectra provide evidence of the presence of macromolecules and aggregated compounds. Different DOM structural composition was found depending on the collection source of subtropical wetlands and revealed specific signatures. Those signatures revealed different sources of DOM, such as microbial, higher plants, oxidized, and combustion-derived compounds [7]. Humic substances induce resonance peaks in the aromatic and aliphatic region. The degree of humification of humic-like substances was related to the proportion of aliphatic protons to aromatic protons. Small organic compounds likely originating from microbes were also measured in DOM [56]. Changes in the NMR of DOM with environmental contaminants give insight on the reactivity of DOM with contaminant or between the many dissolved organic compounds. The NMR spectra of humic acids with fluoroquinolone antibiotics in aqueous alkaline solution show the weak interactions formed by monitoring changes of binding sites related to changes in proton chemical shift of the contaminant signals upon addition of humic acids and the potential of DOM specific components to be used in pollution remediation [57]. Characteristic 2D ^{1}H—^{13}C HSQC of DOM contains aliphatic cross peaks in the δ_H 0.6–1.6/δ_C 10–40 ppm with strong methyl signature as identified by DEPT-HSQC, CRAM is located in the δ_H 1.6–3.2/δ_C 25–55 ppm region and compounds containing methoxy chemical environment are found in the δ_H 3.0–4.5/δ_C 50–65 ppm region [58]. The carbohydrate region is divided into nonanomeric and anomeric carbons. In addition to aromatic structures cross peaks in the δ_H 5–9/δ_C 105–140 ppm region, cross peaks in the δ_H 5–6.5/δ_C 115–135 ppm region were tentatively assigned to materials derived from linear terpenoids MDLT with signals arising from conjugated and isolated carbon-carbon double bonds [52]. In the same study, DOM composition varied with the activity of microorganisms. Changes in the DOM structures were found during cyanobacterial bloom by comparing NMR spectra before and after the bloom. Cyanobacteria perform photosynthesis and have a high metabolic activity resulting in an increase of concentration of small molecules, mainly carbohydrates in the DOM composition. A decrease of carbohydrate due to the uptake of those compounds by heterotrophic organisms was also observed. The change in the DOM spectral signature during and after the algal bloom season was also accompanied with increasing the abundance of carboxylic acids, peptides, and esters. The absence of aliphatic lipids indicated the absence of algal extracts. Increased peak intensity in the functionalized aliphatic region was attributed to the formation of CRAM substances from microbial degradation of algae and it was the main component of organic matter after the algal bloom.

Solution-state NMR spectroscopy of organic matter in air and rainwater

Characteristic of atmospheric organic matter (organic aerosols)

Atmospheric aerosols are emitted directly or are produced by gas-to-particles conversion through chemical transformations. Low-volatility vapors can condense either on aerosols already present in the atmosphere (heterogeneous condensation) or can homogeneously condense to form an aerosol (homogeneous condensation). Water molecules condense mostly on existing aerosols, while inorganic and organic acids can condense both ways. Condensation and coagulation occur during the aging of particles, as well as photooxidation, inducing the formation of aerosols of size range up to 1 μm of diameter including ultrafine (particles with diameter < 0.1 μm) and fine particles (particles with diameter < 2.5 μm). Coarse particles (particles with diameter > 2.5 μm) are mainly emitted from mechanical processes such as wind-blown soil dust or road dust, sea spray, erosion of epicuticular plant waxes, and volcanic eruptions. Therefore, the nature, size, and composition of aerosols are correlated with origin (natural or anthropogenic) and formation mechanisms. The removal mechanism from the atmosphere (sink mechanism) also depends on the size and mass of particles. Fine particles are mainly removed from the atmosphere by wet deposition when precipitation occurs, while coarse particles are removed from the atmosphere by dry deposition (or sedimentation) [59].

Organic aerosols (OAs) are the organic fraction of atmospheric aerosols. Primary organic aerosols (POAs) are emitted directly by terrestrial sources, while secondary organic aerosols (SOAs) are chemically transformed during their stay in the atmosphere from volatile organic compound (VOCs) precursors or aging of POAs [2]. POA is mainly emitted by biomass burning or combustion of fossil fuel. ^{1}H NMR analysis has shown that freshly emitted POA is transformed fairly rapidly undergoing oxidation reaction in the atmosphere and remains suspended in the air from days to months. There is a growing interest in studying the chemical composition of water-soluble organic compounds as organic contaminants can interact with the respiratory tract through inhalation and translocate into the blood, interact with organ tissues or induce systemic inflammation, and other undesirable biological responses and health outcomes. Particulate matter has been established as a carcinogen by inducing DNA damages, while, recent in vitro studies also suggested that particulate matter induces epigenetic alterations [60, 61]. Multidimensional NMR spectroscopy has helped to characterize most of the organic aerosols that are not possible to be characterized with ultrasensitive methods, such as GC-MS and LC-MS, and with less preparation steps [62].

NMR sample preparation

The methodological framework developed for the analysis of SOM and DOM by NMR has been initially implemented to determine the chemical profiles of the water-soluble organic carbon (WSOC) fraction by NMR, including similar solvent systems followed by purification/isolation techniques. Humic-like substances (HULIS) have been hypothesized to be major components of organic airborne particulate matter, with a structure resembling a humic acid fraction of humic substances. The NMR spectrum of HULIS isolated from standard reference air dust NIST 1648 was published by Havers et al. in 1996 [63]. Low-molecular-size organic acids, diacids, oxoacids, and polyacids have been characterized in D_2O extracts of urban aerosols [64]. WSOC extraction typically includes simple sonication in water that is sufficient to recover hydrophilic substances. Extraction with 0.1 M NaOH allows for the recovery of hydrophobic compounds not readily extracted with water [65].

The extract may be subsequently fractionated to basic/neutral, acidic, or polyacidic species by anion-exchange chromatography [66]. The first fraction eluted is composed of neutral and basic compounds with strong NMR signals of proton in α-position to a hydroxyl (RCH_2—OH) or methoxy (RCH_2—OCH_3) group. Further analysis includes the determination of the normalized abundance of nonexchangeable protons from the integration of regions of functional groups and the comparison of the relative abundances of functional groups in different samples. It was shown that fog samples contained a higher proportion of protons in α-position to a hydroxyl and lower content of purely aliphatic groups than aerosols. Overlapping signals remain unresolved at the time, requiring the application of 2D homonuclear and heteronuclear experiments to identify individual compounds. The second fraction is composed of mono- and dicarboxylic acids, and the third fraction of polycarboxylic acids with a strong unsaturated carbon signature. The latter also included convoluted peaks and lumps at higher fields that are characteristic of the presence of HULIS. This study also showed that the monocarboxylic fraction presented highly oxidized species formed in the atmosphere.

NMR characterization

The chemical profiles of different organic aerosols types have been characterized in order to obtain the NMR spectral fingerprints of organic aerosol sources in an attempt to improve source reconciliation and apportionment, and/or uniquely identify potentially tracers and biomarkers and links to environmental and human health outcomes [61]. Typical 2D ^1H—^1H DQF COSY NMR spectra of organic aerosol types show cross peaks in the F2: 0.8–2.5 ppm/F1: 0.8–2.5 ppm region attributed to the aliphatic chains of lignin, α,ω-dicarboxylic acids, and neutral amino acids in biomass burning and biogenic aerosol [8] (Fig. 2.3). Cross peaks in the F2/F1: δ_H 0.89/1.33 ppm, δ_H 1.55/0.86 ppm, δ_H 1.33/1.54 ppm, δ_H 1.79/1.29, δ_H 1.97/1.33 ppm, δ_H 2.56/0.99, and δ_H 2.56/1.55 regions in traffic exhausts were associated with a branched aliphatic alcohol, ether, or ester compounds and aliphatic amines. Resonances of branched methyl groups in alpha position of aliphatic or aromatic carboxylic acids (δ_H F2/F1 2.4–3.2 ppm/0.9–1.4 ppm) were found only in wood burning. The abundance of resonance in the δ_H F2/F1 1.8–3.3 ppm/1.8–3.3 ppm region in road dust and traffic (and absence in soil) corroborated the contamination of soil particles by traffic exhausts. Resonances typically associated with alcohols, deoxy sugars, ethers, and esters with the OH—CH_2—CH_2—CH(OH)-R structure, amino-acids, hydroxylcarboxylic acid (HOOC—CH(OH)—CH_2-R), or acetate derivative (R-CO-—O—CH_2—CH_3) were observed in the F2: 2.9–4.5 ppm / F1:0.9–2.5 ppm region for biomass burning aerosol and to a lesser extent for traffic exhausts, road dust, and soil particles. Resonance signals in the F2: 3.2–4.7 ppm / F1: 3.2–4.7 ppm region was indicative of polyol compounds including microbial carbohydrates, sugar alcohols, and anhydrosugars, polyfunctionalized aliphatic structure attached to aromatic rings in HULIS compounds and polyhydroxylated carboxylic acids (HOOC—$(CHOH)_n$—CH_2OH) in biomass burning aerosols [67]. The strong abundance of CH_2 and CH in the anomeric position of cyclic carbohydrates found in sucrose, glucose, and trehalose in pollen and soil particles was observed in the F2: 4.9–5.4 ppm /

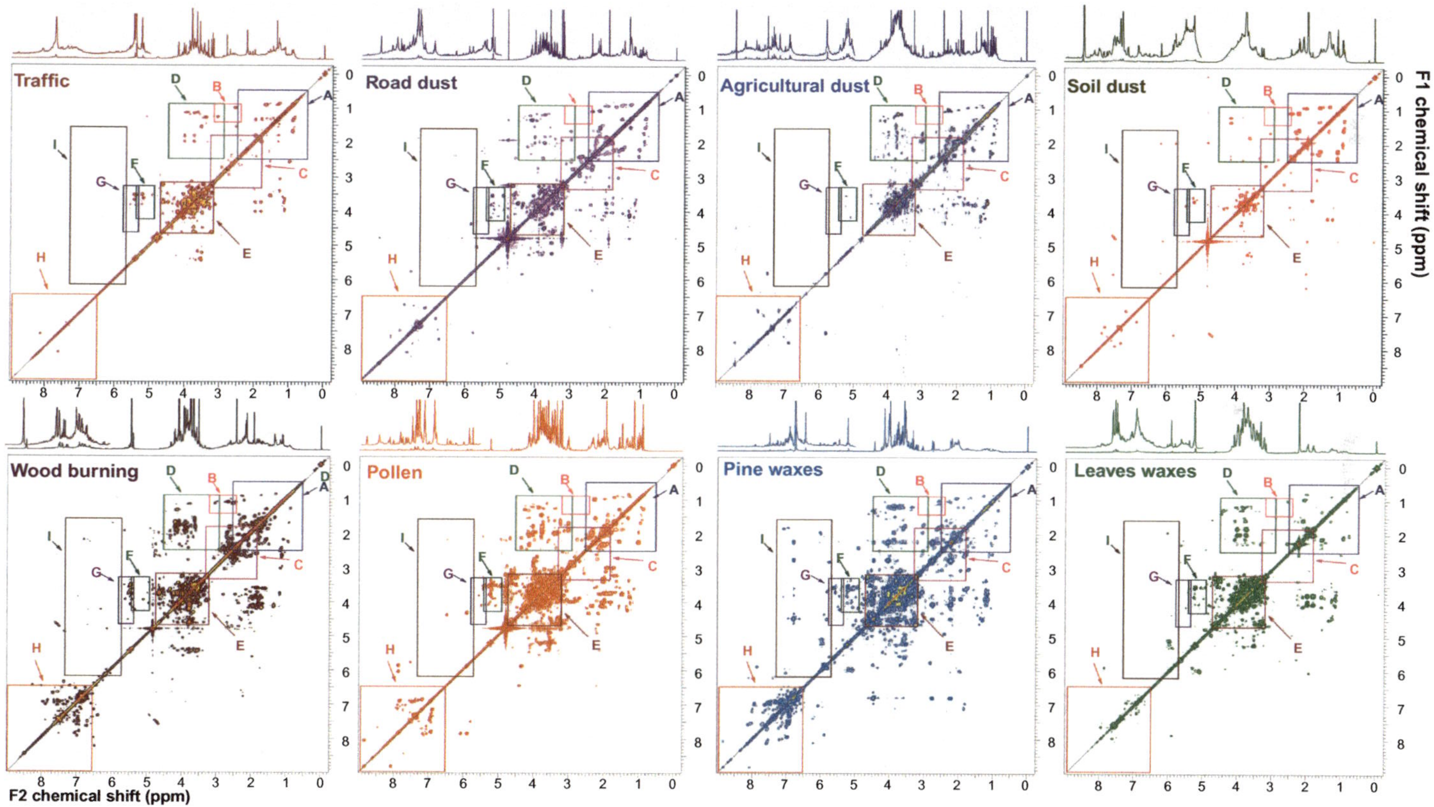

Fig. 2.3 2D ^1H—^1H DQF-COSY NMR spectra for traffic exhausts, road dust, agricultural dust, soil dust, and pine waxes, leaf waxes, pollen, and biomass burning, and cross peak regions.
With permission from M.C. Chalbot, I. Kavouras, Nuclear magnetic resonance characterization of water soluble organic carbon of atmospheric aerosol types, Nat. Prod. Commun. 14 (5) (2019), https:/doi.org/10.1177/1934578X19849972

F1: 3.3–4.3 ppm region. Levoglucosan, a tracer of biomass burning was previously observed in ^{1}H NMR. Signals in the F2: 6.5–9.2 ppm/F1: 6.5–9.2 ppm region may be associated with aliphatic protons attached to aromatic rings, aromatic amino-acids, hydroxybenzoic acid, and phenyl rings with aliphatic carboxylic acids.

SOA formation involves the oxidation of anthropogenic but also natural precursors (hemiterpenes, monoterpenes, and sesquiterpenes) by ozone (O_3), hydroxyl and nitrate (OH and NO_3, respectively) radicals. NMR analysis of SOA may provide insights on the formation of individual compounds but more importantly, NMR is suitable for kinetic studies to elucidate the formation mechanisms of SOA. The oxidation of α- and β-pinene leads to the formation of pinonaldehyde, nopinone, cis- and trans-pinonic, nonpinonic, and pinic acids [68, 69]. Methyl-tetrols have been identified by NMR as photooxidation products from isoprene precursor and are present in the H—C—O region, contributing to the higher proportion of acyl groups, but also to methyl signals in the aliphatic region of the NMR spectra [70].

Comparison of cloud water, wet deposition (rain) samples with the WSOC of aerosols showed differences in the organic carbon composition that could be related to the impact of reactions occurring in the atmosphere. The ^{1}H NMR spectra of cloud water and rain contained a higher proportion of saturated aliphatic structures in both H—C and H—C—O regions than in those of aerosol WSOC, but a lower proportion of aliphatic compounds containing C=O and C=C functional groups, and a small fraction of polycarboxylic acid compounds [71]. The organic composition of rainwater shows distinct regions that correspond to aromatic, carbohydrate, CRAM, and MDLT regions that were defined in DOM ^{1}H NMR spectra. Seasonal variation of the organic composition of rainwater has also been described and showed an increase in the CRAM region due to enhanced photochemistry in summer months and, thus, an increase in the formation of highly oxidized species [72].

Processing of environmental NMR data

NMR spectral analysis software

Chemical compounds and metabolites in SOM, DOM, aerosol, and precipitation may be identified by analyzing ^{1}H and ^{13}C NMR spectra using reference spectra from authenticated compounds or in environmental and biological media [65]. Today, databases with reference NMR spectra are available, including the biological magnetic resonance data bank (BMRB) [66] and the human metabolome database (HMDB) [67]. The BMRB NMR peak query allows for the molecular identification based on the resonances of characteristic groups, while HMDB may be used to evaluate potential molecular structures of the unknown compound. Tools such as spectral database search, custom-made resonance database development, and NMR spectra prediction have already been integrated into NMR data processing software. In addition, software modules were developed to analyze the variability of NMR spectra and identify common patterns and link them to sample features. These include Chenomx and Bruker's AMIX, used for the analysis of NMR spectra in biological specimens and precipitation samples [65]. The Bruker's AMIX software encompasses all steps in NMR signal processing, including raw data filtering and adjustment, integration, bucketing, factor analysis, and metabolites identification or matching. Metaboanalyst (Version 4.0), an online software tool, includes a broad range of statistical tools to analyze NMR spectra [68].

Functional analysis

Solution-state NMR data processing by nontargeted methods are applied to compare the chemical composition of samples collected from different sources/environmental matrices or study seasonal variations, changes occurring after exposures to contaminant or stressors, or pattern identification. When studying organic aerosols, 1D-NMR data gives valuable information for their source apportionment by integrating defined functional group regions on the one hand and statistical factor analysis by binning on the other hand. Fig. 2.4 shows several data processing methods and statistical analyses of environmental samples. In functional group analysis, spectra are integrated for each region of functional group (aliphatic, oxygenated unsaturated aliphatics, oxygenated saturated, olefinic, and aromatic data). Alkoxy (H—C—C=O) and acyl (H—C—O) contributions to the total of aliphatics (H—C—C= +H—C—O+H—C—C) give a diagram showing aerosol source and aging [73, 74]. The organic hydrogen concentrations calculated for each region are transformed into carbon concentration using H/C ratios of 0.4 for aromatic carbons (Ar—C), 1.1 for hydroxylated saturated (H—C—O), and 2 for aliphatics (saturated H—C and unsaturated H—C—C=). Unsaturated oxygenated aliphatic H—C—C=O was estimated by subtracting aromatic C=C contributions to the H—C—C= region, using an H/C ratio for benzylic groups (H—C—Ar) of 1/6. Biomass burning, SOA, aged SOA, and marine aerosols were the first regions well characterized for fine particles identified by Decesari et al. [73] (Fig. 2.4A). Subsequent studies characterized additional regions identified as an urban aerosol, fresh biomass burning, and pollen [15, 75].

Factorial analysis

The factorial untargeted analysis aims to identify common features among a large number of NMR spectra and relate them to specific characteristics of the samples. It relies on the assumption that the majority of the variability in NMR spectra may be explained by a limited number of variables. The first step for untargeted analysis is reconstructing the spectra using bucketing that creates bins of the same ppm range (for example 0.04 ppm) to reduce the number of variables (e.g., 400 bins) of a spectrum as compared to the total number of points collected while computing NMR spectra (up to 35 K). Intelligent bucketing is used to recognize the start and end of a peak, adjust for individual bin size and, therefore, reducing artifacts associated with improper integration of peaks. An important requirement is that all NMR spectra used in the analysis must be aligned so the binning correspond exactly to the same signals for each spectrum. Subsequently, the binned NMR spectra may be normalized (as needed) and analyzed by factorial methods including principal components analysis

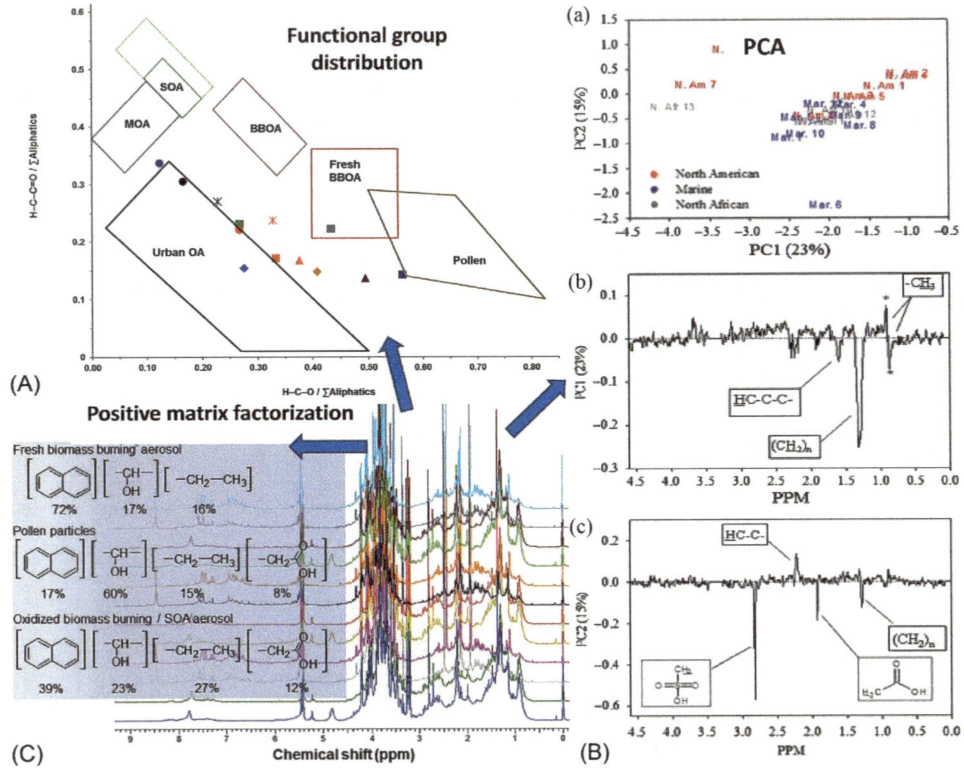

Fig. 2.4 Fingerprint recognition using typical methods. (A) functional group distribution derived from the integration of regions of NMR spectrum, (B) PCA applied to NMR binning data: on top, the score plot of PC1 vs PC2, and PC1 and PC2 variable loading vs the chemical shift, and (C) positive matrix factorization (PMF) derived from bucketing NMR data allow for the quantification of sources that contribute to variability of NMR data. PMF enable to extract factors and their respective contribution to the organic carbon composition.

(A) Reproduced with permission from R.M.B.O. Duarte, J.T.V. Matos, A.S. Paula, S.P. Lopes, et al., Structural signatures of water-soluble organic aerosols in contrasting environments in South America and Western Europe, Environ. Pollut. 227 (2017) 513–525. (B) Reproduced with permission from A.S. Wozniak, R.U. Shelley, S.D. McElhenie, W.M. Landing, et al., Aerosol water soluble organic matter characteristics over the North Atlantic Ocean: Implications for iron-binding ligands and iron solubility, Mar. Chem. 173 (2015) 162–172. (C) Reproduced with permission from M.C. Chalbot, P. Chitranshi, G. Gamboa da Costa, E. Pollock, et al., Characterization of water-soluble organic matter in urban aerosol by 1H-NMR spectroscopy, Atmos. Environ. 128 (2016) 235–245.

(PCA), partial-least square-discriminant analysis (PLS-DA), and positive matrix factorization (PMF). These methods do not require previous knowledge of factors (i.e., characteristic/property) that may explain the variability of the database [76]. In PCA, the dataset is reduced to a limited subset of principal components (PC) that explain most of the dataset variability (>90%). The associations of NMR bins with each PC (loadings) allow for the chemical characterization of PC. Moreover, samples that have similar chemical composition cluster together in PCA score plots. For example, soil samples with high antibiotic activity are separated from samples with low antibiotic activity. Recent PCA of soils extracts taken at different locations and/or different land use have shown that SOM depends largely on land use (i.e., remnant vs managed soils), rather than location [34, 77]. Fig. 2.4B (top) shows the PCA score plot for atmospheric samples and sample separation as a function of the sample location [78]. Plotting the PC scores vs the chemical shift in ppm allows to identify the signals resonances (and hence the functional groups/compounds) that are responsible for the variability of the signals.

PMF was successfully applied to determine the fingerprints of organic aerosols from NMR data and calculate the contribution of sources or types of organic aerosols (Fig. 2.4C) [15, 79, 80]. Analysis of ^{1}H NMR spectra revealed the abundance of background and aged marine organic aerosol and urban polluted aerosol in a marine remote location, brought by air masses transport. Factorial analysis was also applied on NMR data to identify the sources of biogenic SOA in a boreal forest, in Finland, resolving four distinct profiles as follows: "Glycols," "HULIS"-containing factor (anthropogenic contributions + levoglucosan, "Amines" factor (diethyl and dimethyl amines and biogenic marine species), and biogenic "Terpene-SOA like" factor originated from a polluted environment [81]. PMF analysis of urban aerosol in the Midwest US indicated the prevalence of fresh and aged biomass burning emissions in winter and summer, respectively [15, 82]. A strong carbohydrate signature for spring aerosol was assigned to pollen and biogenic aerosol. The three types of organic carbon identified by NMR spectroscopy explained the majority of organic carbon measured in the same region. The finding of these studies shows the relevance of NMR data analysis as a method for organic aerosols source apportionment.

Challenges in environmental NMR metabolomics

Environmental metabolomics may also be applied to study the effect of environmental contaminants in human studies. Data must be examined in both 1D and 2D, in order to identify metabolites or biomarkers. This implies that most of the peaks have to be present not only in the 1D but also in the 2D to improve confidence in identifying the right compound. Atmospheric aerosols or rainwater contain compounds in which chemical shifts correspond to metabolites present in biological fluids. A large diversity of compounds exists in organic aerosols such as carbohydrate, amino acids, and low-molecular-weight carboxylic acids, compounds for which some but not all the signals can be found because of overlap of multiple signals from multiple molecules. The building of databases of standard compounds must eliminate identification errors. Aromatic compounds in soils and dissolved organic matter are formed mainly from the decomposition, oxidation, or burning of plant materials, while in aerosols they are emitted from biomass burning of plants and trees or resuspension of soils.

Recent advances in environmental NMR

New techniques have been designed to better characterize and understand the structural composition of environmental matrices, plus the advection of new research areas has led to the significant transformation of the NMR instrument. Comprehensive multiphase NMR (CMP-NMR) has been designed for in situ analysis of organic aerosols transformation, and in vivo metabolomics. CMP-NMR was used to analyze the soil organic matter in the liquid, gel, and solid state inside the same NMR instrument [9]. Application to the soil is the most interesting because of the soil multiphase scale and the possibility to study soil based on the size of single molecules and aggregates. Indeed, CMP-NMR allows the identification of the structure of molecules present in a soil sample, use diffusion-based NMR to analyze the colloid formation issued from the aggregation of organic matter, and their association with clay. In this way, one can obtain the whole picture of the reactivity of the SOM component with water or minerals (SiO_2, Al_2O_3, MgO) or soil contaminants [9].

In situ NMR photoreactor was designed to directly monitor products formed during the irradiation of organic compounds having broad interest [83]. The field of application is extremely large as photochemistry in different phases applies for SOM, DOM, and organic aerosols. Quantitative studies with changes in the intensity of signals in NMR spectra allow for monitoring the kinetics of the appearance or disappearance of products. NMR methods are also useful to monitor soil remediation efficacity as well as water quality.

The field of in vivo environmental metabolomics arose from the need to monitor changes in the metabolisms of microorganisms and organisms of small size following or during environmental exposures in real time. This analytical method has limitations due to the experimental design that is still in the preparation phase and requires the transformation of the NMR instrument and the necessity to reduce stress for the living organisms in the experimental setup. Pumps are used to oxygenize and clean the matrix, thus allowing a run of samples containing a low concentration of organic compounds/metabolites and thus lower sensitivity [84]. Better results have been obtained by ^{13}C enrichment to quantify the metabolites of interest.

References

[1] A.J. Simpson, D.J. McNally, M.J. Simpson, NMR spectroscopy in environmental research: from molecular interactions to global processes, Prog. Nucl. Magn. Reson. Spectrosc. 58 (3) (2011) 97–175.

[2] U. Pöschl, M. Shiraiwa, Multiphase chemistry at the atmosphere–biosphere interface influencing climate and public health in the Anthropocene, Chem. Rev. 115 (10) (2015) 4440–4475.

[3] X. Zhang, Z. Li, X. Nie, M. Huang, et al., The role of dissolved organic matter in soil organic carbon stability under water erosion, Ecol. Indic. 102 (2019) 724–733.

[4] T. Kirchgeorg, S. Schüpbach, N. Kehrwald, D.B. McWethy, et al., Method for the determination of specific molecular markers of biomass burning in lake sediments, Org. Geochem. 71 (2014) 1–6.
[5] S. Killops, V. Killops, Long-term fate of organic matter in the geosphere, in: S. Killops, V. Killops (Eds.), Introduction to Organic Geochemistry, Blackwell Publishing Ltd., 2015, pp. 117–165.
[6] S. Liu, Y. Zhu, F. Wu, W. Meng, et al., Characterization of plant-derived carbon and phosphorus in lakes by sequential fractionation and NMR spectroscopy, Sci. Total Environ. 566-567 (2016) 1398–1409.
[7] N. Hertkorn, M. Harir, K.M. Cawley, P. Schmitt-Kopplin, et al., Molecular characterization of dissolved organic matter from subtropical wetlands: a comparative study through the analysis of optical properties, NMR and FTICR/MS, Biogeosciences 13 (8) (2016) 2257–2277.
[7a] N.S. Santini, M.F. Adame, R.H. Nolan, Y. Miquelajauregui, D. Pinero, A. Mastretta-Yanes, A.P Cuervo-Robayo, D. Eamus, Storage of organic carbon in the soil of Mexican temperate forests, Forest Ecol. Manag. 446 (2019) 115–125.
[8] M.C. Chalbot, I. Kavouras, Nuclear magnetic resonance characterization of water soluble organic carbon of atmospheric aerosol types. Nat. Prod. Commun. 14 (5) (2019). https://doi.org/10.1177/1934578X19849972.
[9] D. Courtier-Murias, H. Farooq, H. Masoom, A. Botana, et al., Comprehensive multiphase NMR spectroscopy: basic experimental approaches to differentiate phases in heterogeneous samples, J. Magn. Reson. 217 (2012) 61–76.
[10] R. Mylotte, A. Sutrisno, H. Farooq, H. Masoom, et al., Insights into the composition of recalcitrant organic matter from estuarine sediments using NMR spectroscopy, Org. Geochem. 98 (2016) 155–165.
[11] A.J. Simpson, S.A. Brown, PURGE NMR: effective and easy solvent suppression, J. Magn. Reson. 175 (2005) 340–346.
[12] Y. Li, M. Harir, M. Lucio, B. Kanawati, et al., Proposed guidelines for solid phase extraction of Suwannee River dissolved organic matter, Anal. Chem. 88 (13) (2016) 6680–6688.
[13] T.D.W. Claridge, (Chapter 10). Diffusion NMR spectroscopy, in: T.D.W. Claridge (Ed.), High-Resolution NMR Techniques in Organic Chemistry, third ed., Elsevier, Boston, 2016, , pp. 381–419.
[14] N. Hertkorn, M. Harir, B.P. Koch, B. Michalke, et al., High-field NMR spectroscopy and FTICR mass spectrometry: powerful discovery tools for the molecular level characterization of marine dissolved organic matter, Biogeosciences 10 (3) (2013) 1583–1624.
[15] M.C. Chalbot, P. Chitranshi, G. Gamboa da Costa, E. Pollock, et al., Characterization of water-soluble organic matter in urban aerosol by 1H-NMR spectroscopy, Atmos. Environ. 128 (2016) 235–245.
[16] M.C. Chalbot, S.V. Pirela, L. Schifman, V. Kasaraneni, et al., Synergistic effects of engineered nanoparticles and organics released from laser printers using nano-enabled toners: potential health implications from exposures to the emitted organic aerosol, Environ. Sci. Nano 4 (11) (2017) 2144–2156.
[17] M.C. Chalbot, I.G. Kavouras, Nuclear magnetic resonance spectroscopy for determining the functional content of organic aerosols: a review, Environ. Pollut. 191 (2014) 232–249.
[18] Ē. Kupče, W. Hiller, Clean adiabatic TOCSYs, Magn. Reson. Chem. 39 (5) (2001) 231–235.
[19] H. Ma, C.M. Pedersen, Q. Zhao, Z. Lyu, et al., NMR analysis of the Fischer-Tropsch wastewater: combination of 1D selective gradient TOCSY, 2D DOSY and qNMR, Anal. Chim. Acta 1066 (2019) 21–27.

[20] L.N. Soucémarianadin, B. Erhagen, M.B. Nilsson, M.G. Öquist, et al., Two dimensional NMR spectroscopy for molecular characterization of soil organic matter: application to boreal soils and litter, Org. Geochem. 113 (2017) 184–195.
[21] S.S. Yekta, M. Hedenström, B.H. Svensson, I. Sundgren, et al., Molecular characterization of particulate organic matter in full scale anaerobic digesters: an NMR spectroscopy study, Sci. Total Environ. (2019).
[22] A.J. Simpson, W.L. Kingery, P.G. Hatcher, The identification of plant derived structures in humic materials using three-dimensional NMR spectroscopy, Environ. Sci. Technol. 37 (2) (2003) 337–342.
[23] T. Parella, Pulse program catalogue: 1D and 2D NMR experiments, in: Topspin v3.0 NMR Guide, 2010.
[24] A.J. Simpson, M.J. Simpson, R. Soong, Environmental nuclear magnetic resonance spectroscopy: an overview and a primer, Anal. Chem. 90 (1) (2018) 628–639.
[25] N. Hertkorn, A. Permin, I. Perminova, D. Kovalevskii, et al., Comparative analysis of partial structures of a peat humic and fulvic acid using one- and two-dimensional nuclear magnetic resonance spectroscopy, J. Environ. Qual. 31 (2) (2002) 375–387.
[26] R.M.B.O. Duarte, A.M.S. Silva, A.C. Duarte, Two-dimensional NMR studies of water-soluble organic matter in atmospheric aerosols, Environ. Sci. Technol. 42 (22) (2008) 8224–8230.
[27] J. Ralph, L.L. Landucci, NMR of lignins, in: Lignins and Lignans: Advances in Chemistry, Taylor and Francis, Boca Raton, FL, 2010, , pp. 137–244.
[28] Y. Li, M. Harir, J. Uhl, B. Kanawati, et al., How representative are dissolved organic matter (DOM) extracts? A comprehensive study of sorbent selectivity for DOM isolation, Water Res. 116 (2017) 316–323.
[29] A.J. Simpson, Determining the molecular weight, aggregation, structures and interactions of natural organic matter using diffusion ordered spectroscopy, Magn. Reson. Chem. 40 (13) (2002) S72–S82.
[30] J. Kikuchi, K. Ito, Y. Date, Environmental metabolomics with data science for investigating ecosystem homeostasis, Prog. Nucl. Magn. Reson. Spectrosc. 104 (2018) 56–88.
[31] V.V. Krishnamurthy, D.J. Russell, C.E. Hadden, G.E. Martin, 2J,3J-HMBC: A new long-range Heteronuclear shift correlation technique capable of differentiating 2JCH from 3JCH correlations to protonated carbons, J. Magn. Reson. 146 (1) (2000) 232–239.
[32] P.M. Huang, A.G. Hardie, Formation mechanisms of humic substances in the environment, in: N. Senesi, B. Xing, P.M. Huang (Eds.), Biophysico-Chemical Processes Involving Natural Nonliving Organic Matter in Environmental Systems, John Wiley and Sons, Inc., 2009, pp. 41–109.
[33] X. Feng, K.M. Hills, A.J. Simpson, J.K. Whalen, et al., The role of biodegradation and photo-oxidation in the transformation of terrigenous organic matter, Org. Geochem. 42 (3) (2011) 262–274.
[34] S. Rochfort, V. Ezernieks, P. Mele, M. Kitching, NMR metabolomics for soil analysis provide complementary, orthogonal data to MIR and traditional soil chemistry approaches—a land use study, Magn. Reson. Chem. 53 (9) (2015) 719–725.
[35] J.N. James, C.D. Gross, P. Dwivedi, T. Myers, et al., Land use change alters the radiocarbon age and composition of soil and water-soluble organic matter in the Brazilian Cerrado, Geoderma 345 (2019) 38–50.
[36] P. Sharma, Y. Laor, M. Raviv, S. Medina, et al., Compositional characteristics of organic matter and its water-extractable components across a profile of organically managed soil, Geoderma 286 (2017) 73–82.

[37] A.J. Simpson, M.J. Simpson, E. Smith, B.P. Kelleher, Microbially derived inputs to soil organic matter: are current estimates too low? Environ. Sci. Technol. 41 (23) (2007) 8070–8076.
[38] R.R. Chang, R. Mylotte, M.H. Hayes, R. McLnerney, et al., A comparison of the compositional differences between humic fractions isolated by the IHSS and exhaustive extraction procedures, Naturwissenschaften 101 (3) (2014) 197–209.
[39] A.J. Simpson, G. Song, E. Smith, B. Lam, et al., Unraveling the structural components of soil Humin by use of solution-state nuclear magnetic resonance spectroscopy, Environ. Sci. Technol. 41 (3) (2007) 876–883.
[40] A.J. Simpson, M.J. Simpson, Nuclear magnetic resonance analysis of natural organic matter, in: N. Senesi, B. Xing, P.M. Huang (Eds.), Biophysico-Chemical Processes Involving Natural Nonliving Organic Matter in Environmental Systems, 2009, , pp. 589–650.
[41] K.M. Holtman, H.-m. Chang, J.F. Kadla, Solution-state nuclear magnetic resonance study of the similarities between milled wood lignin and cellulolytic enzyme lignin, J. Agric. Food Chem. 52 (4) (2004) 720–726.
[42] T. Komatsu, J. Kikuchi, Comprehensive signal assignment of 13C-Labeled lignocellulose using multidimensional solution NMR and 13C chemical shift comparison with solid-state NMR, Anal. Chem. 85 (18) (2013) 8857–8865.
[43] M. Derrien, S.R. Brogi, R. Gonçalves-Araujo, Characterization of aquatic organic matter: assessment, perspectives and research priorities, Water Res. 163 (2019) 114908.
[44] J.J. Cole, Y.T. Prairie, N.F. Caraco, W.H. McDowell, et al., Plumbing the global carbon cycle: integrating inland waters into the terrestrial carbon budget, Ecosystems 10 (1) (2007) 172–185.
[45] M.H. Chantigny, Dissolved and water-extractable organic matter in soils: a review on the influence of land use and management practices, Geoderma 113 (3) (2003) 357–380.
[46] E.U. Hobley, L.C. Zoor, H.R. Shrestha, L.T. Bennett, et al., Prescribed fire affects the concentration and aromaticity of soluble soil organic matter in forest soils, Geoderma 341 (2019) 138–147.
[47] J. Lu, A.C. Chang, L. Wu, Distinguishing sources of groundwater nitrate by 1H NMR of dissolved organic matter, Environ. Pollut. 132 (2) (2004) 365–374.
[48] P.J. Mitchell, A.J. Simpson, R. Soong, A. Oren, et al., Solution-state NMR investigation of the sorptive fractionation of dissolved organic matter by alkaline mineral soils, Environ. Chem. 10 (4) (2013) 333–340.
[49] A. Sardana, B. Cottrell, D. Soulsby, T.N. Aziz, Dissolved organic matter processing and photoreactivity in a wastewater treatment constructed wetland, Sci. Total Environ. 648 (2019) 923–934.
[50] K. Mopper, A. Stubbins, J.D. Ritchie, H.M. Bialk, et al., Advanced instrumental approaches for characterization of marine dissolved organic matter: extraction techniques, mass spectrometry, and nuclear magnetic resonance spectroscopy, Chem. Rev. 107 (2) (2007) 419–442.
[51] H.-C. Kim, M.-J. Yu, I. Han, Multi-method study of the characteristic chemical nature of aquatic humic substances isolated from the Han River, Korea, Appl. Geochem. 21 (7) (2006) 1226–1239.
[52] B. Lam, A. Baer, M. Alaee, B. Lefebvre, et al., Major structural components in freshwater dissolved organic matter, Environ. Sci. Technol. 41 (24) (2007) 8240–8247.
[53] G.C. Woods, M.J. Simpson, P.J. Koerner, A. Napoli, et al., HILIC-NMR: toward the identification of individual molecular components in dissolved organic matter, Environ. Sci. Technol. 45 (9) (2011) 3880–3886.

[54] N. Hertkorn, R. Benner, M. Frommberger, P. Schmitt-Kopplin, et al., Characterization of a major refractory component of marine dissolved organic matter, Geochim. Cosmochim. Acta 70 (12) (2006) 2990–3010.
[55] J.-J. Wang, M.J. Lafrenière, S.F. Lamoureux, A.J. Simpson, et al., Differences in riverine and pond water dissolved organic matter composition and sources in Canadian high Arctic watersheds affected by active layer detachments, Environ. Sci. Technol. 52 (3) (2018) 1062–1071.
[56] B.G. Pautler, G.C. Woods, A. Dubnick, A.J. Simpson, et al., Molecular characterization of dissolved organic matter in glacial ice: coupling natural abundance 1H NMR and fluorescence spectroscopy, Environ. Sci. Technol. 46 (7) (2012) 3753–3761.
[57] X. Zhao, Z. Hu, X. Yang, X. Cai, et al., Noncovalent interactions between fluoroquinolone antibiotics with dissolved organic matter: a 1H NMR binding site study and multi-spectroscopic methods, Environ. Pollut. 248 (2019) 815–822.
[58] F. Zhang, M. Harir, F. Moritz, J. Zhang, et al., Molecular and structural characterization of dissolved organic matter during and post cyanobacterial bloom in Taihu by combination of NMR spectroscopy and FTICR mass spectrometry, Water Res. 57 (2014) 280–294.
[59] M.C. Chalbot, I.G. Kavouras, Concentrations and physicochemical properties of ambient aerosol in the northern hemisphere, in: Air Pollution and Pollutants, Academy Publish, 2012.
[60] I.R. Miousse, M.C. Chalbot, N. Aykin-Burns, X. Wang, et al., Epigenetic alterations induced by ambient particulate matter in mouse macrophages, Environ. Mol. Mutagen. 55 (5) (2014) 428–435.
[61] I.R. Miousse, M.C. Chalbot, R. Pathak, X. Lu, et al., In vitro toxicity and epigenotoxicity of different types of ambient particulate matter, Toxicol. Sci. 148 (2) (2015) 473–487.
[62] B. Nozière, M. Kalberer, M. Claeys, J. Allan, et al., The molecular identification of organic compounds in the atmosphere: state of the art and challenges, Chem. Rev. 115 (10) (2015) 3919–3983.
[63] N. Havers, P. Burba, J. Lambert, D. Klockow, Spectroscopic characterization of humic-like substances in airborne particulate matter, J. Atmos. Chem. 29 (1) (1998) 45–54.
[64] Y. Suzuki, M. Kawakami, K. Akasaka, 1H NMR application for characterizing water-soluble organic compounds in urban atmospheric particles, Environ. Sci. Technol. 35 (13) (2001) 2656–2664.
[65] S.P. Lopes, J.T.V. Matos, A.M.S. Silva, A.C. Duarte, et al., 1H NMR studies of water- and alkaline-soluble organic matter from fine urban atmospheric aerosols, Atmos. Environ. 119 (2015) 374–380.
[66] S. Decesari, M.C. Facchini, S. Fuzzi, E. Tagliavini, Characterization of water-soluble organic compounds in atmospheric aerosol: a new approach, J. Geophys. Res. Atmos. 105 (D1) (2000) 1481–1489.
[67] B. Graham, O.L. Mayol-Bracero, P. Guyon, G.C. Roberts, et al., Water-soluble organic compounds in biomass burning aerosols over Amazonia 1. Characterization by NMR and GC-MS, J. Geophys. Res. Atmos. 107 (D20) (2002). LBA 14-11-LBA 14-16.
[68] I.G. Kavouras, N. Mihalopoulos, E.G. Stephanou, Formation of atmospheric particles from organic acids produced by forests, Nature 395 (6703) (1998) 683–686.
[69] M. Claeys, R. Szmigielski, I. Kourtchev, P. Van der Veken, et al., Hydroxydicarboxylic acids: markers for secondary organic aerosol from the photooxidation of α-pinene, Environ. Sci. Technol. 41 (5) (2007) 1628–1634.
[70] N. Zanca, A.T. Lambe, P. Massoli, M. Paglione, et al., Characterizing source fingerprints and ageing processes in laboratory-generated secondary organic aerosols using proton-nuclear magnetic resonance (1H-NMR) analysis and HPLC HULIS determination, Atmos. Chem. Phys. 17 (17) (2017) 10405–10421.

[71] S. Decesari, M.C. Facchini, S. Fuzzi, G.B. McFiggans, et al., The water-soluble organic component of size-segregated aerosol, cloud water and wet depositions from Jeju Island during ACE-Asia, Atmos. Environ. 39 (2) (2005) 211–222.
[72] B.A. Cottrell, M. Gonsior, L.M. Isabelle, W. Luo, et al., A regional study of the seasonal variation in the molecular composition of rainwater, Atmos. Environ. 77 (2013) 588–597.
[73] S. Decesari, M. Mircea, F. Cavalli, S. Fuzzi, et al., Source attribution of water-soluble organic aerosol by nuclear magnetic resonance spectroscopy, Environ. Sci. Technol. 41 (7) (2007) 2479–2484.
[74] M. Brege, M. Paglione, S. Gilardoni, S. Decesari, et al., Molecular insights on aging and aqueous-phase processing from ambient biomass burning emissions-influenced Po Valley fog and aerosol, Atmos. Chem. Phys. 18 (17) (2018) 13197–13214.
[75] R.M.B.O. Duarte, J.T.V. Matos, A.S. Paula, S.P. Lopes, et al., Structural signatures of water-soluble organic aerosols in contrasting environments in South America and Western Europe, Environ. Pollut. 227 (2017) 513–525.
[76] J. Lever, M. Krzywinski, N. Altman, Principal component analysis, Nat. Methods 14 (2017) 641.
[77] C.W. Johns, A.B. Lee, T.I. Springer, E.N. Rosskopf, et al., Using NMR-based metabolomics to monitor the biochemical composition of agricultural soils: a pilot study, Eur. J. Soil Biol. 83 (2017) 98–105.
[78] A.S. Wozniak, R.U. Shelley, S.D. McElhenie, W.M. Landing, et al., Aerosol water soluble organic matter characteristics over the North Atlantic Ocean: Implications for iron-binding ligands and iron solubility, Mar. Chem. 173 (2015) 162–172.
[79] S. Decesari, J. Allan, C. Plass-Duelmer, B.J. Williams, et al., Measurements of the aerosol chemical composition and mixing state in the Po Valley using multiple spectroscopic techniques, Atmos. Chem. Phys. 14 (22) (2014) 12109–12132.
[80] S. Decesari, E. Finessi, M. Rinaldi, M. Paglione, et al., Primary and secondary marine organic aerosols over the North Atlantic Ocean during the MAP experiment, J. Geophys. Res. Atmos. 116 (D22) (2011).
[81] E. Finessi, R.T. Lidster, F. Whiting, T. Elliott, et al., Improving the quantification of secondary organic aerosol using a microflow reactor coupled to HPLC-MS and NMR to manufacture ad hoc calibration standards, Anal. Chem. 86 (22) (2014) 11238–11245.
[82] M.C. Chalbot, B. McElroy, I.G. Kavouras, Sources, trends and regional impacts of fine particulate matter in southern Mississippi valley: significance of emissions from sources in the Gulf of Mexico coast, Atmos. Chem. Phys. 13 (7) (2013) 3721–3732.
[83] L. Bliumkin, R. Dutta Majumdar, R. Soong, A. Adamo, et al., Development of an in situ NMR photoreactor to study environmental photochemistry, Environm. Sci. Technol. 50 (11) (2016) 5506–5516.
[84] M. Bastawrous, A. Jenne, M. Tabatabaei Anaraki, A.J. Simpson, In-vivo NMR spectroscopy: a powerful and complimentary tool for understanding environmental toxicity, Meta 8 (2) (2018) 35.

Advanced two-dimensional solid-state NMR spectroscopy and its application in environmental sciences

3

Xi Chen[a], Jennifer Mejia[b], Wenying Chu[b], Jingdong Mao[b], Sophia Viar[c], Hongjian Gao[a]
[a]Anhui Province Key Laboratory of Farmland Ecological Conservation and Pollution Prevention, School of Resources and Environment, Anhui Agricultural University, Hefei, China, [b]Department of Chemistry and Biochemistry, Old Dominion University, Norfolk, VA, United States, [c]2505 Tiswood Court, Chesapeake, VA, United States

Introduction to nuclear magnetic resonance

Overview

Nuclear magnetic resonance (NMR) spectroscopy is a spectroscopic technique that probes the intrinsic spin properties of atomic nuclei [1]. In order to collect an NMR spectrum, the sample is placed in a magnetic field and then excited using selected radiofrequency electromagnetic radiation. NMR has far-reaching applications and can be used to characterize molecular structures, as well as investigate molecular dynamics and interactions. In environmental science, it has been used extensively to characterize a variety of environmental samples, such as plant litter [2, 3], peats [4], humic substances [5], fossil fuels [6], soil organic matter [7, 8], marine sediment [9], aerosols [10], and also to study the interactions of environmental media with pollutants [11].

There are two major kinds of NMR, liquid- and solid-state. Liquid-state NMR has limitations in that it can only probe soluble samples. Many environmental samples, such as natural organic matter (NOM), are insoluble or contain poorly soluble fractions, rendering solution-state NMR useless. For example, the characterization of humin, biochar, and coal is often difficult when using just solution-state. This restriction can be overcome by resorting to solid-state NMR. Solid-state NMR is nondestructive and allows for structural characterization on bulk or solid samples without extraction [12, 13]. Recent technical developments have made it an indispensable tool for chemical analysis, structural elucidation, and the study of dynamics in organic, inorganic, and biological systems [12, 14–18].

In this chapter, we will review some advanced two-dimensional (2D) solid-state NMR techniques and their applications for environmental research. We will highlight a few recent developments including new correlation experiments and spectral-editing

Multidimensional Analytical Techniques in Environmental Research. https://doi.org/10.1016/B978-0-12-818896-5.00003-X

techniques that provide carbon functional group, domain, heterogeneity, and connectivity information. Also, to assist in the understanding of the potential applications of solid-state NMR, a short review of NMR basics is provided.

NMR theory

NMR theory is typically described using quantum mechanics because it provides an accurate, complete understanding of NMR spectroscopy. Here, for practical purposes, we use the simpler vector model to explain the basics of NMR, although it is important to note that the vector model can be only applied to uncoupled spins [19].

As shown later, some nuclei possess spin angular momentum, which leads to a nuclear spin magnetic moment. The consequence of this is that these nuclei can generate small magnetic fields, like tiny bar magnets. When the sample (a collection of nuclei) is placed in a strong magnetic field, an interaction between the nuclear spin magnetic moments and the applied external magnetic field occurs. The energy of this interaction varies with the angle between the magnetic moments and the applied external field. For magnetic moments aligned with the external field with the angle zero, the energy is lowest, whereas for those opposed to the field the energy is highest. Therefore, one would expect that the energy of all the spins in the sample is minimal if all the individual magnetic moments are aligned with the magnetic field. However, due to disruption by thermal motion of the molecules, not all the individual magnetic moments can adopt the minimal energy. The energy of the interactions between nuclear magnetic moments is much smaller than that of thermal motion so that across the sample, the arrangement of the nuclear magnetic moments is almost random. Nevertheless, there is a slight preference for magnetic moments to be parallel to the magnetic field. If averaged over the whole sample, there will be a net magnetization aligned along the magnetic field. Due to this net magnetization, we say that the sample is magnetized and there is magnetization along the direction of the external magnetic field represented by a bulk magnetization vector (Fig. 3.1). Once the bulk magnetization is formed and reaches an equilibrium state, it does not change with time in magnitude and direction. However, if we tip the magnetization away from the z-axis so that the magnetization vector makes a certain α angle to the z-axis, the magnetization vector starts to rotate around the direction of the external field forming a cone with a constant angle (Fig. 3.2). This kind of rotation is called free precession and rotates with the frequency ω. The frequency of this rotation, the Larmor frequency, is dependent on the external magnetic field B_0 and γ, the gyromagnetic ratio, unique to the nuclei under observation (Eq. 3.1) [19, 20].

$$\omega = -\gamma B_0 \left(\text{rad s}^{-1}\right) \tag{3.1}$$

In order to tip the magnetization away from z-axis toward the xy-plane, it is necessary to apply a small radiofrequency irradiation at or near Larmor frequency along the x-axis. That is, the radiofrequency irradiation frequency is resonant with the Larmor frequency. Then, the magnetization is no longer along the z-axis but moves toward the xy-plane, meanwhile precessing about the external field. To detect the signal, a small

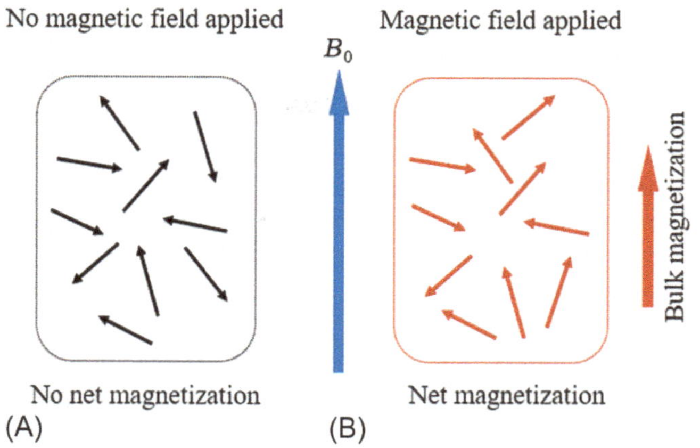

Fig. 3.1 The effect of magnetization on an NMR sample using the vector model. (A) With no magnetic field applied, the nuclei magnetic dipoles are oriented randomly. (B) When an external magnetic is field applied, a net magnetization occurs along the *z* direction.

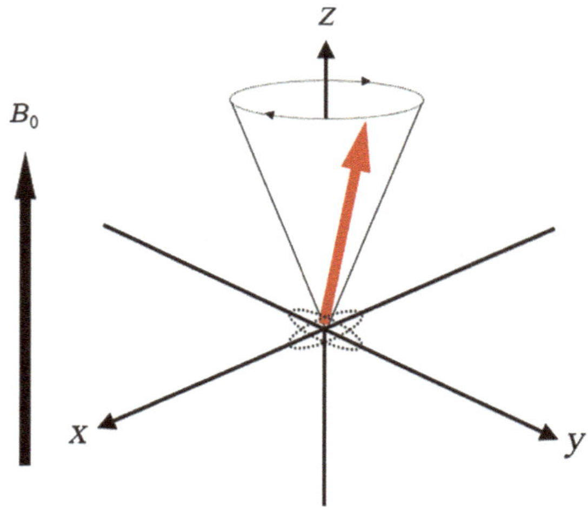

Fig. 3.2 Larmor precession.

coil, mounted in the *xy*-plane is around the sample. As the magnetization precesses and cuts the coil, a current is induced. Note that this coil is also used to generate radio-frequency pulses. This current is amplified and recorded, which is recognized as the NMR signal, or free induction decay (FID). After Fourier transformation, the time-domained FID becomes the frequency-domained NMR spectrum [19].

Solid-state NMR

Solid-state NMR is different from solution NMR in nuclear spin interactions. Anisotropic interactions, such as chemical shift anisotropy, dipole-dipole coupling, and quadrupolar coupling are all orientation dependent and present in solids [20]. These interactions result in significant line broadening in solid-state NMR spectra. In solid-state NMR, one often deals with powder samples that consist of many crystallites with random orientations. For such samples, the NMR spectra are due to the statistical distribution of all possible crystallite orientations and much broader than those of solution samples since different molecular orientations in the samples lead to different spectral frequencies. Dipole-dipole couplings in solids also result in additional line broadening. Furthermore, for nuclei with spin >½, the dominant line broadening source is the quadrupolar nuclei. Finally, anisotropy from the bulk magnetic susceptibility of solids further contributes to some line broadening.

In contrast, for solution-state NMR, these anisotropic interactions are averaged due to rapid molecular tumbling. Therefore, solid-state NMR spectra, without any special manipulations like magic-angle spinning, are broad [20]. The lack of resolution in these kinds of NMR spectra obscures significant structural information that can be extracted. Therefore, unlike solution-state NMR, solid-state NMR requires additional techniques to obtain high-resolution spectra.

The primary techniques used in solid-state NMR to obtain high-resolution spectra are magic-angle spinning (MAS), cross-polarization, and high-power decoupling. The primary function of magic-angle spinning is to (1) reduce or completely remove chemical shift anisotropy and (2) assist in the removal of heteronuclear dipolar couplings. With very high spinning speeds achievable due to technological developments, it is also used to remove the effects of homonuclear dipolar coupling and to narrow spectral lines due to quadrupolar coupling. As mentioned above, rapid molecular tumblings in solution NMR average the orientation-dependent interactions and the use of MAS is to achieve the same result for solids that lack these motions. In this technique, samples are put into rotors and spun at high spinning speeds at a specific angle, 54.74°, which is called the magic angle (Fig. 3.3) [20].

The molecular orientation is dependent on $3\cos^2\theta - 1$, where θ is the angle used to describe the orientation of the spin interaction tensor. The average of $3\cos^2\theta - 1$, $<3\cos^2\theta - 1>$ is proportional to $3\cos^2\theta_R - 1$, where θ_R is the angle between the rotor (the sample) and the applied magnetic field, B_0. If we set θ_R to 54.74°, $3\cos^2\theta_R - 1$ will be zero, causing $<3\cos^2\theta - 1>$ to be zero as well. If the spinning speed of the sample in

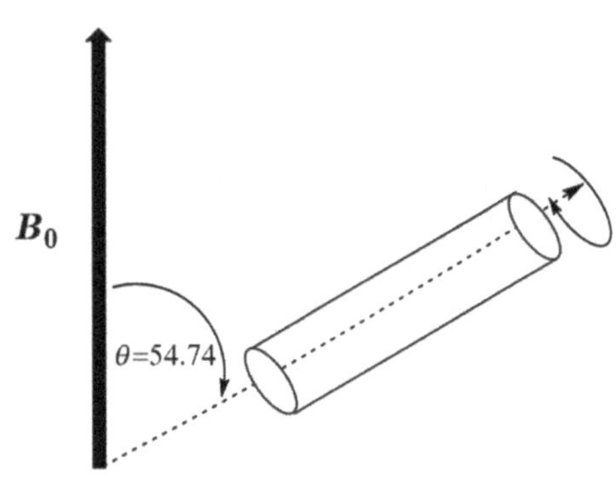

Fig. 3.3 MAS schematic.

the rotor is rapid enough for θ to be averaged compared to the interaction anisotropy, then the interaction anisotropy will be averaged to zero. Spinning at slower rates, particularly at those that are less than what eliminates the chemical shift anisotropy, leads to the appearance of a set of spinning sidebands. There are a few choices available to suppress these spinning sidebands. One of them is the insertion of a total suppression of sidebands (TOSS) pulse sequence before detection at low spinning speeds. The second major solid-state technique, cross-polarization, is primarily employed to enhance the signal intensity of dilute nuclei such as ^{13}C via cross-polarization with ^{1}H. The poor signal-to-noise ratios observed in ^{13}C spectra are due to the low natural abundance of the ^{13}C isotope as well as its low gyromagnetic ratio, which can lead to very long relaxation times. This ultimately results in longer recycle delays used and long measuring time. To overcome these challenges, the cross-polarization technique was developed. The polarization process capitalizes on the favorable features of ^{1}H isotopes (or any naturally abundant nuclei). The gyromagnetic ratio of ^{1}H isotopes is 4 times that of ^{13}C and the natural abundance of ^{1}H is 99.98% compared to the 1.07% for ^{13}C. During the polarization transfer, after an initial $90°_x$ pulse that moves ^{1}H atoms into the *xy*-plane, the protons are spin locked to keep them there. Then, the dilute nuclei (e.g., ^{13}C) and ^{1}H nuclei are subjected to pulses with frequencies set to meet the Hartmann-Hahn condition. It is during this period (or contact time) that a transfer of ^{1}H magnetization to the ^{13}C occurs as a result of the dipolar coupling between the two nuclei. Finally, the dilute nuclei now have sensitivity and relaxation times closer to that of the abundant species. The third technique, high-power heteronuclear decoupling, is used to remove the dipolar couplings between ^{1}H and ^{13}C nuclei during the detection. The above three techniques, combined, are the most frequently employed techniques and are collectively referred to as cross-polarization/magic-angle spinning (CP/MAS) (Fig. 3.4).

2D solid-state NMR spectroscopy

Basics of 2D NMR

Recent developments in 2D solid-state NMR techniques have greatly advanced the possibilities for structural identification of complex organic materials in nature [17, 18, 21]. Compared to the information available from using one-dimensional (1D) solid-state NMR experiments, 2D solid-state NMR techniques provide superior

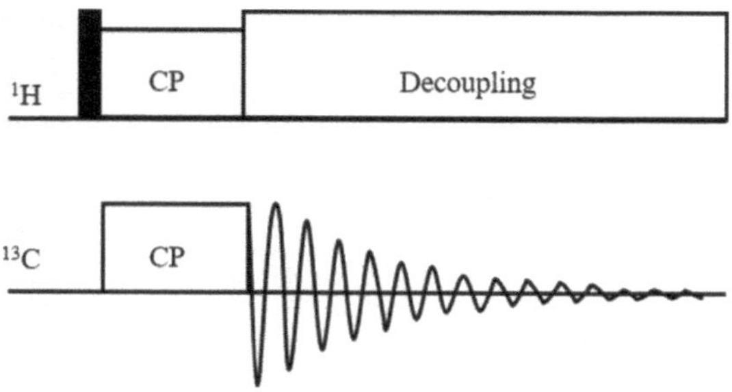

Fig. 3.4 Pulse sequence of a CP/MAS NMR experiment.

structural details for the complex compounds in environmental research [22, 23]. The diversity of chemical species and broad lines observed in the spectra of complex organic materials make detailed analyses difficult by using 1D solid-state ^{13}C NMR techniques alone [23, 24]. For example, some 1D ^{13}C solid-state NMR spectra usually have broad and heavily overlapped bands in the aromatic (93–150 ppm) region that cannot be clearly distinguished. Utilizing 2D solid-state NMR techniques can greatly mitigate such ambiguity in peak assignments as well as provide extra structural information never achievable in 1D spectra of environmental samples [10, 11].

The general scheme for 2D NMR spectroscopy (Fig. 3.5) consists of four distinct periods: (1) preparation, (2) evolution, (3) mixing, and (4) detection [25, 26]. Acquisition of 2D NMR spectra first begins with bringing the nuclear spin system to a well-defined initial state during the preparation period. During this period, magnetization returns toward thermal equilibrium and a single RF pulse, or a combination of pulses, create a transient state of the spin system. Coherence, both observable and unobservable, developed in the preparation period evolves during the evolution period, t_1, but is not detected. Note that t_1 is not a fixed period but is systematically incremented by Δt. After this evolution period where the spin system has evolved with frequency ω_1 under the influence of various nuclear spin interactions, comes the mixing period. Here, unobservable coherence present at the end of the evolution period t_1 will be converted into observable signals following RF pulses. After the mixing period, the observable resonance signal is detected as a function of the second time variable, t_2. Now time-domain functions depending on both t_1 and t_2 are generated. Fourier transformation with respect to t_2 and then t_1 results in a 2D spectrum with two frequency axes, ω_1 and ω_2.

The above describes the general scheme for 2D NMR, but the interactions determining the time evolution of the spin system can be manipulated in many ways, for example, using homo- or heteronuclear decoupling, MAS, or polarization transfer [23, 25, 26]. Especially, when these techniques are combined with various spectral-editing methods, 2D NMR spectra have great potential for characterizing organic matter or other complex molecules at the molecular level.

Advanced 2D solid-state NMR techniques

There are two main types of 2D NMR spectroscopy: homonuclear and heteronuclear correlation. Both involve a magnetization transfer from one type of nuclei to another. Homonuclear correlation involves the observation of the transfer between like nuclei, for example, ^{1}H-^{1}H. In contrast, heteronuclear experiments seek to observe the transfer between different nuclei such as: ^{1}H and ^{13}C or ^{13}C and ^{15}N.

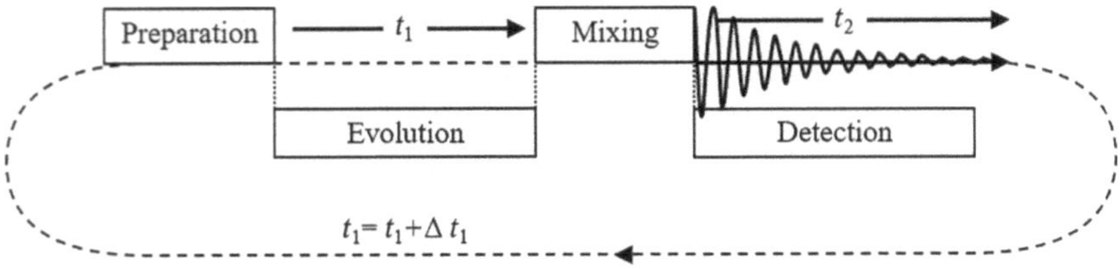

Fig. 3.5 Schematic representation for the generic 2D NMR experiment.

2D ^1H-^{13}C heteronuclear correlation (HETCOR) solid-state NMR

The 2D ^{1}H-^{13}C heteronuclear correlation (HETCOR) NMR is arguably the most important 2D technique for advanced structural elucidation of environmental samples. It provides useful information on the structure of complex organic materials via the transfer of magnetization from proton to carbon nuclei. ^{1}H-^{13}C HETCOR can be used for: (1) clearer determination of functional groups, (2) measurements of connectivity of specific functional groups, and (3) identification of heterogeneities and domains. Therefore, it is used for studying a broad variety of samples such as coal, soil organic matter, proteins, and multicomponent polymers [16, 21, 23, 27–29]. ^{1}H-^{13}C heteronuclear correlation in solids is based on dipolar-coupling-driven magnetization transfer and different schemes have been developed for examining various samples [30].

HETCOR for identification of functional groups

The correlation of ^{1}H and ^{13}C chemical shifts in solid-state NMR has been extensively used in the analysis of complex natural organic materials [23, 27, 31]. Unfortunately, the poor resolution in solids limits the structural details that can be obtained by traditional HETCOR experiments [23, 32]. Since its original version was developed for solid-state NMR, extensive research and additional enhancements have been made to improve the HETCOR technique [16, 21].

Using routine CP and/or insertion of ^{1}H spin diffusion combined with various spectral-editing techniques, this technique offers a variety of analytical options. Spectral-editing techniques such as dipolar dephasing, ^{13}C T_2 (spin-spin relaxation time) and ^{13}C chemical shift anisotropy (CSA) filters enable experimenters to simplify spectra and identify more specific functional groups (Table 3.1) [21]. For example, using a CSA filter, anomeric (O-C-O) signals can be separated from those of aromatic compounds [33]. Another example involves clever manipulations of spins to isolate CH_2 group signals [34].

HETCOR for connectivity

In addition to identification of specific functional groups, ^{1}H-^{13}C 2D HETCOR can provide details regarding functional group connectivity. While routine 2D HETCOR does identify the nearest protons for carbons, combining it with various spectral-editing techniques can show connectivity and proximities of particular functional groups more clearly [32]. For example, 2D HETCOR with 40-μs dipolar dephasing and 1-ms CP provides connectivity information on nonprotonated carbons and mobile carbons, like CCH_3 and OCH_3 [35]. In addition, adding a proton spin-diffusion period to the abovementioned 2D HETCOR experiment can further extend the observed long-range correlations to more distant nuclei [23, 32].

HETCOR for domains and heterogeneities

^{1}H-^{13}C 2D HETCOR experiments with ^{1}H spin diffusion can be used to identify domains and heterogeneities in environmental samples. This knowledge is useful for understanding the complex formation mechanisms associated with the environmental samples. Generally, the "domain" refers to the structural unit that is constructed by similar kinds of chemical compounds, such as lignin, with diameters

Table 3.1 Functional groups of sp^2 and sp^3 hybridized carbons with corresponding NMR techniques.

Functional groups	Formula	Technique
sp^2 hybridized carbons		
Aldehyde	RCHO	^{13}C CP-TOSS with short CP
Ketone	R(C=O)R′	^{13}C CP-TOSS with 40-μs dipolar dephasing
Carboxylic acid	RCOOH	2D HETCOR with OH/NH selection combined with SUPER and SPIDER
Carboxylate ion	$RCOO^-$	SUPER and SPIDER
Ester of carboxylic acid	RCOOR	SUPER and SPIDER
Amide	RCON	SUPER and SPIDER
Phenol	ArOH	2D HETCOR with OH/NH suppression
Aromatic ester	ArOOR	2D HETCOR
Aromatic ether	$ArOCH_3$	2D HETCOR
Aromatic amine	ArNR	SPIDER
Anilide	ArNHCOR	SPIDER and SPIDER with dipolar dephasing
Protonated aromatics	Ar C-H	Dipolar dephasing and CH selection
Nonprotonated aromatics	Ar C-C	Dipolar dephasing and CH selection
Exomethylenes	$CH_2=C$	CH_2 selection
Bridgehead carbons	Ar C=C	Long-range dipolar dephasing
Heterocyclic N		SPIDER
sp^3 hybridized carbons		
Protonated anomerics	OC(H, R′)O	^{13}C CSA filter with short CP
Nonprotonated anomerics	OC(R, R′)O	^{13}C CSA filter with dipolar dephasing
OCH of carbohydrate	OCH	CH selection
OCH2 of carbohydrate	OCH_2	CH_2 selection
Alkyl ether	ROR′	2D HETCOR with OH/NH suppression and SUPER
Alcohol	ROH	2D HETCOR with OH/NH suppression and SUPER
Methoxyl	OCH_3	Dipolar dephasing
N-alkyl	NC	Dipolar dephasing, CH selection and SPIDER
Alkyl CH	CCH	CH selection
Alkyl CH2	CCH_2	CH_2 selection
Alkyl CH3	CCH_3	Dipolar dephasing
Alkyl quaternary carbon	CC_q	Dipolar dephasing
Crystalline methylene	$-(CH_2)_n-$	^{13}C DP-MAS and ^{13}C DP-MAS with a short recycle delay, ^1H spin diffusion
Amorphous methylene	$-(CH_2)_n-$	^{13}C DP-MAS and ^{13}C DP-MAS with a short recycle delay, ^1H spin diffusion

larger than 5 nm [16, 25]. In contrast, if the diameter of a structural unit is lower than 5 nm, it can be defined as "heterogeneity" [16, 25]. According to the characteristics of spin diffusion, the equilibration time of magnetization is short for small domains and heterogeneities while it is longer for larger domains [16]. Hence, the signals from the domains of organic materials can be observed by ^{1}H-^{13}C 2D HETCOR with different ^{1}H spin-diffusion times. For example, ^{1}H-^{13}C 2D HETCOR spectrum (Fig. 3.6A) with spin diffusion of $t_m = 10\,\mu s$ only shows one bond correlations between aromatic protons and aromatic carbons [21]. Increasing spin-diffusion time to 200 μs (Fig. 3.6B), extends connectivity and is visualized in the enhanced cross-peak region, where correlations between many types of protons and carbons are seen [21].

2D solid-state ^{13}C-^{13}C correlation NMR

2D solid-state ^{13}C-^{13}C correlation NMR is very helpful in the observation of the carbon skeleton of organic materials [17, 22]. It is used to determine carbon atom connectivity via ^{13}C-^{13}C dipolar coupling. Due to the low natural abundance of ^{13}C (1.11%), however, the applicability of 2D ^{13}C-^{13}C correlation techniques is often limited in the environmental sciences. Hence, isotopic labeling or enrichment strategies are necessary in 2D ^{13}C-^{13}C correlation solid-state experiments. In addition to isotopic labeling, the recent development of dynamic nuclear polarization (DNP) also offers significant signal enhancements in MAS experiments and provides

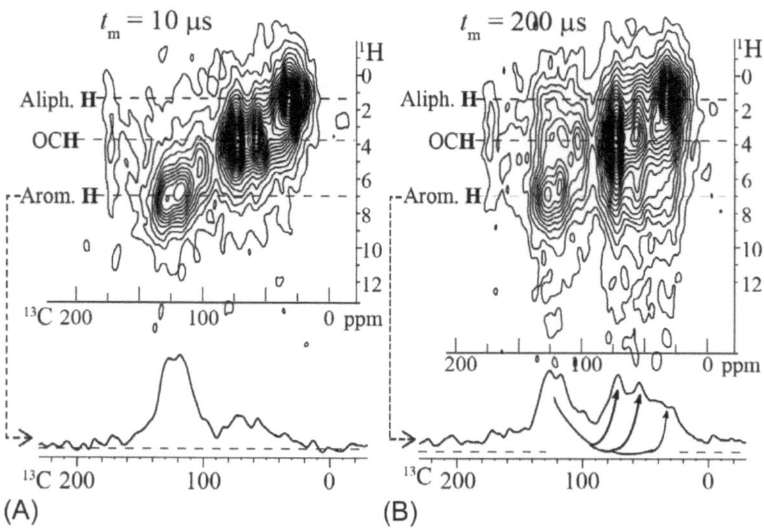

Fig. 3.6 2D HETCOR spectra with different ^{1}H spin-diffusion times, (A) $t_m = 10\,\mu s$ and (B) $t_m = 200\,\mu s$, for a humic acid sample.
Reproduced with permission (A) from J. Mao, et al., Advanced solid-state NMR spectroscopy of natural organic matter, Prog. Nucl. Magn. Reson. Spectrosc. 100 (2017) 17-51; (B) from J. Mao, N. Chen, X. Cao, Characterization of humic substances by advanced solid state NMR spectroscopy: demonstration of a systematic approach, Org. Geochem. 42 (2011) 891-902.

opportunities for 2D ^{13}C-^{13}C correlation solid-state experiments in natural abundance organic matter [36].

One of the most frequently used 2D ^{13}C-^{13}C correlation techniques is 2D double-quantum (DQ)/single-quantum (SQ) ^{13}C-^{13}C NMR. The core of DQ/SQ sequences is the classic Incredible Natural Abundance DoublE QUAntum Transfer Experiment (INADEQUATE). The INADEQUATE experiment is commonly used to obtain information regarding directly bonded ^{13}C atoms. To date, only a few examples of solid-state 2D DQ/SQ ^{13}C-^{13}C correlation experiments have been reported in the environmental science field [17, 22, 37]. Compared to SQ/SQ-type experiments, DQ/SQ simplifies analysis of spectra so that the undesirable diagonal signals from uncorrelated magnetization are significantly reduced in intensity [38]. This reduction of the diagonal signal increases resolution and makes the spectra easier to analyze because it is less crowded [39]. Specifically, the cross-peaks near the diagonal ridge are simpler and more easily identified. Hence, 2D DQ/SQ ^{13}C-^{13}C correlation is particularly valuable for identification of C-C pairs in organic matter.

Fig. 3.7 shows two ^{13}C-^{13}C correlation NMR spectra of ^{13}C glucose samples carbonized by treatment with fuming sulfuric acid [22]. In Fig. 3.7A, there is an intense diagonal peak in the aromatic region (100–168 ppm), which makes the cross-peaks near the diagonal ridge difficult to identify. The 2D DQ/SQ ^{13}C-^{13}C correlation NMR spectrum of the same sample seen in Fig. 3.7B shows that the immense diagonal ridge is removed with the use of the DQ filter. From this added clarity, we can identify the strong cross-peak near 130 ppm to be bridgehead carbons of fused aromatic rings. Overall, the 2D DQ/SQ ^{13}C-^{13}C correlation experiment is a useful method for characterizing the chemical structure of organic materials, especially for aromatic-enriched materials.

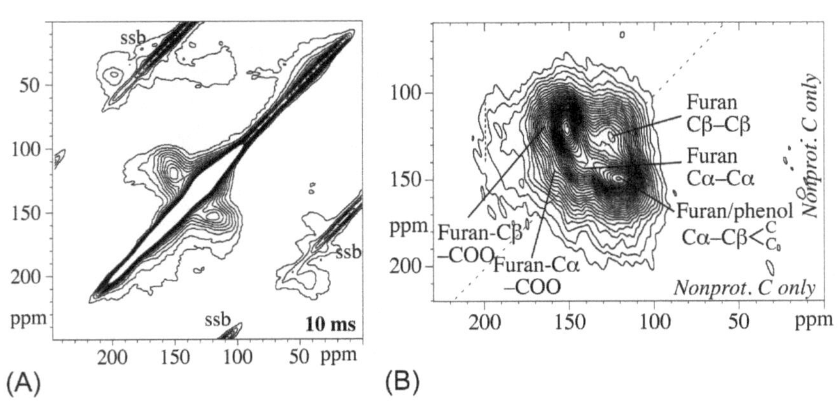

Fig. 3.7 (A) The 2D SQ/SQ ^{13}C-^{13}C NMR spectrum without spectral editing of a ^{13}C labeled glucose sample charred by fuming sulfuric acid, with a spin-diffusion time of 10 ms and (B) a sheared dipolar-dephased DQ/SQ NMR spectrum of the same sample.
Reproduced with permission from R.L. Johnson, et al., Spectrally edited 2D ^{13}C-^{13}C NMR spectra without diagonal ridge for characterizing ^{13}C-enriched low-temperature carbon materials, J. Magn. Reson. 234 (2013) 112-124.

Summary and outlook

NMR techniques have been extensively applied in the field of environmental science, which lead to thousands of publications. However, most of them only have employed simple ^{13}C or ^{15}N CP/MAS NMR techniques. While advanced 2D solid-state NMR techniques are extensively used in biomolecular NMR, they have so far been severely underexploited in the environment research, mainly due to a lack of expertise and appropriate techniques as well as instruments and spectroscopists dedicated to environmental research. Compared with simple 1D NMR spectra, advanced 2D solid-state NMR spectra have great potential to elucidate the chemical structures of environmental samples at the molecular level, especially combined with the isotopic labeling method and spectral-editing techniques. The structural information obtained from advanced solid-state NMR will help clarify the biogeochemistry processes in various environmental systems.

References

[1] R. Pohmann, Physical basics of NMR, in: In Vivo NMR Imaging, Springer, 2011, pp. 3–21.

[2] G. Bonanomi, et al., Linking bacterial and eukaryotic microbiota to litter chemistry: Combining next generation sequencing with ^{13}C CPMAS NMR spectroscopy, Soil Biol. Biochem. 129 (2019) 110–121.

[3] X. Wang, et al., Structural convergence of maize and wheat straw during two-year decomposition under different climate conditions, Environ. Sci. Technol. 46 (13) (2012) 7159–7165.

[4] P.G. Hatcher, D.C. Waggoner, H. Chen, Evidence for the existence of humic acids in peat soils based on solid-state ^{13}C NMR, J. Environ. Qual. 48 (6) (2019) 1571–1577.

[5] I. Kostenko, E. Abakumov, Characterization of humic acids in mountainous meadow soils and burozems of the crimea using ^{13}C-NMR, Eurasian Soil Sci. 51 (12) (2018) 1411–1418.

[6] N. Hao, et al., Review of NMR characterization of pyrolysis oils, Energy Fuel 30 (9) (2016) 6863–6880.

[7] X. Fang, et al., Quantitative ^{13}C NMR of whole and fractionated Iowa Mollisols for assessment of organic matter composition, Geochim. Cosmochim. Acta 74 (2) (2010) 584–598.

[8] N.J. Mathers, Z. Xu, Solid-state ^{13}C NMR spectroscopy: characterization of soil organic matter under two contrasting residue management regimes in a 2-year-old pine plantation of subtropical Australia, Geoderma 114 (1-2) (2003) 19–31.

[9] Y. Gélinas, J.A. Baldock, J.I. Hedges, Demineralization of marine and freshwater sediments for CP/MAS ^{13}C NMR analysis, Org. Geochem. 32 (5) (2001) 677–693.

[10] R.M. Duarte, A.C. Duarte, Unraveling the structural features of organic aerosols by NMR spectroscopy: a review, Magn. Reson. Chem. 53 (9) (2015) 658–666.

[11] P. Mazzei, A. Piccolo, Interactions between natural organic matter and organic pollutants as revealed by NMR spectroscopy, Magn. Reson. Chem. 53 (9) (2015) 667–678.

[12] I. Kögel-Knabner, ^{13}C and ^{15}N NMR spectroscopy as a tool in soil organic matter studies, Geoderma 80 (3-4) (1997) 243–270.

[13] S. Chukov, E. Lodygin, E. Abakumov, Application of ^{13}C NMR spectroscopy to the study of soil organic matter: a review of publications, Eurasian Soil Sci. 51 (8) (2018) 889–900.

[14] B. Blümich, et al., Solid-State NMR II: Inorganic Matter, vol. 31, Springer Science & Business Media, 2013.

[15] D.D. Laws, H.M.L. Bitter, A. Jerschow, Solid-state NMR spectroscopic methods in chemistry, Angew. Chem. Int. Ed. 41 (17) (2002) 3096–3129.
[16] J. Mao, et al., Advanced solid-state NMR spectroscopy of natural organic matter, Prog. Nucl. Magn. Reson. Spectrosc. 100 (2017) 17–51.
[17] X. Cao, et al., A major step in opening the black box of high-molecular-weight dissolved organic nitrogen by isotopic labeling of synechococcus and multibond Two-Dimensional NMR, Anal. Chem. 89 (22) (2017) 11990–11998.
[18] K. Schmidt-Rohr, J.-D. Mao, D. Olk, Nitrogen-bonded aromatics in soil organic matter and their implications for a yield decline in intensive rice cropping, Proc. Natl. Acad. Sci. 101 (17) (2004) 6351–6354.
[19] J. Keeler, Understanding NMR Spectroscopy, John Wiley & Sons, 2011.
[20] M.J. Duer, Solid state NMR spectroscopy: principles and applications, John Wiley & Sons, 2008.
[21] J. Mao, N. Chen, X. Cao, Characterization of humic substances by advanced solid state NMR spectroscopy: demonstration of a systematic approach, Org. Geochem. 42 (8) (2011) 891–902.
[22] R.L. Johnson, et al., Spectrally edited 2D ^{13}C-^{13}C NMR spectra without diagonal ridge for characterizing ^{13}C-enriched low-temperature carbon materials, J. Magn. Reson. 234 (2013) 112–124.
[23] J.-D. Mao, B. Xing, K. Schmidt-Rohr, New structural information on a humic acid from two-dimensional ^{1}H-^{13}C correlation solid-state nuclear magnetic resonance, Environ. Sci. Technol. 35 (10) (2001) 1928–1934.
[24] J.-D. Mao, K. Schmidt-Rohr, Accurate quantification of aromaticity and nonprotonated aromatic carbon fraction in natural organic matter by ^{13}C solid-state nuclear magnetic resonance, Environ. Sci. Technol. 38 (9) (2004) 2680–2684.
[25] K. Schmidt-Rohr, H.W. Spiess, Multidimensional Solid-State NMR and Polymers, Elsevier, 2012.
[26] B. Blümich, H.W. Spiess, Two-dimensional solid-state NMR spectroscopy: new possibilities for the investigation of the structure and dynamics of solid polymers [New Analytical Methods (38)], Angew. Chem. Int. Ed. Engl. 27 (12) (1988) 1655–1672.
[27] B.-J. Van Rossum, H. Förster, H. De Groot, High-field and high-speed CP-MAS ^{13}C NMR heteronuclear dipolar-correlation spectroscopy of solids with frequency-switched Lee–Goldburg homonuclear decoupling, J. Magn. Reson. 124 (1997) 516–519.
[28] K.W. Zilm, G.G. Webb, ^{13}C proton shift correlation spectroscopy of a whole coal, Fuel 65 (5) (1986) 721–724.
[29] M.A. Wilson, et al., ^{1}H CRAMPS NMR derived hydrogen distributions in various coal macerals, Org. Geochem. 20 (7) (1993) 985–999.
[30] A. Lesage, et al., Carbon – proton chemical shift correlation in solid-state NMR by through-bond multiple-quantum spectroscopy, J. Am. Chem. Soc. 120 (50) (1998) 13194–13201.
[31] H. Kono, Two-dimensional magic angle spinning NMR investigation of naturally occurring chitins: Precise ^{1}H and ^{13}C resonance assignment of α-and β-chitin, Biopolymers 75 (3) (2004) 255–263.
[32] X. Yao, K. Schmidt-Rohr, M. Hong, Medium- and long-distance ^{1}H–^{13}C heteronuclear correlation NMR in solids, J. Magn. Reson. 149 (2001) 139–143.
[33] J.-D. Mao, K. Schmidt-Rohr, Separation of aromatic-carbon ^{13}C NMR signals from di-oxygenated alkyl bands by a chemical-shift-anisotropy filter, Solid State Nucl. Magn. Reson. 26 (1) (2004) 36–45.
[34] J.-D. Mao, K. Schmidt-Rohr, Methylene spectral editing in solid-state ^{13}C NMR by three-spin coherence selection, J. Magn. Reson. 176 (1) (2005) 1–6.

[35] J. Mao, et al., Characterization of a nitrogen-rich fulvic acid and its precursor algae from solid state NMR, Org. Geochem. 38 (8) (2007) 1277–1292.
[36] T. Maly, et al., Dynamic nuclear polarization at high magnetic fields, J. Chem. Phys. 128 (5) (2008) 02B611.
[37] G.P. Holland, et al., Quantifying the fraction of glycine and alanine in β-sheet and helical conformations in spider dragline silk using solid-state NMR, Chem. Commun. 43 (2008) 5568–5570.
[38] T. Kobayashi, I.I. Slowing, M. Pruski, Measuring long-range ^{13}C–^{13}C correlations on a surface under natural abundance using dynamic nuclear polarization-enhanced solid-state nuclear magnetic resonance, J. Phys. Chem. C 121 (44) (2017) 24687–24691.
[39] Y. Nishiyama, et al., Modern Methods in Solid-State NMR: A Practitioner's Guide, Royal Society of Chemistry, 2018.

High-resolution mass spectrometry strategies for the investigation of dissolved organic matter

4

Jeffrey A. Hawkes[a], *William Kew*[b]
[a]Uppsala University, Uppsala, Sweden, [b]Environmental Molecular Sciences Laboratory, Pacific Northwest National Laboratory, Richland, WA, United States

Motivation

High-resolution mass spectrometry (HRMS) reveals an unparalleled chemical complexity in environmental samples, and an incredible amount of data can be gathered from single samples. Unfortunately, without analytical standards, the data is not quantitative and cannot be assigned to specific structures, so the exact nature of natural mixtures like dissolved organic matter (DOM) is unknown. Current research of DOM using HRMS aims to either probe the nature of such mixtures or to explore how DOM mixtures change during environmental processing. In this chapter, we will discuss the elements of both research trends.

We begin by exploring the environmental reasons for the ubiquitous feature diversity found and then will discuss analytical strategies that aim to improve our understanding of environmental samples and the way that they are processed in nature. This work is intended as a reference for aquatic scientists who want to better understand the analytical obstacles presented by complex mixture HRMS analysis as well as some of the recent analytical strategies that have been developed to deal with these problems.

Environmental production and processing of DOM with regards to individual analytes

Due to the complexity of DOM and typically low analyte concentrations, it is often difficult to trace the origins of individual components [1]. Many analytes or signals (e.g., NMR features, fluorescence signals, HRMS peaks) seem to be interrelated and are often assumed to have a similar source, history, and fate. In aquatic systems, it is common to find references to "allochthonous," or "terrestrial," and "autochthonous," or "aquatic," organic matter, meaning produced previously and transported to the location or produced locally, respectively. Other studies take a more functional approach and define groups as "recalcitrant" and "fresh." "Terrestrial" DOM tends to have more phenolic, aromatic character than fresh DOM and is therefore highly light absorbent. Its source is generally considered to be plant structure material such as lignin, which has been processed and solubilized in soils before washing into the aquatic system through soil

Multidimensional Analytical Techniques in Environmental Research. https://doi.org/10.1016/B978-0-12-818896-5.00004-1

pore waters. "Autochthonous" organic matter is formed via primary production and local processing of the waste products of primary producers. Other "fresh" sources of DOM may include aerial sources to aquatic systems such as leaf litter, pollen, and dust (i.e., wet deposition), although these are clearly allochthonous.

Generally, biological processes use enzyme-mediated reactions to produce relatively large abundances of a small number of compounds (metabolites), while geochemical processes are less specific and tend to produce many more randomized structures. Consider, for example, the production of glucose via the Calvin cycle, which requires specific nucleotides/cofactors (ATP and NADH) and produces a single product (glyceraldehyde-3-phosphate) and compare with the Maillard browning reaction, which is a purely chemical process that takes a reducing sugar (e.g., glucose) and an amino acid and forms a complex mixture of products after application of heat [2]. For this reason, it can be expected that "fresh" organic matter has less molecular diversity than geochemically processed DOM.

Labile DOM may be cycled so rapidly that its steady-state concentration is low, despite participating in a large portion of the total active carbon biogeochemistry in the system. Confounding this problem are analytical issues. It is possible, even likely, that due to its more hydrophilic nature [3], fresh organic matter is harder to isolate from water [4], which means that studies which employ solubility-based isolation methods like solid-phase extraction are often biased toward processed, less reactive DOM. Further still, freshly produced compounds may be less compatible with electrospray ionization (ESI) (see "Ionization and optimizing sample transfer to the mass spectrometer" section) due to a lower abundance of readily ionized groups (e.g., more sugars and fewer organic acids), higher mass (proteins and polysaccharides), and less hydrophobic character (worse ESI spray character). ESI-HRMS and SPE thus combine with the biogeochemical fact of low concentration to form a situation where fresh DOM is less well characterized and understood than relatively inactive, stable DOM.

The few studies that investigate fresh DOM with MS find that it does indeed have a lower isobaric diversity and a less Gaussian distribution, indicating a less randomized and processed nature [4–6]. By inference, the Gaussian and highly patterned distributions found in processed DOM are controlled by combinatorial limits of molecular compositions, and the averaging of an exponentially increasing number of isomers [6–8]. Furthermore, the overall distributions may also be a factor of ionization efficiency and response based on structural functionality. Hertkorn et al. [9] conducted a comprehensive, philosophical analysis of the isobaric, isomeric, and structural diversity of DOM. They report that the number of chemically sensible isomers at 178 Da ranges from 10^4 to 10^8 depending on the molecular formula. Additionally, they note that increasing oxygen saturation at first increases the number of isomers, then eventually decreases it, as oxygen only takes two chemical bonds. As a result, the diversity of isomers is highest in the central area of the van Krevelen diagram (near to $H/C = 1$ and $O/C = 0.4$). It is probably no coincidence that this area of van Krevelen space contains the largest peak intensities as measured by ESI-HRMS—the isomeric diversity is greatest here, and so even though each isomer may have a vanishingly low concentration [1], the accumulated concentration is higher. It is debated whether functional diversity is as large as isomeric diversity—it may be that similar core structures are present with randomly arranged but similar functional groups [7, 10, 11], and similar core structures may be present in diverse environments due to their chemical stability [6, 8].

HRMS strategies for the investigation of DOM

In an idealized case, an organic analyte in a water sample would be completely extracted and concentrated from the sample without alteration, it would ionize efficiently (give a large response) with a known and reproducible sensitivity (signal/abundance) and it would have no interferences from other analytes with the same mass. Unfortunately, each of these components of the analytical pipeline has multiple complications that do not allow such confidence in analyte determination.

Sample preparation

Seldom can environmental samples be analyzed by MS directly [12], without some preparation, due to the problematic interference of salt and the typically low concentration of DOM. Recent developments in sample preparation have been reviewed comprehensively [13, 14], and here we will focus on how the most common approaches influence the results gathered by HRMS.

Most methods simply aim to maximize the total amount of organic carbon recovered [3, 15–17]. The most common techniques for isolating DOM involve acidification of the sample in order to protonate carboxylic acids, facilitating their immobilization onto hydrophobic resins, polymers or silica phases, and the most common sorbent used is Agilent PPL due to the relative ease and low cost of the method [15]. DEAE cellulose and DEAE sepharose [16, 18], along with weak anion exchange (WAX) resins are also useful, because they can be used at ambient pH, minimizing modification of the sample and also allowing long-term passive sampling [19].

A selection of 24 SPE cartridges was tested for extraction efficiency and proton environment as observed by nuclear magnetic resonance and obtained O/C and H/C ratio by FT-ICR MS using a freshwater and marine sample [20]. The study clearly demonstrated that the chemistry of DOM isolates can be tailored to suit research goals in cases where certain compound classes are required (e.g., weak anions can be extracted by weak anion exchangers; polar sorbents preferentially extract low oxygen species). The study showed that while PPL gave the highest recovery for extraction of marine DOM, different sorbents may be useful for different research questions. Studies that aim to investigate hydrophobic naphthenic acids can selectively target these compounds using hydrophobic polymers [21].

Raeke et al. [3] found that PPL extraction at pH 2 only had a minor effect on the results of ESI(−)-MS in environmental mixtures, despite only recovering 61–66% of total carbon. This suggests that PPL has similar selectivity to the available analytes as negative mode ESI. The same study found that carbohydrates and other polar compounds were poorly extracted, particularly compounds with log D (octanol/water partition coefficient at pH 2) lower than −1. Further tests found that a large portion of environmental material was extracted but not eluted (i.e., irreversibly retained) [22]. This effect can even be observed by eye in very colored samples, as the frit on top of the cartridge often stains yellow-brown during extraction. This indicates that some irreversible adsorption occurs, possibly by filtration of colloidal DOM aggregates rather than bonding.

Because carboxylic acids constitute a large proportion of aquatic DOM, hydrophobic resin techniques (e.g., PPL) are quite successful in recovering a majority of DOM but are highly selective and poor at extracting more labile and bioactive components such as amino acids (and presumably peptides) and free sugars or polysaccharides [3]. This leads to important and drastic limitations for studies that aim to investigate these types of molecules. Any purification method is inherently biased and selective to some portion of the rich diversity of DOM and this must be considered when planning the study and evaluating the results. This is especially important in the context of dissolved organic nitrogen or phosphorous, which are not selectively targeted with current protocols.

Because there is little ESI(−) ionizable material in the whole water sample that is not also found in the PPL method extract, the ESI(−) ionizable portion of DOM is likely to be quantitatively constrained at the upper end by the typical extraction efficiency on PPL (~50%) in aquatic systems and could be somewhat lower [23, 24]. It may also be considered that methods which use methanol for elution of retained material (e.g., C18, PPL, as opposed to DEAE) can inadvertently lead to methylation (methyl esterification) of carboxylic acids in the sample, and this may explain the apparent loss of the most carboxylated analytes [17, 25, 26]. Samples prepared in methanol for MS analysis may also suffer the same fate although this is not proven to occur on the typical sample preparation timescales [27]. Sequential elution from sorbents can yield useful results similar to preparative chromatography ("Formula assignment and distinguishing isobars" section), as the material is fractioned into chemically defined portions [22, 28–31]. Sample preparation is not a solved problem and although the PPL SPE technique is cheap and convenient, its limitations should always be kept in mind.

Ionization and optimizing sample transfer to the mass spectrometer

Ionization techniques

A fundamental stage in MS is the generation of ions, and the choice of ionization technique substantially influences the resulting mass spectrum. In complex mixture analysis, especially DOM characterization, negative mode ESI dominates for several reasons. It is exceptionally well suited to ionizing polar carboxylic acids that seem to constitute a major portion of DOM, and it is a soft technique generating thousands of singly charged, deprotonated molecular ions across a large mass range [32] so the spectra are readily interpreted. Additionally, it is the standard source that is sold with commercial instruments. However, ESI—like all ionization techniques—is biased in which species it ionizes and how efficiently it ionizes them.

In ESI, the solution is vaporized, ejecting sequentially smaller droplets, and eventually desolvated ions or neutral molecules. ESI preferentially ionizes polar species [33–35]. The complex ESI mechanism leads to a challenging prediction of the extent to which gas-phase ions are produced, called the "ionization efficiency" [33, 36] or "monomer ion yield" [24]. Ionization efficiency spans orders of magnitude and is further complicated by ionization suppression in which ions "compete" for charge with each other and solvent and additive molecules. Ionization efficiency is thus affected by solvent composition and other matrix effects (protic vs aprotic, polarity, pH, other additive concentrations, and overall sample concentration) [35], meaning that it is not a well-predicted property. The effect of aspects such as solvent composition, concentration, and pH have been investigated to some extent in complex organic mixtures [37–39].

Ultimately, not every analyte in the spray is ionized and drawn into the mass spectrometer to the same extent, and the unknowable ionization efficiency for unknown analytes means that ESI-MS is not quantitative. The intensity of each peak in the spectrum is thus correlated to the final ion abundance in the mass analyzer (with ion abundances also moderated by the ion optics leading up the analyzer) but has no necessary relationship with the analyte abundance in solution (concentration). Approximate relative comparison—of the same ions—between samples requires a matched matrix. This nonquantitative property has an obvious problematic effect on the interpretation of MS data from complex mixtures, as the most intense ions can skew statistical or visualization efforts without appropriate normalization. Additionally, some analytes are so poorly ionized that their peak is below the limit of detection of the analyzer and they are not considered in the resulting data.

Adducted ions—where the charge is introduced with a, for example, sodium or potassium (positive mode) or chloride (negative) ion—can be produced in ESI, further complicating the spectra. This is typically more problematic in positive mode for DOM analysis, thus further encouraging negative mode analysis. Complementary ionization techniques, including atmospheric pressure photoionization and chemical ionization (APPI and APCI), ionize less polar species through photon or electron capture mechanisms and are less susceptible to interference from the salts or fatty acids which plague ESI spectra [40, 41]. However, APPI and APCI spectra are complicated by the formation of molecular radical ions and may require higher performance mass analyzers to resolve and correctly identify these species.

These solution state techniques can be complemented by solid-phase ionization, such as matrix-assisted laser desorption/ionization and laser desorption ionization (MALDI, LDI), where one (or more) lasers ablate and ionize molecules from a solid surface. This approach has had some recent interest [27, 42] although its advantages are more apparent for environmental samples which require extraction—such as soil—rather than DOM, which is or should be, inherently soluble. That said, analysis of DOM standards such as SRFA and SRNOM by LDI yields complementary spectra and largely different compound identification than by ESI. The mechanisms and analysis of LDI spectra of environmental samples are still being actively pursued by the community. Another technique that uses deposited material is paperspray ionization (PSI, PSCI), however, ions produced are similar to those of ESI and APCI. One reported advantage of this technique is the lowered sensitivity to salt [43], which may reduce the amount of sample clean up required.

Alternatively, a gas-phase method—proton transfer reaction (PTR)-MS—has been used to analyze the volatile and semi-volatile components of seawater DOM [44] and successfully used to fingerprint differences between DOM sources based on lower molecular weight compounds (<100 Da) than typically observed by ESI-FTMS [45].

The analytical window is limited by the use of just one ionization technique or polarity, and if more laboratories had access to a wider range of ionization sources, it is likely that more studies would have been conducted with multiple ionization techniques for increased coverage of the mixture [40, 41, 46]. Similar to multiple sample preparation techniques, studies must evaluate the benefit of extra analytical techniques vs the extra time and cost required.

Segmented acquisition

One straightforward method to boost the dynamic range of the mass spectra is to acquire multiple regions of the mass spectrum and stitch them together [47–49]. Use of a quadrupole mass filter can remove all but a narrow range of masses (10–100 *m/z*). This, coupled with an increase in the ion accumulation time, increases the signal-to-noise within that mass region. This can be repeated across the mass ranges of interest to produce a pseudo-broadband mass spectrum, but with a substantially higher dynamic range than typically achieved through normal broadband acquisition. Note that ion accumulation time has been shown to affect relative ion abundances in complex mixtures [50]. The increased sensitivity can be used to characterize DOM in samples with very low carbon concentrations [51]. This approach was recently applied, along with absorption mode processing, in petroleomics to yield over 200,000 unique molecular formula assignments in a single sample [49]. Mass window isolation can also be used to exclude high abundance contaminants like fatty acids, which ionize exceptionally well in negative mode ESI and can obscure mass spectra.

Investigating complex mixture diversity

Formula assignment and distinguishing isobars

Complex environmental mixtures have both isobaric and isomeric diversity. Isobars are defined as ions that have the same nominal mass but a different molecular formula (and a different exact mass), while isomers have the same formula and exact mass, but a different structure. Isobars can theoretically be resolved into different peaks by a mass spectrometer while isomers cannot. In complex mixtures such as DOM, both isobaric and isomeric complexity present problems to the analyst. Isobars have different masses because each atom has a slightly different "mass defect," which results from the nuclear binding energy (dependent on the specific number of protons and neutrons). Due to these slight differences, CH_4 and O deviate slightly in mass from the nominal 16 Da by 36.4 mDa. The number of isobaric peaks found at any nominal mass is dependent on two factors: the diversity of the sample (i.e., the number of ionizable isobars above the detection limit) and the resolving power of the mass spectrometer (how narrow the peaks are). The isobaric complexity of the sample is unknown and could theoretically be extremely high when an increasing number of atom possibilities are considered [9, 40]. Hundreds of peaks per nominal mass have been observed in complex natural mixtures [49], but theoretical possibilities can be constrained after considering some chemical factors like solubility, ion type, and knowledge of the sample such as heteroatom (e.g., S, P) content.

It is difficult to know in advance what resolving power is required. Usually, researchers simply use the highest resolution instrument available, but there may be reasons to decrease resolving power due to easier access to an instrument [52] or increased frequency of transients at a lower resolution, meaning more transients (better signal-to-noise) or better coupling with chromatography [53].

Resolving the CH_4 vs O mass split (36.4 mDa) and C_4 vs O_3 mass split (21.1 mDa) are essential for the analysis of aquatic samples because these substitutions are so common in the carboxylic acids that dominate the sample's ionizable material. This requires a resolving power of at least 30,000 at 200–600 Da, which is in the range of some modern time of flight analyzers and is easily achieved by all commercially produced Orbitrap and Fourier transform—ion cyclotron resonance (FT-ICR) analyzers (Fig. 4.1). The higher the resolving power, coupled with high mass measurement accuracy, the greater the confidence in formula assignment. Because CHO species dominate negative mode ESI-HRMS spectra, high resolving powers may not be necessary to find major geochemical trends [52, 54], but more diverse samples would require higher resolution, meaning that prior knowledge of the sample is important for determining the required analytical technique.

Various strategies for formula assignment have been reported [55–60]. Broadly speaking, most tools utilize exact mass accuracy, within a tolerance threshold, matching to a precompiled database of molecular formulas based on elemental constraints, coupled with homologous series and isotope peak matching. Unambiguous assignments are often resolved by a preference for simplicity with heteroatomic counts (fewer NSP atoms is deemed more likely). These tools have mostly been developed and investigated in the context of negative mode ESI and often presume CHO compounds the most prevalent. Formula assignment is not a solved problem, and new strategies that utilize the quality of the data, identify more isotopologues, and which report confidence of assignments are needed in the community.

Separation methods for investigating complex mixture diversity

Preparative chromatography

Preparative chromatography, like sequential SPE, can be used to simplify complex environmental samples [11, 61–67]. This technique has been shown to be very useful, as fraction collectors can be programmed to collect multiple replicates of the same fractions [62], greatly increasing the abundance of analytes in certain polarity fractions [65, 66]. These portions of the material can then be reconstituted for ESI-HRMS [65], or deposited for paper spray or LDI. The advantage of this strategy is that ionization suppression is greatly alleviated, allowing for observation of analytes that were previously suppressed [30, 68, 69]. This helps with gaining a more complete coverage of the sample's diversity (assuming the anticipated diversity exists), but if only one

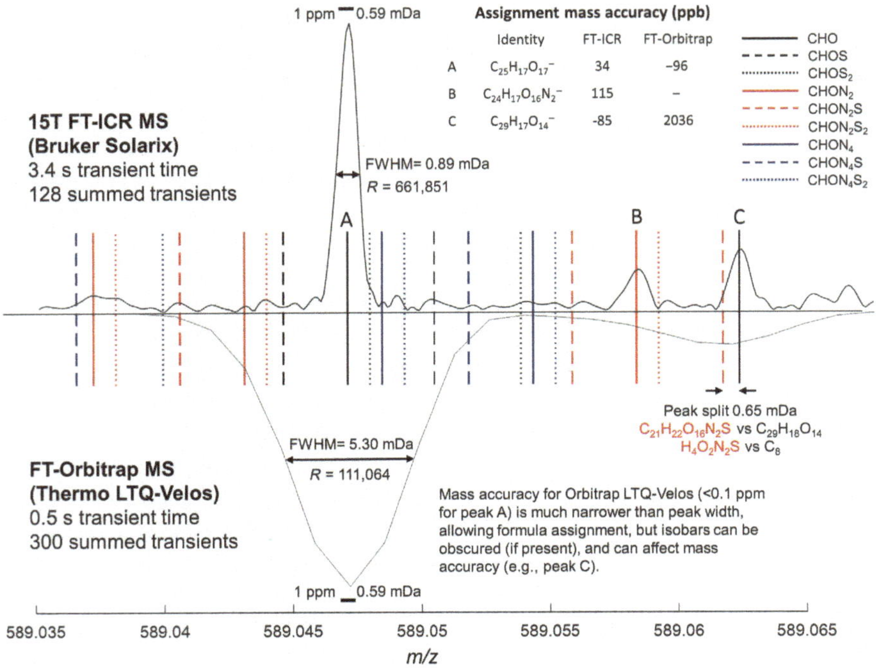

Fig. 4.1 Mass spectra generated for Elliot Soil Fulvic Acid in negative mode by two instruments with ~6× difference in resolving power. A small mass window (30 mDa) is shown, with possible formula assignment locations indicated by vertical lines, allowing for reasonable atomic ratio constraints (e.g., $H/C < 2.2$), up to four nitrogen atoms and two sulfur atoms, as indicated in the legend. Two CHO peaks should be assigned, and one $CHON_2$ peak was resolved by the FT-ICR and not by the Orbitrap. The possibility for false assignment increases as resolving power decreases or sample diversity increases. After calibration, FT-ICR MS can typically assign formulas to better than 200 ppb accuracy. FT-Orbitraps can also routinely achieve <500 ppb accuracy, but obscured isobars can affect the apex position and accuracy, as in Peak C of this figure.

ionization technique is used to measure the isolated fractions, there is still substantial bias in the analysis. The dissolution of fractions in the same solvent is a key benefit of fraction collection over online chromatography, as ionization methods are highly sensitive to solvent and buffer composition. Additionally, each fraction can be analyzed better by summing multiple transients, while online chromatography can lead to low signal to noise, as transients (which usually take at least 1 s) are usually slow compared with chromatographic separation [11].

The clear disadvantage of this approach is the large demand on time and materials. A lot of extra equipment is required—HPLC equipment, fraction collectors, vacuum drying apparatus, along with a requirement for a larger amount of DOM material. The time required can increase to a great extent as well, as the DOM is typically separated into at least four fractions [66], or even hundreds of fractions in some studies led by Stenson [11, 61, 67]. This tends to restrict preparative chromatography approaches to the detailed study of a small sample set, and thus is not suited to studies across large geographical regions or timescales.

Studies to date have shown that isobaric complexity is maintained after fractionation, but also that some molecular peaks were unique to one particular fraction, hinting at the possibility of isolating individual components of the mixture through further separation [66, 67, 69]. More comprehensive attempts to isolate individual molecules from complex environmental mixtures have been made by taking smaller chromatographic slices and via the second dimension of chromatography [11, 61–63]. Various studies have undertaken the detailed mass spectrometric analysis of such highly fractionated material, reporting a decrease in isobaric complexity but not usually finding that individual compounds with simple fragmentation patterns could be isolated [11, 61]. Brown et al. (2016) reported a decrease in isomeric complexity for the most polar compounds in SRFA compared to the more hydrophobic fractions [61]. They first used a long reversed-phase HPLC method to separate 100 fractions, subjecting an early and late fraction to a second, orthogonal separation based on charge and polarity. The hydrophilic compounds that eluted early on column 1 have the highest oxygen content, and therefore should have the lowest isomeric diversity (following Hertkorn et al. [9]), while the oxygen-poor, late-eluting material has higher theoretical isomeric diversity. Brown et al. seemed to confirm this theory using low resolution tandem MS on the second column's eluent, finding increased feature diversity in the chromatograms of individual fragments although the feature resolution remained poor. Capley et al. [11] found that passing the most hydrophilic material over a second column with an immobilized metal (nickel) showed very little retention of material, and suggest that this indicates that the structures are small and rigid [11]. They also used fragmentation of masses 341 and 339 in the late-eluting material to find that the core structures also had a mass difference of 2, demonstrating common neutral losses, (see "Fragmentation and structural elucidation" section) with the mass difference of two hydrogens a feature of the core structure.

Offline chromatographic fractionation has also been used in petroleomics using different techniques such as normal phase chromatography [30] and size exclusion chromatography [24]. These complex mixtures, like DOM, are inherently inseparable by chromatography, but these approaches are useful for adding functional information to the sample (e.g., how basic the nitrogen-containing compounds are and therefore whether they are pyridinic [30]) or giving access to higher molecular mass material (>1000 Da) that is otherwise not observed because of ion suppression [24, 30, 68]. These studies used the actual signal intensity and quantity of material present in each fraction to assess the "monomer ion yield" of the material in each fraction. This is a useful concept that may be greatly exploited in future chromatography studies in order to investigate which chromatographic (e.g., polarity) fractions of the material have the highest ionization potential and therefore have a great effect on HRMS results [70].

Chromatography, electrophoresis, and ion mobility to "separate" isomers

Direct online coupling of high-pressure liquid chromatography or gas chromatography to mass spectrometry (LCMS/GCMS) may seem to be an obvious strategy for DOM fingerprinting considering their widespread use in fields like metabolomics and contamination research. However, these topics often have a targeted approach, or relatively little diversity (in the case of metabolomics), due to the freshness of the material, which has not been chemically processed. The lower diversity makes coupled chromatography or electrophoresis work well because analytes give well-resolved peaks that can be easily distinguished from noise and separated from other compounds that may suppress ionization or interfere with the signal (i.e., isomers).

In very diverse, chemically processed mixtures such as DOM, this is not the case, and material elutes in humps, or "unresolved complex mixtures (UCMs)." Signal-to-noise ratio is sacrificed because the number of transients that can be summed is limited by the fact that the eluent cannot easily be stopped for analysis without causing diffusion and band broadening. Additionally, variable ion count generation through an LC run is not compatible with fixed ion accumulation time detectors (e.g., Bruker SolariX FT-ICR systems), and the result is variable ion populations in the ICR cell, which results in space charge effects, peak splitting and/or coalescence, and frequency shifts. This can be solved by the use of automated gain control (AGC) available on Orbitrap instruments (Thermo Fisher). The chromatographic timescale does not match the analytical timescale well, and the solvent cannot be optimized for ionization without diluting the eluent with an additional flow of organic solvent although this could be used to add ionization modifiers (e.g., acid/base).

Despite these drawbacks, and even though chromatographic peaks cannot be resolved, online chromatography can give important advantages for analysis. A wider coverage of material is detected due to the alleviation of ionization suppression [53, 70] and sample contaminants are conveniently removed from most transients. Additionally, the measured material has added chemical information related to the separation—for example, its octanol:water partitioning coefficient, in the case of reversed-phase chromatography [71], or its apparent size or charge density in the case of size exclusion chromatography [23, 72, 73] or electrophoresis [69]. Transient analysis frequency can be increased using quadrupolar detection (on four detector plates instead of two) in FT-ICR [53, 74], or by decreasing resolution, in cases where this can be sacrificed because the analyzer's power exceeds the mixture's diversity [61, 75]. It may also be the case that isobaric interferences are overcome by the separation itself, allowing lower resolution analysis, as in the case presented by Byer et al. [76], where molecules differing by C_3 vs SH_4 (3.4 mDa) were resolved using 2D gas chromatography rather than by mass alone. Furthermore, unlike preparative fraction collection methods, online chromatography can be conducted on a large number of samples in a short time. Hawkes et al. [77] used a 30-min reversed-phase method to analyze 74 headwater stream samples, 4 SRFA controls, and 2 blanks in less than 48 h, which would be challenging to match with a manual (user-operated) direct infusion approach.

Online chromatography of DOM has provided several recent insights to long-standing debates in biogeochemistry. Reversed-phase chromatography combined with ESI-HRMS has demonstrated that isomeric diversity exceeds tens or possibly hundreds of isomers per exact mass, by showing that extracted ion currents have almost no peak features, even after additional separation steps [8, 53, 63, 70]. Size exclusion chromatography-HRMS has cast doubt on the popular theory that apparently large molecular weight material is composed of loose aggregates of carboxylic acid monomers [23, 72, 78]. Size exclusion combined with NMR also found that larger material did not contain similar proton environments as smaller material—but was rather more alcoholic and aromatic in nature, indicating lignin and polysaccharide material, as opposed to smaller carboxylic rich alicyclic material (CRAM) and material derived from linear terpenoids (MDLT) [62]. These assignments correspond well with the expected ionization efficiency and mass to charge ratio of these components, and it is likely that larger polysaccharides and lignin are not detected by ESI-MS. It is unknown how abundant this pool of ESI-invisible material is, but it should be considered that this material is not separated or assigned in direct infusion or reversed-phase chromatography methods [23].

Another recently exploited technique is coupled ion mobility spectrometry (IMS)-mass spectrometry in which ions are separated inside the mass spectrometer based on their cross-sectional area by moving (IMS) or trapping (TIMS) them in an electric field counter to the flow of a buffer gas. These separations take place over the course of several milliseconds, so a very fast mass analyzer is typically required to detect the emerging ions. For this reason, only recently has IMS-MS been used in the analysis of complex mixtures where formula assignment is required, as time-of-flight (TOF) mass spectrometry technology has reached the point where it is possible to achieve resolving power > 30,000 at hundreds of scans per second. Trapped-ion mobility separation (TIMS) can also be coupled to FT-ICR MS and this gives much improved coverage in DOM mixtures compared with current TIMS-TOF capability [79].

Complex mixtures of isomers emerge from IMS drift cells in humps [22, 79–81] as we have come to expect, but the alicyclic, carboxyl rich material that is typical for DOM can quite effectively be separated from other classes of compounds that may otherwise be difficult to separate by mass, including anthropogenic contaminants such as endocrine-disrupting hormones [82]. Furthermore, Benigni et al. [82] found that the number of mass peaks identified at a large range of nominal masses increased several times after adding this ion separation step, presumably as a result of decreased isobaric complexity after ion separation. Several papers have attempted to constrain the number of isomers that contribute to a molecular formula by fitting idealized chromatographic peaks into a mobility profile [79, 80, 83] (Fig. 4.2), and this concept has also been used to demonstrate decreasing isomeric diversity after a chemical process [84] and to demonstrate high isomeric diversity after chromatography [8]. Caution should be taken in this approach as peak resolution or capacity is low for IMS, it assumes idealized peak profiles and therefore can only constrain the lower limit of isomeric diversity as unresolved peaks are not accounted for [83]. A further benefit of IMS-MS is that it can be used to monitor the formation of doubly charged ions (which most analysts try to avoid), as these travel faster through the drift tube [85].

It is also possible to determine the collisional cross section of ions directly in Fourier transform mass spectrometry [86, 87]. As ions collide with neutral background gas, their motion is perturbed and their induced signal attenuated. Application of short-time Fourier transforms, to study the ion damping profile, allows for the calculation of CCS of ions in the analyzer cell. This method requires calibration standards and a robust model and the use of short-time Fourier transforms reduces

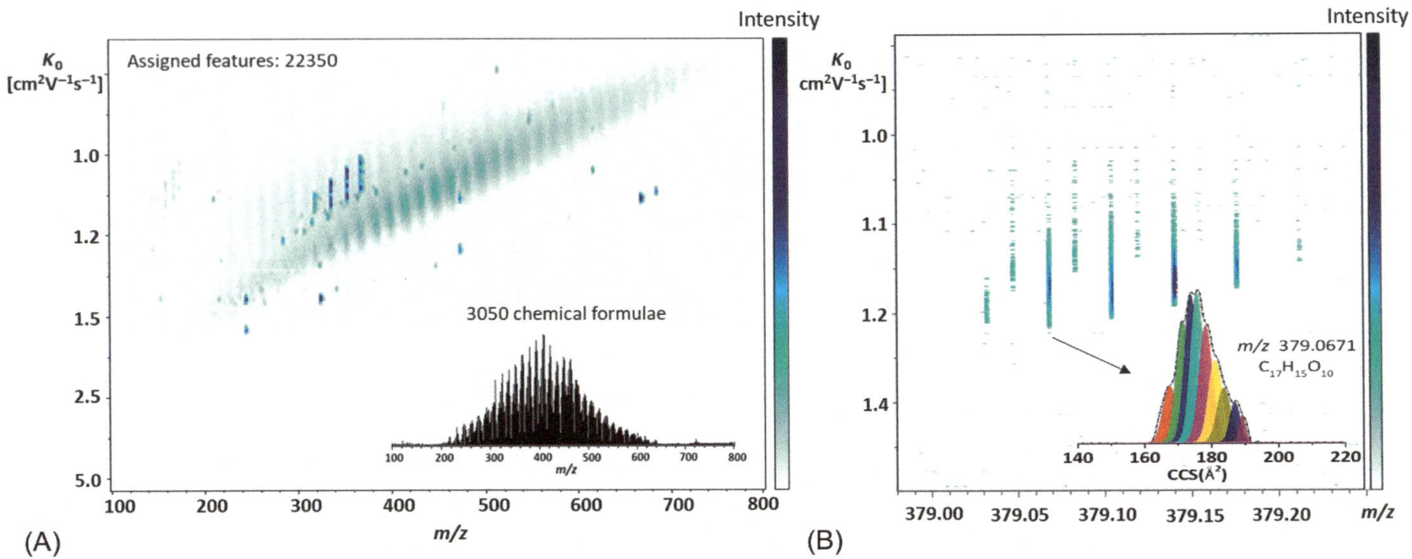

Fig. 4.2 (A) Two-dimensional TIMS-FT-ICR MS plot and MS projection (inset) of SRFA standard. (B) Expansion of (A) around *m/z* 379 showing IMS deconvolution of up to 9 isomers by SAME algorithm for *m/z* 379.0671 ($C_{17}H_{15}O_{10}$, mass error $<$ 0.1 ppm).
Credit: Dennys Leyva and Francisco Fernandez Lima (Florida International University).

the mass-domain resolution achieved. The comparability of these measurements to those made with dedicated ion mobility spectrometers requires further attention, but this approach may allow for CCS determination on existing FTICR and FT Orbitrap hardware.

MS methods for investigating complex mixture diversity

Increasing mass resolving power to increase isobaric resolution

Resolving power is key to correctly identifying species in a complex mixture; isobaric compounds which are not resolved will appear as broadened peaks with an apex between the real *m/z* of either compound, potentially leading to incorrect assignments (Fig. 4.1). Thus, any means to increase resolving power represent important advances in the field. Broadly speaking, these include hardware upgrades (larger magnets and newer instruments), signal detection changes (longer transients or quadrupolar detection of higher order harmonic signals), and signal processing developments (absorption mode processing).

Development of two ultrahigh field FT-ICR systems at 21 T at the National High Magnetic Field Laboratory (NHMFL) and Pacific Northwest National Laboratory has yielded substantial improvements in resolving powers, as well as numbers and errors of molecular formula assignments, for complex mixture characterization [75, 88]. Resolving powers in excess of 1 million (at 367 *m/z*) for 3-s transients were achieved in magnitude mode, and 1.7 million in absorption mode for SRFA, allowing for resolution of sub-miliDalton splits between chemical species across the typical mass range of DOM. Such instrumental upgrades, however, are often prohibitively expensive and require substantial planning, more cryogens, and more lab space. Furthermore, the improvements can be modest—the theoretical resolving power enhancement from a 12 to 21 T system would only be 1.75×.

Absorption mode processing has been routinely applied in NMR for decades but is only recently finding use in FT-ICR. The delay has historically been a lack of computational resources and algorithms to calculate the phase correction, and thus FTMS data is traditionally presented in magnitude mode, where absorption mode can yield a twofold increase in resolving power. Custom hardware, such as the Predator system at the NHMFL, has achieved absorption mode processing as mentioned before. However, the software has also been developed allowing for the calculation of phase correction from normally acquired data on commercial instruments [89, 90]. This allows for a "free" enhancement of resolving power, and thus also mass accuracy, either increasing the utility of lower field instruments [91] or for improved coupling to online chromatography.

Recently, commercial instrumentation has become available which includes additional detect electrodes on the ICR cell—so-called quadrupolar detection—allowing for increased resolving power in the same measurement time [92]. This is achieved through measuring the multiples of the reduced cyclotron frequency, with appropriate ICR cell design and application of tuned electric fields. These two-omega (for twice the frequency) measurements have a twofold boost in resolving power. Four-omega measurements have been reported at 21 T [93], but the very high frequencies of the low mass ions bring difficulties in digitization and signal acquisition of lower masses (below 600 m/z) and were therefore not suitable for DOM characterization. As with absorption mode processing, this method can be applied to lower field instruments to achieve results comparable to those from larger magnetic fields [94] or be used to achieve high resolution with shorter transient length, and thus apply online chromatography of NOM or petroleum to FT-ICR [53, 74].

Fragmentation and structural elucidation

Structures and molecular functionality are elucidated in fragmentation studies, most commonly with collision-induced dissociation (CID) by breaking ions apart with a buffer gas and calculating which neutral losses of molecules can occur to leave the measured stable ion behind [95]. Some mechanisms are fairly straightforward, such as loss of CO_2 from a carboxylic acid group, while others can require more complex pathways like ring closures, as in the case of loss of H_2O from two adjacent carboxylic acids, leaving a ring anhydride [11, 95]. Phenol OH groups are stable and do not fragment easily as a water loss, while aliphatic OH groups can fragment with a nearby proton as H_2O [95]. As CO_2 and H_2O losses absolutely dominate the fragmentation patterns of DOM peaks, it follows that carboxylic acid groups are very common structures [7, 8, 10, 11, 96]. Larger neutral losses indicating amide, ether or glycoside bonds are not observed, suggesting that these structures are less common in the ions produced by the sample preparation and instrumental parameters [8, 10]. Several neutral losses of CO_2 can be observed from DOM precursor ions (Fig. 4.1), so it is likely that the majority of the oxygen in typical ESI(−) ionizable DOM is, in fact, carboxylic oxygen (two per group) [10]. Alternatives to CID are photon-based fragmentation methods, such as ultraviolet photodissociation (UVPD) and infrared multiphoton dissociation (IRMPD), which may complement CID fragmentation and be more structurally selective, although their use for NOM characterization seems to be limited so far [59].

There is a well-documented problem with fragmentation in complex mixtures that if the isolation window contains several peaks, it can be difficult to elucidate which precursor peak or peaks a charged fragment originates from [7, 8, 11, 80, 95]. This can be overcome using "in-cell" isolation via selective or shaped radiofrequency pulses, which allow for the isolation of a single mass peak for fragmentation [10, 61]. Witt et al. [10] compared the fragmentation patterns of peaks that differ in CH_4 vs O at a single nominal mass, finding similar backbone masses after fragmentation of the labile carboxylic acid groups. They argue that the structural diversity of DOM is not as high as the isobaric diversity due to the common backbone structures and similar chemistry of functional groups. Further MS_n experiments in which ions from one fragmentation are trapped and fragmented again, also show that sequential losses of CO_2 and H_2O can lead to common "backbone" structures, potentially revealing the origins of the material [10, 11]. The result that carboxylic acids surround a large, alicyclic backbone with some degree of unsaturation also corresponds well with NMR data and extrapolation of the average double bond equivalence of DOM molecular formulas [97, 98].

The fragmentation pattern of DOM peaks is highly similar in different aquatic settings and biogeochemical constraints may lead to similar molecular mixtures in every environment [6]. The number of structural isomers must be very high, at least approaching hundreds per molecular formula otherwise, the fragmentation patterns would not reach such an averaged response [7]. The fact that this averaged, similar pattern remains after chromatographic separation (Fig. 4.3) suggests that structural diversity is extremely high [8]—but as pointed out by Witt et al. [10], functional diversity may be substantially lower.

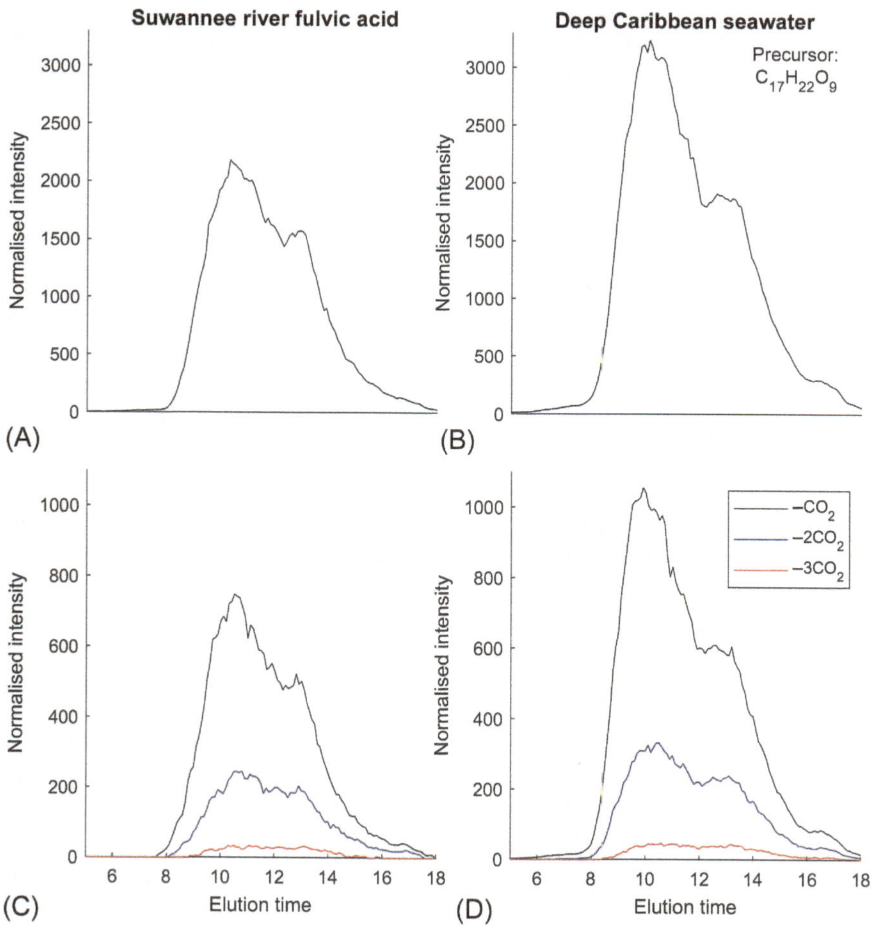

Fig. 4.3 Extracted ion currents of a reversed-phase chromatographic separation and ESI-FT-Orbitrap MS analysis of SRFA (left) and deep Caribbean seawater (right). (A and B) precursor ion $C_{17}H_{22}O_9$ with no collision-induced dissociation (CID) energy, (C and D) neutral losses of 1–3 CO_2 molecules from the precursor in (A) and (B) with 27 V normalized CID energy. The lack of feature diversity in the chromatograms (both precursors and fragments) suggests there is extremely high isomeric diversity. The similarity between the samples suggests that structures and functionality are similar across aquatic environments.

Complexation, derivatization, and exchange techniques

Along with fragmentation, there are several tools available to the analyst to probe the structural and functional chemistry of complex environmental mixtures using HRMS. Unlike NMR, or FTIR, which cannot resolve individual components of DOM directly, the information derived can be attributed to individual molecular formulas, demonstrating the nature and diversity of thousands of individual components of the mixture, rather than the bulk sample.

Derivatization methods include esterification of carboxylic acid groups to deuterated methyl esters, allowing enumeration of carboxylic acid groups [25]. Sulfur functionality has been investigated using selective derivatization methods [99] and these strategies could be extended to other atoms such as nitrogen and phosphorus. Labile oxygen atoms in carboxylic and alcohol groups naturally exchange with oxygen in solvating water, allowing determination of the number of nonlabile oxygen atoms in molecular formulas by exchanging ^{18}O for natural ^{16}O using $H_2^{18}O$ [100]. Carrying out the reaction at different temperatures and reaction times can lead to information about how labile different oxygen atoms are and how many nonlabile (e.g., ether and ketone) oxygen atoms are present. The H/D exchange is a similar process but can be carried out in the ESI spray for carboxylic and alcohol protons [100, 101], or after long reaction can be used to probe the structure of the backbone of the molecule by exchange with aromatic protons [100, 102]. H/D exchange can also be carried out inside the mass spectrometer with a deuterated reaction gas [103].

The main complication with all of these techniques is the difficulty in distinguishing incomplete reactions from isomeric diversity [11]. It was shown that a $[H_3PO_4]_4$ cluster with 11 labile hydrogens gives an almost Gaussian distribution of H/D exchanges [101] with a maximum at 11, indicating some degree of incomplete reaction. This can be confused by actual structural diversity in complex mixtures, where some isomers have more labile hydrogens than others, leading to an additional distribution of product ions. This drawback is very difficult to overcome and the results of derivatization or exchange are usually discussed in terms of the maximum number of exchanges, rather than attempting to unravel this complexity [102]. Alternatively, the techniques can be used to investigate peaks with a low number of total atoms that may be exchanged [104].

It is interesting to note that certain exchanges that are usually considered undesirable artifacts of ionization techniques such as ESI can be harnessed to gain structural information. For example, usually it is considered a concern that storing DOM samples in methanol may lead to methylation of carboxylic acid groups [17, 26], but it may actually be utilized to count the carboxylic acid groups [25]. Similarly, primary amines like the additive ammonia may form imines in the spray with ketone or aldehyde groups in positive ESI, which can be used to detect this functionality [105]. Further still, base cations are usually considered a nuisance in positive mode, particularly sodium, because of their abundance and the ease with which they form adducts with carboxylic acids [106], but they can be used to form interesting and useful complexes in certain cases using a "ligand fishing" strategy [107].

Stenson [108] used metal ions (Be^{2+}, Mn^{2+}, Cr^{3+}) to investigate the peaks that emerged in the mass spectrum of SRFA due to complexation of humic acids with these base cations, which were specifically chosen because the complexes would be placed at even nominal masses, making them easily stand out. The study showed that these ions form hundreds or thousands of complexes either in solution or during ESI spray. Waska et al. [109] have investigated the possibility of using a similar method to investigate iron- and copper-binding ligands in seawater DOM. They found that isobaric complexity in seawater DOM made the assignment of Fe and Cu complexes extremely challenging, even after further fractionation of the sample by different SPE protocols, but did successfully characterize the siderophore ferrioxamine B with this method, along with other known ligands [110]. They recommend HPLC fractionation of metal ligand complexes, and this approach was recently successfully used alongside HPLC-ICP-MS to detect iron-binding siderophores at concentrations as low as 1 pM in seawater [111]. Duncan et al. [112] demonstrated that divalent barium ions form strong complexes with carboxylic acid groups, strengthening the C—COOH bond and therefore limiting fragmentation. This is useful in the characterization of naphthenic acids, which only have one carboxylic acid group, as fragmentation can then lead to more structurally diagnostic product ions. Boron, conversely, can form complexes with diols and triols, as demonstrated for model compounds [107]. This paper demonstrated that sugar-like formulas and lignin compounds in soil fulvic acid could bind boron with a similar ligand fishing approach to Stenson's.

Visualization and data handling techniques

Visualization

HRMS data containing thousands of features is inherently difficult to visualize and it is therefore typical to present the data and compare samples in a variety of figure types that reduce dimensionality. The most common visualization is the van Krevelen (VK) diagram in which elemental ratios (typically H/C, O/C) are plotted for each formula (Fig. 4.4A). This diagram was introduced for HRMS visualization at the beginning of the millennium [114] and has been used extensively since to show how environmental trends correlate with apparent chemistry of the ionized material [26, 77, 115, 116], to compare samples [117, 118], to show how geochemical transformations affect samples [119] and as an interactive tool for exploring data [57, 120]. More common in petroleomics and HRMS analysis of naphthenic acids is the double bond equivalence (DBE) vs C or DBE vs O diagram (Fig. 4.4D; [121]). Problematically, DBE is independent from O and S, thus does not account for the common C=O double bonds in DOM. Koch and Dittmar proposed the alternative "aromaticity index" in 2006 to allow for a better characterization of the aromaticity of DOM molecules [122].

Two-dimensional scatter plots with thousands of data points suffer from overlap biasing the visualization. It can be effective to plot density, not points, to more fairly represent the underlying data or to visualize the data interactively [120]. Alternatively, additional diagrams can encode further information, such as other heteroatomic content or DBE, which is preferable to three-dimensional plots, which are impossible to effectively show in 2D space (i.e., printed) (Fig. 4.4E).

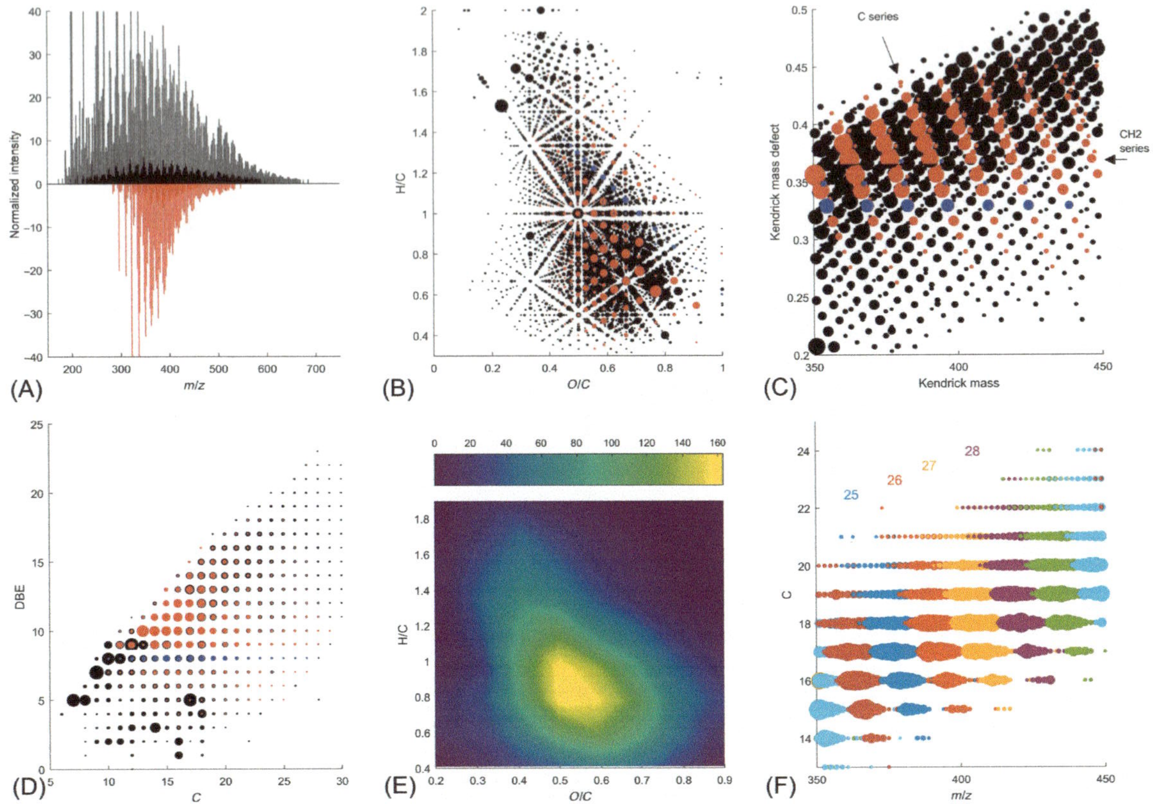

Fig. 4.4 A selection of visualization methods, displaying FT-ICR-MS data of Elliot Soil Fulvic Acid (ESFA) collected on a 15T Bruker Solarix instrument. (A) Reconstructed mass spectrum comparison (B) van Krevelen diagram, (C) Kendrick plot, (D) DBE vs C diagram, (E) van Krevelen density plot, (F) C vs *m/z* diagram. In (B), (C), (D), and (F), intensity is displayed as point size. In (A), all assigned peaks are shown as black lines, and peaks with 10 oxygen atoms are shown in the negative direction in red for comparison. In (B–D), all points are colored black, overlaid in red are formulas with 10 oxygen atoms, and overlaid in blue are formulas with 10 oxygen atoms and double bond equivalence (DBE) of 8. In (E), the density (color) shows the number of peaks in each bin and intensity is not considered. In (F), the sum of carbon and oxygen are shown in color, and for some "islands" this sum is indicated as a number above the island after [113].

The Kendrick plot is an older tool that is extremely useful for identifying series of compounds that are related by some sort of homologous series [60, 123, 124]. The traditional Kendrick conversion is used to transform the mass of $^{12}CH_2$ (14.01565 Da) to be exactly 14 Da. Each homologous series of compounds that differ by a CH_2 unit (i.e., a link in an aliphatic chain), now differ by a Kendrick mass of exactly 14 and have an identical Kendrick Mass Defect (the decimal after the exact mass). Such a series can be grouped together in a Kendrick diagram (Fig. 4.4C), while they are harder to spot in a VK, and overlay on top of many other points in a DBE vs C diagram (Fig. 4.4A and B). As the main use of Kendrick plots is to aid molecular formula assignment, their utility is now limited since automatic formula assignment tools exist for DOM [55–57]. Further transformation of the data into subsequent mass defect families can allow an extreme degree of data reduction—Roach et al. [125] showed that a crude oil sample with 13,000 peaks could be reduced to just 25 families (25 points), when considered as Kendrick-like networks of CH_2, H_2, and O homologues. This type of network approach has also been used to provide impressive 3D molecular maps of HRMS data and the homologous connections that link formulas together via so-called "transformation pathways" [126–128].

There are numerous other options for displaying data that can be useful depending on the research question. Other atomic ratios can be plotted, using different heteroatoms (N, S, P, Cl, Na) and the formulas that contain specific combinations of these, or other atoms can be extracted from the data as compound classes (e.g., the SO_4 class, containing one S and four O). Reemtsma [113] examined the information gained by plotting carbon number vs m/z and grouping compounds by the sum of carbon and oxygen (Fig. 4.4F). Within each "island" there are series at the same nominal mass for which a CH_2 substitutes oxygen and a double bond, meaning that the compounds to the top are aliphatic, O limited, and the bottom is the most aromatic and carboxylic rich.

HRMS resolves thousands of peaks in DOM and displaying this data in one figure can be useful, though is challenging to achieve without compromise. In the case of the traditional H/C, O/C van Krevelen diagram, mass, intensity, and other heteroatom content (e.g., N, S, and P) information is lost. This is before chromatographic data and intersample differences are even considered. DBE vs C diagrams do indicate mass via C and hydrogen deficiency via DBE, but lack specific information about further heteroatom content (O, N, S, P), and have much more overlap of points. They are therefore often used in compound class diagrams, where the single classes are shown (e.g., O_2) in their own diagram. This limited heteroatom content information means that they are not commonly used in DOM studies but are extremely popular in petroleomics, where heteroatom content is limited.

VK diagrams are often chosen to present HRMS data in biogeochemical studies because the diagram lends itself to the division of the points into groups according to arbitrary boundaries in the atomic ratios of H/C and O/C. Because molecular classes also fall within defined boundaries (e.g., lipids are nearly saturated with hydrogen and contain few oxygen atoms), it is conceptually useful to divide the VK diagram into these defined molecular areas and therefore annotate the peaks found by HRMS according to their corresponding compound class (e.g., lipid-like peaks). This type of data analysis typically includes combinations of the following: lipids, peptides/

proteins, sugars, lignin, tannin, polyphenols, condensed aromatics, black carbon. Such "assignments" should be well constrained and accompanied with strong caveats. The removal of dimensions from the data risks poor annotations in many cases. For example, peptides should necessarily contain nitrogen (at least one per 100 Da, depending on the amino acid residues). The "peptide-like" region should, therefore, include an N/mass dimension constraint as well as H/C and O/C dimension constraints, but this is rarely discussed and is not apparent from the VK diagram [129]. Multiple N containing formulas (i.e., peptides) are also very difficult to resolve from CHO peaks (Fig. 4.1). Even after satisfying these formula constraints, it is impossible to confirm these assignments from mass spectral peaks alone, without isotopic fine structure and/or fragmentation confirmation. Visualization techniques reduce data complexity for easier consideration, but also hide important information about the signals and assignments. In the context of environmental fingerprinting, therefore, visualizations serve a very useful purpose, but their limitations must be considered when interrogating the data.

Statistical treatment of data

Many HRMS DOM studies investigate the changing nature of DOM over environmental gradients in space and time. In these cases, changing peak intensities are often correlated with environmental factors such as radiocarbon age [130], inorganic element concentrations [77], mean annual precipitation [115], or even bacterial community DNA composition [131]. This type of analysis requires large sample sets (e.g., 74 in Ref. [77], 120 in Ref. [115], and 137 in Ref. [130]) and strict criteria for correlation in order to avoid "p-hacking," as some fraction of thousands of mass peaks are likely to have random trends with at least one environmental factor. A Pearson's Rho of 0.001 was chosen in Ref. [77], and 0.01 was chosen in Ref. [130]. Kellerman et al. [115] instead used Spearman's rank correlation to acknowledge that trends may not be linear and only used peaks that were present in 90% of the samples above a certain intensity threshold. They also performed an analysis of the false discovery rate in order to assess the level of confidence required to consider significance, leading them to a value of 0.027.

These types of correlation analysis require very reliable peak intensities, as it is assumed that an increasing or decreasing ion abundance is environmentally significant, rather than due to variability (random or otherwise) in the analysis. Samples should be run in random order and ideally, replicates are analyzed to evaluate the magnitude of peak intensity change required to signify genuine change [132–134], but under the burden of constraint on instrument time, analysis of a larger number of samples is often preferred over replicate analysis. The reliability of peak intensity values is seldom discussed [132, 135, 136] and this should be addressed moving forward. Riedel and Dittmar [48] pointed out that analyzing triplicates with 1/3 the number of scans takes the same amount of time and allows an evaluation of artifact and noise peaks (those which are present in only one or two of the replicates) and this approach may be considered as a way to also evaluate peak variability without losing the capability of running a high number of samples. We found that peak intensity variability was below 5% for five sequential replicate analyses of SRFA by FT-Orbitrap MS and five analysis dispersed throughout a sample set for 15 T FT-ICR MS (Fig. 4.5A and B). This <5% variability was only found for peaks with a normalized intensity >10% of the maximum (Obitrap) and >5% of the maximum (FT-ICR). The lower intensity peaks had larger assignment errors and poorer reproducibility for both analyzers, and this should be factored into statistical analysis when considering if environmental trends are significant.

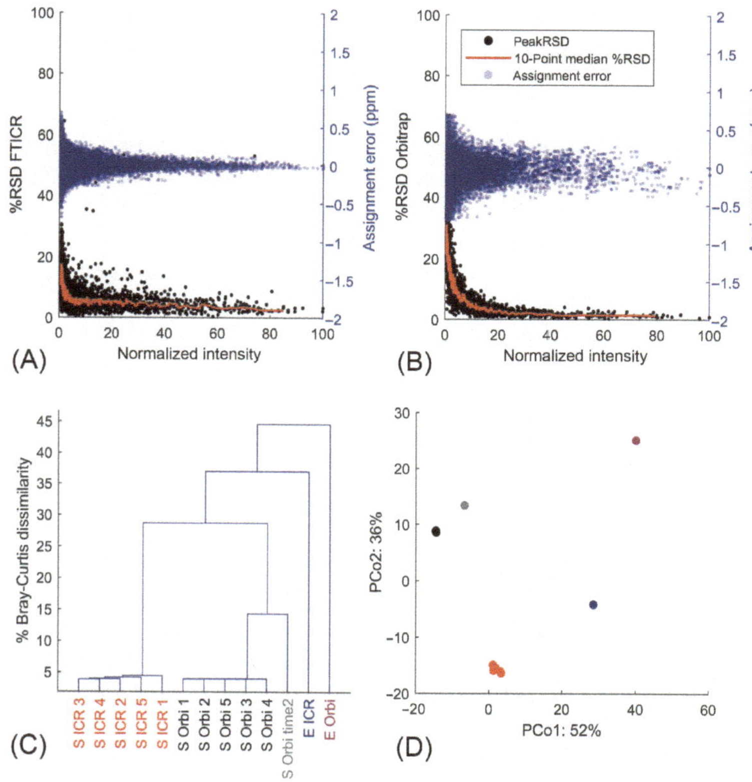

Fig. 4.5 (A and B) Relative standard deviation (%RSD) of normalized peak intensities for peaks that were present in all five replicates of SRFA measured by FT-ICR MS (A) and FT-Orbitrap MS (B), also shown are assignment errors for each replicate. (C) Hierarchical cluster analysis of Bray-Curtis dissimilarities for five replicates of SRFA (labeled S) measured by FT-ICR-MS (labeled ICR) and FT-Orbitrap MS (labeled Orbi). These are compared with SRFA measured by FT-Orbitrap MS 8 months previously and a different sample, Elliot soil fulvic acid (labeled E) measured once on each instrument. (D) Principal coordinate analysis (PCoA) diagram reproducing the dissimilarities between the samples shown in (C) (same color code). PCo1 and PCo2 are shown. Replicate analysis typically yields Bray-Curtis dissimilarities (BCD) of ~5%, while analysis of the same sample after several months may lead to BCD of 15%. Different analyzers can differ more (e.g., 30% for these examples) and different samples differ greatly (40–45% for SRFA vs ESFA).

The peaks that are found to correlate with environmental variables are usually collected together and presented with the typical visualization tools such as VK diagrams. A potential confirmation of the reliability of trends is that molecular formulas with similar chemistry (e.g., occupying a similar region of the VK diagram) follow the same biogeochemical patterns. This has been used on several occasions to show how DOM changes in character as it degrades in nature [130, 136–138], or has been used to find peaks that can be used as tracers for terrestrial input of DOM to the deep ocean [139]. The trends of individual peak intensities are therefore quite useful, but over the last decade, it has been recognized that these types of large datasets are very well suited to multivariate statistical analysis, where the data are reduced to a small number of factors, particularly because so many peaks clearly follow the same trends, indicating a great deal of redundancy in the full dataset.

A typical multivariate approach is principal component analysis (PCA), which can be conducted directly on the formula assigned and aligned sample intensity data, giving scores for the samples and loadings for each formula or mass, indicating the contribution of each formula to each principal component (PC). Often it is found that very few PCs (1–3) explain a majority (more than 50%) of all data variability, even for a fairly large number of diverse samples [77, 140, 141]. Typically, the first or second PC is found to correlate with aromaticity or oxygen:carbon ratio, leading to broad shifts (top to bottom or left to right) in VK space. These broad trends are a further result of the extreme isomeric diversity of the samples, which lead to a variety of reactivity within each molecular formula [142] and a consequent buffering of the peak intensity to the chemical change. If only half of the isomers are removed, the molecular formula's peak will decrease by $\sim$50% (disregarding isomer-specific ionization efficiency). This is the same observed result as the case where every isomer is half removed, and it bears remembering that neither case can be proven.

As an alternative to principal component analysis, sample peak intensity lists can be compared using distance metrics like Bray-Curtis Dissimilarity or Jaccard dissimilarity, which allow numerous pairwise sample distances to be reconstructed into principal coordinates via nonclassical multidimensional scaling, or to be distributed into groups using hierarchical cluster analysis (HCA; Fig. 4.5). Clustering analysis and principal coordinate modeling of intersample dissimilarities usually show that replicate analysis of the same sample are highly similar (usually more than 95% similar), while different samples can vary by 10% to almost 100% (completely different, almost no shared peaks) depending on the distance metric [52, 77, 141, 143]. This type of analysis can be conducted on presence/absence data rather than using peak intensities, which can be useful in cases where peak variability between replicates in not assessed or is unknown [5, 115, 141, 144]. Molecular formulas that correlate with hierarchical clusters, principal components (as loadings), or principal coordinate scores are often mapped onto VK diagrams to demonstrate how DOM composition can be deconstructed and clustered by the multidimensional factor analysis [135, 141], finding geochemical trends in the data and sometimes "molecular indicator species" which are indicative of certain regions or processes [5].

These multivariate approaches are almost always conducted on normalized data (where peak intensities are scaled to sum to a common value, and sometimes scaled to the standard deviation across the dataset [135]) for mathematical reasons and also to account for analytical factors that do not allow strict quantification like ionization suppression, but this often presents a problem when it comes to interpreting the data. In cases where DOC concentrations vary greatly, it can be confusing to describe trends in normalized data that go counter to the concentration gradient. For example, if a peak intensity increases in normalized data while DOC decreases, it is unclear whether this is because the sum of the analytes with that molecular formula increase are present at similar concentrations, or simply decrease less than the other analytes. The lack of understanding about the quantity of ionizable material in each sample thus stifles scientific understanding of actual (rather than relative) environmental trends. Because of variable ionization potential, great caution should be taken in correlating HRMS data with data from other techniques such as excitation emission matrices or NMR [135, 145–147], as it is possible that different analytes in the mixture are determined with the different techniques [23], making correlations coincidental, albeit related to similar geochemical trends.

Concluding remarks

There are now a wide range of HRMS techniques available for analysis of environmental samples, including tailored sample preparation techniques, ionization methods, coupling with different chromatographic phases, derivatization methods and other MS separation and structural elucidation methods. In the coming years, we hope that methods will be selected based on research questions, rather than research questions being designed around the methods and data available. The analytical window of HRMS techniques will continue to be an important concern in data interpretation, but it is equally important to remember that this window can be modified to suit the needs of the study.

An important question facing researchers is the cost/benefit analysis of running multiple samples vs multiple techniques (analytical windows) vs replicates. The conclusions drawn from results must consider the exact nature of the analytical window(s) used, including sample handling and preparation, ionization, detection limits, and data processing (e.g., formula assignment). Method validation should become commonplace and best practice should include the use of model compounds to test the analytical window(s). Additionally, studies should explicitly consider the reliability of intersample differences in comparison to analytical variability.

For environmental fingerprinting, it must be recognized that the data represents the ions in the mass analyzer, not the molecules in the raw sample, and it is of utmost importance to prepare and measure the samples as consistently as possible. Care should be taken to analyze large sample sets in one batch to minimize analytical variability and drift in signal of reference samples should always be monitored. Replicates, quality controls, and blanks remain invaluable and underused. As with all science, reproducibility and repeatability are key to ensure validity. Therefore, it should be common practice to make all raw and processed data and all data processing scripts and routines publicly available. This will also reduce the amount of reinventing performed, allowing scientists to focus their efforts on new problems.

References

[1] J.K. Geuer, B. Krock, T. Leefmann, B.P. Koch, Quantification, extractability and stability of dissolved domoic acid within marine dissolved organic matter, Mar. Chem. 215 (2019), 103669. https://doi.org/10.1016/j.marchem.2019.103669.

[2] D. Hemmler, C. Roullier-gall, J.W. Marshall, M. Rychlik, J. Andrew, P. Schmitt-kopplin, Evolution of complex Maillard chemical reactions, resolved in time, Sci. Rep. 7 (2017), https://doi.org/10.1038/s41598-017-03691-z.

[3] J. Raeke, O.J. Lechtenfeld, M. Wagner, P. Herzsprung, T. Reemtsma, Selectivity of solid phase extraction of freshwater dissolved organic matter and its effect on ultrahigh resolution mass spectra, Environ, Sci. Process. Impacts 18 (2016) 918–927, https://doi.org/10.1039/c6em00200e.

[4] M. Gonsior, L.C. Powers, E. Williams, A. Place, F. Chen, A. Ruf, N. Hertkorn, P. Schmitt-kopplin, The chemodiversity of algal dissolved organic matter from lysed *Microcystis aeruginosa* cells and its ability to form disinfection by-products during chlorination, Water Res, 155 (2019) 300–309, https://doi.org/10.1016/j.watres.2019.02.030.

[5] E.B. Kujawinski, K. Longnecker, N.V. Blough, R. Del Vecchio, L. Finlay, J.B. Kitner, S. J. Giovannoni, Identification of possible source markers in marine dissolved organic matter using ultrahigh resolution mass spectrometry, Geochim. Cosmochim. Acta 73 (2009) 4384–4399, https://doi.org/10.1016/j.gca.2009.04.033.

[6] M. Zark, T. Dittmar, Universal molecular structures in natural dissolved organic matter, Nat. Commun. 9 (2018) 1–8, https://doi.org/10.1038/s41467-018-05665-9.

[7] M. Zark, J. Christoffers, T. Dittmar, Molecular properties of deep-sea dissolved organic matter are predictable by the central limit theorem: Evidence from tandem FT-ICR-MS, Mar. Chem. 191 (2017) 9–15, https://doi.org/10.1016/j.marchem.2017.02.005.

[8] J.A. Hawkes, C. Patriarca, P.J.R. Sjöberg, L.J. Tranvik, J. Bergquist, Extreme isomeric complexity of dissolved organic matter found across aquatic environments, Limnol. Oceanogr. Lett. 3 (2018) 21–30, https://doi.org/10.1002/lol2.10064.

[9] N. Hertkorn, C. Ruecker, M. Meringer, R. Gugisch, M. Frommberger, E.M. Perdue, M. Witt, P. Schmitt-Kopplin, High-precision frequency measurements: indispensable tools at the core of the molecular-level analysis of complex systems, Anal. Bioanal. Chem. 389 (2007) 1311–1327, https://doi.org/10.1007/s00216-007-1577-4.

[10] M. Witt, J. Fuchser, B.P. Koch, Fragmentation studies of fulvic acids using collision induced dissociation Fourier transform ion cyclotron resonance mass spectrometry, Anal. Chem. 81 (2009) 2688–2694, https://doi.org/10.1021/ac802624s.

[11] E.N. Capley, J.D. Tipton, A.G. Marshall, A.C. Stenson, Chromatographic reduction of isobaric and isomeric complexity of fulvic acids to enable multistage tandem mass spectral characterization, Anal. Chem. 82 (2010) 8194–8202, https://doi.org/10.1021/ac1016216.

[12] R.L. Sleighter, P.G. Hatcher, Molecular characterization of dissolved organic matter (DOM) along a river to ocean transect of the lower Chesapeake Bay by ultrahigh resolution electrospray ionization Fourier transform ion cyclotron resonance mass spectrometry, Mar. Chem. 110 (2008) 140–152, https://doi.org/10.1016/j.marchem.2008.04.008.

[13] E.C. Minor, M.M. Swenson, B.M. Mattson, A.R. Oyler, Structural characterization of dissolved organic matter: a review of current techniques for isolation and analysis, Environ. Sci. Process. Impacts 16 (2014) 2064–2079, https://doi.org/10.1039/c4em00062e.

[14] S. Sandron, A. Rojas, R. Wilson, N.W. Davies, P.R. Haddad, R.A. Shellie, P.N. Nesterenko, B.P. Kelleher, B. Paull, Chromatographic methods for the isolation, separation and characterisation of dissolved organic matter, Environ. Sci. Process. Impacts 17 (2015) 1531–1567, https://doi.org/10.1039/c5em00223k.

[15] T. Dittmar, B. Koch, N. Hertkorn, G. Kattner, A simple and efficient method for the solid-phase extraction of dissolved organic matter (SPE-DOM) from seawater, Limnol. Oceanogr. Methods 6 (2008) 230–235. https://doi.org/10.4319/lom.2008.6.230.

[16] I.V. Perminova, I.V. Dubinenkov, A.S. Kononikhin, A.I. Konstantinov, A.Y. Zherebker, M.A. Andzhushev, V.A. Lebedev, E. Bulygina, R.M. Holmes, Y.I. Kostyukevich, I.A. Popov, E.N. Nikolaev, Molecular mapping of sorbent selectivities with respect to isolation of arctic dissolved organic matter as measured by Fourier transform mass spectrometry, Environ. Sci. Technol. 48 (2014) 7461–7468, https://doi.org/10.1021/es5015423.

[17] Y. Li, M. Harir, M. Lucio, B. Kanawati, K. Smirnov, R. Flerus, B.P. Koch, P. Schmitt-Kopplin, N. Hertkorn, Proposed guidelines for solid phase extraction of Suwannee River dissolved organic matter, Anal. Chem. 88 (2016) 6680–6688, https://doi.org/10.1021/acs.analchem.5b04501.

[18] M. Kida, H. Sato, A. Okumura, H. Igarashi, N. Fujitake, Introduction of DEAE Sepharose for isolation of dissolved organic matter, Limnology 20 (2019) 153–162, https://doi.org/10.1007/s10201-018-0561-3.

[19] B. Lam, A.J. Simpson, Passive sampler for dissolved organic matter in freshwater environments, Anal. Chem. 78 (2006) 8194–8199, https://doi.org/10.1021/ac0608523.

[20] Y. Li, M. Harir, J. Uhl, B. Kanawati, M. Lucio, K.S. Smirnov, B.P. Koch, P. Schmitt-Kopplin, N. Hertkorn, How representative are dissolved organic matter (DOM) extracts? A comprehensive study of sorbent selectivity for DOM isolation, Water Res. 116 (2017) 316–323, https://doi.org/10.1016/j.watres.2017.03.038.

[21] K. Zhang, A.S. Pereira, J.W. Martin, Estimates of octanol-water partitioning for thousands of dissolved organic species in oil sands process-affected water, Environ. Sci. Technol. 49 (2015) 8907–8913, https://doi.org/10.1021/acs.est.5b01656.

[22] Y. Gao, W. Wang, C. He, Z. Fang, Y. Zhang, Q. Shi, Fractionation and molecular characterization of natural organic matter (NOM) by solid-phase extraction followed by FT-ICR MS and ion mobility MS, Anal. Bioanal. Chem. (2019). https://doi.org/10.1007/s00216-019-01943-7.

[23] J.A. Hawkes, P.J.R.R. Sjöberg, J. Bergquist, L.J. Tranvik, Complexity of dissolved organic matter in the molecular size dimension: insights from coupled size exclusion chromatography electrospray ionisation mass spectrometry, Faraday Discuss. 218 (2019) 52–71, https://doi.org/10.1039/c8fd00222c.

[24] J.C. Putman, S. Gutiérrez Sama, C. Barrère-Mangote, R.P. Rodgers, R. Lobinski, A.G. Marshall, B. Bouyssière, P. Giusti, Analysis of petroleum products by gel permeation chromatography coupled online with inductively coupled plasma mass spectrometry and offline with Fourier transform ion cyclotron resonance mass spectrometry, Energy Fuels 32 (2018) 12198–12204, https://doi.org/10.1021/acs.energyfuels.8b02788.

[25] A. Zherebker, Y. Kostyukevich, A. Kononikhin, O. Kharybin, A.I. Konstantinov, K.V. Zaitsev, E. Nikolaev, I.V. Perminova, Enumeration of carboxyl groups carried on individual components of humic systems using deuteromethylation and Fourier transform mass spectrometry, Anal. Bioanal. Chem. 409 (2017) 2477–2488, https://doi.org/10.1007/s00216-017-0197-x.

[26] R. Flerus, B.P. Koch, P. Schmitt-Kopplin, M. Witt, G. Kattner, Molecular level investigation of reactions between dissolved organic matter and extraction solvents using FT-ICR MS, Mar. Chem. 124 (2011) 100–107, https://doi.org/10.1016/j.marchem.2010.12.006.

[27] J.W.T. Blackburn, W. Kew, M.C. Graham, D. Uhrín, Laser desorption/ionization coupled to FTICR mass spectrometry for studies of natural organic matter, Anal. Chem. 89 (2017) 4382–4386, https://doi.org/10.1021/acs.analchem.6b04817.

[28] A. Zherebker, E. Shirshin, O. Kharybin, Y. Kostyukevich, A. Kononikhin, A.I. Konstantinov, D. Volkov, V.A. Roznyatovsky, Y.K. Grishin, I.V. Perminova, E. Nikolaev, Separation of benzoic and unconjugated acidic components of Leonardite humic material using sequential solid-phase extraction at different pH values as revealed by Fourier transform ion cyclotron resonance mass spectrometry and correlation nuclear magnetic resonance spectroscopy, J. Agric. Food Chem. 66 (2018) 12179–12187, https://doi.org/10.1021/acs.jafc.8b04079.

[29] Y. Li, M. Harir, M. Lucio, M. Gonsior, B.P. Koch, P. Schmitt-Kopplin, N. Hertkorn, Comprehensive structure-selective characterization of dissolved organic matter by reducing molecular complexity and increasing analytical dimensions, Water Res. 106 (2016) 477–487, https://doi.org/10.1016/j.watres.2016.10.034.

[30] R.P. Rodgers, M.M. Mapolelo, W.K. Robbins, M.L. Chacón-Patino, J.C. Putman, S.F. Niles, S.M. Rowland, A.G. Marshall, Combating selective ionization in the high resolution mass spectral characterization of complex mixtures, Faraday Discuss. 218 (2019). https://doi.org/10.1039/C9FD00005D.

[31] W. Wang, C. He, Y. Gao, Y. Zhang, Q. Shi, Isolation and characterization of hydrophilic dissolved organic matter in waters by ion exchange solid phase extraction followed by high resolution mass spectrometry, Environ. Chem. Lett. (2019). https://doi.org/10.1007/s10311-019-00898-6.

[32] N.R. Novotny, E.N. Capley, A.C. Stenson, Fact or artifact: the representativeness of ESI-MS for complex natural organic mixtures, J. Mass Spectrom. 49 (2014) 316–326, https://doi.org/10.1002/jms.3345.

[33] T. Henriksen, R.K. Juhler, B. Svensmark, N.B. Cech, The relative influences of acidity and polarity on responsiveness of small organic molecules to analysis with negative ion electrospray ionization mass spectrometry (ESI-MS), J. Am. Soc. Mass Spectrom. 16 (2005) 446–455, https://doi.org/10.1016/j.jasms.2004.11.021.

[34] L. Konermann, E. Ahadi, A.D. Rodriguez, S. Vahidi, Unraveling the mechanism of electrospray ionization, Anal. Chem. 85 (2013) 2–9, https://doi.org/10.1021/ac302789c.

[35] B.A. Huffman, M.L. Poltash, C.A. Hughey, Effect of polar protic and polar aprotic solvents on negative-ion electrospray ionization and chromatographic separation of small acidic molecules, Anal. Chem. 84 (2012) 9942–9950, https://doi.org/10.1021/ac302397b.

[36] A. Kruve, K. Kaupmees, J. Liigand, I. Leito, Negative electrospray ionization via deprotonation: predicting the ionization efficiency, Anal. Chem. 86 (2014) 4822–4830, https://doi.org/10.1021/ac404066v.

[37] K.M. Peru, M.J. Thomas, D.C. Palacio Lozano, D.W. McMartin, J.V. Headley, M.P. Barrow, Characterization of oil sands naphthenic acids by negative-ion electrospray ionization mass spectrometry: influence of acidic versus basic transfer solvent, Chemosphere 222 (2019) 1017–1024, https://doi.org/10.1016/J.CHEMOSPHERE.2019.01.162.

[38] J.V. Headley, K.M. Peru, M.P. Barrow, P.J. Derrick, Characterization of naphthenic acids from athabasca oil sands using electrospray ionization: the significant influence of solvents, Anal. Chem. 79 (2007) 6222–6229, https://doi.org/10.1021/ac070905w.

[39] B.M. Ruddy, C.L. Hendrickson, R.P. Rodgers, A.G. Marshall, Positive ion electrospray ionization suppression in petroleum and complex mixtures, Energy Fuel 32 (2018) 2901–2907, https://doi.org/10.1021/acs.energyfuels.7b03204.

[40] N. Hertkorn, M. Frommberger, M. Witt, B.P. Koch, P. Schmitt-kopplin, E.M. Perdue, Natural organic matter and the event horizon of mass spectrometry, Anal. Chem. 80 (2008) 8908–8919. https://doi.org/10.1021/ac800464g.

[41] W.C. Hockaday, J.M. Purcell, A.G. Marshall, J.A. Baldock, P.G. Hatcher, Electrospray and photoionization mass spectrometry for the characterization of organic matter in

natural waters: a qualitative assessment, Limnol. Oceanogr. Methods 7 (2011) 81–95, https://doi.org/10.4319/lom.2009.7.81.
[42] D. Cao, H. Huang, M. Hu, L. Cui, F. Geng, Z. Rao, H. Niu, Y. Cai, Y. Kang, Comprehensive characterization of natural organic matter by MALDI- and ESI-Fourier transform ion cyclotron resonance mass spectrometry, Anal. Chim. Acta 866 (2015) 48–58, https://doi.org/10.1016/j.aca.2015.01.051.
[43] D. Kim, J. Lee, B. Kim, S. Kim, Optimization and application of paper-based spray ionization mass spectrometry for analysis of natural organic matter, Anal. Chem. 90 (2018) 12027–12034, https://doi.org/10.1021/acs.analchem.8b02668.
[44] S. Kameyama, H. Tanimoto, S. Inomata, U. Tsunogai, A. Ooki, S. Takeda, H. Obata, A. Tsuda, M. Uematsu, High-resolution measurement of multiple volatile organic compounds dissolved in seawater using equilibrator inlet–proton transfer reaction-mass spectrometry (EI–PTR-MS), Mar. Chem. 122 (2010) 59–73, https://doi.org/10.1016/j.marchem.2010.08.003.
[45] D. Materić, M. Peacock, M. Kent, S. Cook, V. Gauci, T. Röckmann, R. Holzinger, Characterisation of the semi-volatile component of dissolved organic matter by thermal desorption—proton transfer reaction—mass spectrometry, Sci. Rep. 7 (2017) 1–8, https://doi.org/10.1038/s41598-017-16256-x.
[46] W. Kew, C.L. Mackay, I. Goodall, D.J. Clarke, D. Uhrín, Complementary ionization techniques for the analysis of scotch whisky by high resolution mass spectrometry, Anal. Chem. 90 (2018) 11265–11272, https://doi.org/10.1021/acs.analchem.8b01446.
[47] A. Gaspar, W. Schrader, Expanding the data depth for the analysis of complex crude oil samples by Fourier transform ion cyclotron resonance mass spectrometry using the spectral stitching method, Rapid Commun. Mass Spectrom. 26 (2012) 1047–1052, https://doi.org/10.1002/rcm.6200.
[48] T. Riedel, T. Dittmar, A method detection limit for the analysis of natural organic matter via Fourier transform ion cyclotron resonance mass spectrometry, Anal. Chem. 86 (2014) 8376–8382, https://doi.org/10.1021/ac501946m.
[49] D.C. Palacio Lozano, R. Gavard, J.P. Arenas-Diaz, M.J. Thomas, D.D. Stranz, E. Mejía-Ospino, A. Guzman, S.E.F. Spencer, D. Rossell, M.P. Barrow, Pushing the analytical limits: new insights into complex mixtures using mass spectra segments of constant ultrahigh resolving power, Chem. Sci. 10 (2019) 6966–6978, https://doi.org/10.1039/C9SC02903F.
[50] D. Cao, J. Lv, F. Geng, Z. Rao, H. Niu, Y. Shi, Y. Cai, Y. Kang, Ion accumulation time dependent molecular characterization of natural organic matter using electrospray ionization-Fourier transform ion cyclotron resonance mass spectrometry, Anal. Chem. 88 (2016) 12210–12218, https://doi.org/10.1021/acs.analchem.6b03198.
[51] R.L. Sleighter, G.A. McKee, P.G. Hatcher, Direct Fourier transform mass spectral analysis of natural waters with low dissolved organic matter, Org. Geochem. 40 (2009) 119–125, https://doi.org/10.1016/j.orggeochem.2008.09.012.
[52] J.A. Hawkes, T. Dittmar, C. Patriarca, L. Tranvik, J. Bergquist, Evaluation of the Orbitrap mass spectrometer for the molecular fingerprinting analysis of natural dissolved organic matter, Anal. Chem. 88 (2016) 7698–7704, https://doi.org/10.1021/acs.analchem.6b01624.
[53] D. Kim, S. Kim, S. Son, M.-J. Jung, S. Kim, Application of Online Liquid Chromatography 7 T FT-ICR Mass Spectrometer Equipped with Quadrupolar Detection for Analysis of Natural Organic Matter, Anal. Chem. (2019). https://doi.org/10.1021/acs.analchem.9b00689.
[54] C. Simon, V.-N. Roth, T. Dittmar, G. Gleixner, Molecular signals of heterogeneous terrestrial environments identified in dissolved organic matter: a comparative analysis of

orbitrap and ion cyclotron resonance mass spectrometers, Front. Earth Sci. 6 (2018) 1–16, https://doi.org/10.3389/feart.2018.00138.
[55] N. Tolić, Y. Liu, A. Liyu, Y. Shen, M.M. Tfaily, E.B. Kujawinski, K. Longnecker, L.-J. Kuo, E.W. Robinson, L. Paša-Tolić, N.J. Hess, Formularity: software for automated formula assignment of natural and other organic matter from ultrahigh-resolution mass spectra, Anal. Chem. 89 (2017) 12659–12665, https://doi.org/10.1021/acs.analchem.7b03318.
[56] E.B. Kujawinski, M.D. Behn, Automated analysis of electrospray ionization Fourier transform ion cyclotron resonance mass spectra of natural organic matter, Anal. Chem. 78 (2006) 4363–4373, https://doi.org/10.1021/ac0600306.
[57] T. Leefmann, S. Frickenhaus, B.P. Koch, UltraMassExplorer: a browser-based application for the evaluation of high-resolution mass spectrometric data, Rapid Commun. Mass Spectrom. 33 (2019) 193–202, https://doi.org/10.1002/rcm.8315.
[58] T. Kind, O. Fiehn, Seven Golden rules for heuristic filtering of molecular formulas obtained by accurate mass spectrometry, BMC Bioinform. 8 (2007) 105, https://doi.org/10.1186/1471-2105-8-105.
[59] A.C. Stenson, A.G. Marshall, W.T. Cooper, Exact masses and chemical formulas of individual Suwannee River fulvic acids from ultrahigh resolution electrospray ionization Fourier transform ion cyclotron resonance mass spectra, Anal. Chem. 75 (2003) 1275–1284, https://doi.org/10.1021/ac026106p.
[60] K. Qian, R.P. Rodgers, C.L. Hendrickson, C.A. Hughey, A.G. Marshall, Kendrick mass defect Spectrum: a compact visual analysis for ultrahigh-resolution broadband mass spectra, Anal. Chem. 73 (2002) 4676–4681, https://doi.org/10.1021/ac010560w.
[61] T.A. Brown, B.A. Jackson, B.J. Bythell, A.C. Stenson, Benefits of multidimensional fractionation for the study and characterization of natural organic matter, J. Chromatogr. A 1470 (2016) 84–96, https://doi.org/10.1016/j.chroma.2016.10.005.
[62] G.C. Woods, M.J. Simpson, P.J. Koerner, A. Napoli, A.J. Simpson, HILIC-NMR: toward the identification of individual molecular components in dissolved organic matter, Environ. Sci. Technol. 45 (2011) 3880–3886, https://doi.org/10.1021/es103425s.
[63] S. Sandron, N.W. Davies, R. Wilson, A.R. Cardona, P.R. Haddad, P.N. Nesterenko, B. Paull, Fractionation of dissolved organic matter on coupled reversed-phase monolithic columns and characterisation using reversed-phase liquid chromatography-high resolution mass spectrometry, Chromatographia 81 (2018) 203–213, https://doi.org/10.1007/s10337-017-3324-0.
[64] S. Sandron, P.N. Nesterenko, M.V. McCaul, B. Kelleher, B. Paull, Normal-phase high-performance counter-current chromatography for the fractionation of dissolved organic matter from a freshwater source, J. Sep. Sci. 37 (2014) 135–142, https://doi.org/10.1002/jssc.201300634.
[65] Z. Liu, R.L. Sleighter, J. Zhong, P.G. Hatcher, The chemical changes of DOM from black waters to coastal marine waters by HPLC combined with ultrahigh resolution mass spectrometry, Estuar. Coast. Shelf Sci. 92 (2011) 205–216, https://doi.org/10.1016/j.ecss.2010.12.030.
[66] B.P. Koch, K.U. Ludwichowski, G. Kattner, T. Dittmar, M. Witt, Advanced characterization of marine dissolved organic matter by combining reversed-phase liquid chromatography and FT-ICR-MS, Mar. Chem. 111 (2008) 233–241, https://doi.org/10.1016/j.marchem.2008.05.008.
[67] A.C. Stenson, Reversed-phase chromatography fractionation tailored to mass spectral characterization of humic substances, Environ. Sci. Technol. 42 (2008) 2060–2065, https://doi.org/10.1021/es7022412.

[68] A. Nyakas, J. Han, K.M. Peru, J.V. Headley, C.H. Borchers, Comprehensive analysis of oil sands processed water by direct-infusion Fourier-transform ion cyclotron resonance mass spectrometry with and without offline UHPLC sample prefractionation, Environ. Sci. Technol. 47 (2013) 4471–4479, https://doi.org/10.1021/es400813s.

[69] A. Gaspar, M. Harir, N. Hertkorn, P. Schmitt-Kopplin, Preparative free-flow electrophoretic offline ESI-Fourier transform ion cyclotron resonance/MS analysis of Suwannee River fulvic acid, Electrophoresis 31 (2010) 2070–2079, https://doi.org/10.1002/elps.200900726.

[70] C. Patriarca, J. Bergquist, P.J.R. Sjöberg, L. Tranvik, J.A. Hawkes, Online HPLC-ESI-HRMS method for the analysis and comparison of different dissolved organic matter samples, Environ. Sci. Technol. 52 (2018) 2091–2099, https://doi.org/10.1021/acs.est.7b04508.

[71] K. Namjesnik-Dejanovic, S.E. Cabaniss, Reverse-phase HPLC method for measuring polarity distributions of natural organic matter, Environ. Sci. Technol. 38 (2004) 1108–1114, https://doi.org/10.1021/es0344157.

[72] A. These, T. Reemtsma, Limitations of electrospray ionization of fulvic and humic acids as visible from size exclusion chromatography with organic carbon and mass spectrometric detection, Anal. Chem. 75 (2003) 6275–6281, https://doi.org/10.1021/ac034399w.

[73] T. Reemtsma, A. These, A. Springer, M. Linscheid, Differences in the molecular composition of fulvic acid size fractions detected by size-exclusion chromatography-on line Fourier transform ion cyclotron resonance (FTICR-) mass spectrometry, Water Res. 42 (2008) 63–72, https://doi.org/10.1016/j.watres.2007.06.063.

[74] M.J. Thomas, E. Collinge, M. Witt, D.C. Palacio Lozano, C.H. Vane, V. Moss-Hayes, M.P. Barrow, Petroleomic depth profiling of Staten Island salt marsh soil: 2ω detection FTICR MS offers a new solution for the analysis of environmental contaminants, Sci. Total Environ. 662 (2019) 852–862, https://doi.org/10.1016/j.scitotenv.2019.01.228.

[75] D.F. Smith, D.C. Podgorski, R.P. Rodgers, G.T. Blakney, C.L. Hendrickson, 21 tesla FT-ICR mass spectrometer for ultrahigh-resolution analysis of complex organic mixtures, Anal. Chem. 90 (2018) 2041–2047, https://doi.org/10.1021/acs.analchem.7b04159.

[76] J.D. Byer, K. Siek, K. Jobst, Distinguishing the C3 vs SH4 mass split by comprehensive two-dimensional gas chromatography-high resolution time-of-flight mass spectrometry, Anal. Chem. 88 (2016) 6101–6104, https://doi.org/10.1021/acs.analchem.6b01137.

[77] J.A. Hawkes, N. Radoman, J. Bergquist, M.B. Wallin, L.J. Tranvik, S. Löfgren, Regional diversity of complex dissolved organic matter across forested hemiboreal headwater streams, Sci. Rep. 8 (2018). https://doi.org/10.1038/s41598-018-34272-3.

[78] T. Reemtsma, A. These, On-line coupling of size exclusion chromatography with electrospray ionization-tandem mass spectrometry for the analysis of aquatic fulvic and humic acids, Anal. Chem. 75 (2003) 1500–1507, https://doi.org/10.1021/ac0261294.

[79] L.V. Tose, P. Benigni, D. Leyva, A. Sundberg, C.E. Ramírez, M.E. Ridgeway, M.A. Park, W. Romão, R. Jaffé, F. Fernandez-Lima, Coupling trapped ion mobility spectrometry to mass spectrometry: trapped ion mobility spectrometry–time-of-flight mass spectrometry versus trapped ion mobility spectrometry–Fourier transform ion cyclotron resonance mass spectrometry, Rapid Commun. Mass Spectrom. 32 (2018) 1287–1295, https://doi.org/10.1002/rcm.8165.

[80] D. Leyva, L. Valadares, J. Porter, J. Wolff, R. Jaffè, F. Fernandez-Lima, Understanding the structural complexity of dissolved organic matter: isomeric diversity, Faraday Discuss. 218 (2019). https://doi.org/10.1039/c8fd00221e.

[81] K. Lu, W.S. Gardner, Z. Liu, Molecular structure characterization of riverine and coastal dissolved organic matter with ion mobility quadrupole time-of-flight LCMS (IM Q-TOF LCMS), Environ. Sci. Technol. 52 (2018) 7182–7191, https://doi.org/10.1021/acs.est.8b00999.

[82] P. Benigni, C.J. Thompson, M.E. Ridgeway, M.A. Park, F. Fernandez-Lima, Targeted high-resolution ion mobility separation coupled to ultrahigh-resolution mass spectrometry of endocrine disruptors in complex mixtures, Anal. Chem. 87 (2015) 4321–4325, https://doi.org/10.1021/ac504866v.
[83] A. Castellanos, P. Benigni, D.R. Hernandez, J.D. Debord, M.E. Ridgeway, M.A. Park, F. Fernandez-Lima, Fast screening of polycyclic aromatic hydrocarbons using trapped ion mobility spectrometry-mass spectrometry, Anal. Methods 6 (2014) 9328–9332, https://doi.org/10.1039/c4ay01655f.
[84] J. Le Maître, M. Hubert-Roux, B. Paupy, S. Marceau, C. Rüger, C. Afonso, P. Giusti, Structural analysis of heavy oil fractions after hydrodenitrogenation by high-resolution tandem mass spectrometry and ion mobility spectrometry, Faraday Discuss. 218 (2019) https://doi.org/10.1039/c8fd00239h.
[85] A. Gaspar, E.V. Kunenkov, R. Lock, M. Desor, I. Perminova, P. Schmitt-Kopplin, Combined utilization of ion mobility and ultra-high-resolution mass spectrometry to identify multiply charged constituents in natural organic matter, Rapid Commun. Mass Spectrom. (2009)https://doi.org/10.1002/rcm.3924.
[86] M. Hu, L. Zhang, S. He, C. Xu, Q. Shi, Collision cross section (CCS) measurement by ion cyclotron resonance mass spectrometry with short-time Fourier transform, Rapid Commun. Mass Spectrom. 32 (2018) 751–761, https://doi.org/10.1002/rcm.8096.
[87] F. Yang, J.E. Voelkel, D.V. Dearden, Collision cross sectional areas from analysis of Fourier transform ion cyclotron resonance line width: a new method for characterizing molecular structure, Anal. Chem. 84 (2012) 4851–4857, https://doi.org/10.1021/ac300379a.
[88] J.B. Shaw, T.-Y. Lin, F.E. Leach, A.V. Tolmachev, N. Tolić, E.W. Robinson, D.W. Koppenaal, L. Paša-Tolić, 21 Tesla Fourier transform ion cyclotron resonance mass spectrometer greatly expands mass spectrometry toolbox, J. Am. Soc. Mass Spectrom. 27 (2016) 1929–1936, https://doi.org/10.1007/s13361-016-1507-9.
[89] F. Xian, C.L. Hendrickson, G.T. Blakney, S.C. Beu, A.G. Marshall, Automated broadband phase correction of Fourier transform ion cyclotron resonance mass spectra, Anal. Chem. 82 (2010) 8807–8812, https://doi.org/10.1021/ac101091w.
[90] Y. Qi, C.J. Thompson, S.L. Van Orden, P.B. O'Connor, Phase correction of Fourier transform ion cyclotron resonance mass spectra using MatLab, J. Am. Soc. Mass Spectrom. 22 (2011) 138–147, https://doi.org/10.1007/s13361-010-0006-7.
[91] Y. Qi, R. Luo, W. Schrader, D.A. Volmer, Application of phase correction to improve the characterization of photooxidation products of lignin using 7 tesla Fourier-transform ion cyclotron resonance mass spectrometry, FACETS 2 (2017) 461–475, https://doi.org/10.1139/facets-2016-0069.
[92] L. Schweihard, M. Lindinger, H.-J. Kluge, Quadrupole-detection FT-ICR mass spectrometry, Int. J. Mass Spectrom. Ion Process. 98 (1990) 25–33, https://doi.org/10.1016/0168-1176(90)85045-4.
[93] J.B. Shaw, M.V. Gorshkov, Q. Wu, L. Paša-Tolić, High speed intact protein characterization using 4X frequency multiplication, ion trap harmonization, and 21 tesla FTICR-MS, Anal. Chem. 90 (2018) 5557–5562, https://doi.org/10.1021/acs.analchem.7b04606.
[94] E. Cho, M. Witt, M. Hur, M.J. Jung, S. Kim, Application of FT-ICR MS equipped with Quadrupole detection for analysis of crude oil, Anal. Chem. 89 (2017) 12101–12107, https://doi.org/10.1021/acs.analchem.7b02644.
[95] J.A. Leenheer, C.E. Rostad, P.M. Gates, E.T. Furlong, I. Ferrer, Molecular resolution and fragmentation of fulvic acid by electrospray ionization/multistage tandem mass spectrometry, Anal. Chem. 73 (2001) 1461–1471, https://doi.org/10.1021/ac0012593.

[96] N. Cortés-Francisco, J. Caixach, Fragmentation studies for the structural characterization of marine dissolved organic matter, Anal. Bioanal. Chem. 407 (2015) 2455–2462, https://doi.org/10.1007/s00216-015-8499-3.
[97] N. Hertkorn, R. Benner, M. Frommberger, P. Schmitt-Kopplin, M. Witt, K. Kaiser, A. Kettrup, J.I. Hedges, Characterization of a major refractory component of marine dissolved organic matter, Geochim. Cosmochim. Acta 70 (2006) 2990–3010, https://doi.org/10.1016/j.gca.2006.03.021.
[98] E. Bae, I.J. Yeo, B. Jeong, Y. Shin, K.H. Shin, S. Kim, Study of double bond equivalents and the numbers of carbon and oxygen atom distribution of dissolved organic matter with negative-mode FT-ICR MS, Anal. Chem. 83 (2011) 4193–4199, https://doi.org/10.1021/ac200464q.
[99] A.M. Pohlabeln, T. Dittmar, Novel insights into the molecular structure of non-volatile marine dissolved organic sulfur, Mar. Chem. 168 (2015) 86–94, https://doi.org/10.1016/j.marchem.2014.10.018.
[100] Y. Kostyukevich, A. Kononikhin, A. Zherebker, I. Popov, I. Perminova, E. Nikolaev, Enumeration of non-labile oxygen atoms in dissolved organic matter by use of 16 O/18 O exchange and Fourier transform ion-cyclotron resonance mass spectrometry, Anal. Bioanal. Chem. 406 (2014) 6655–6664, https://doi.org/10.1007/s00216-014-8097-9.
[101] Y. Kostyukevich, A. Kononikhin, I. Popov, E. Nikolaev, Simple atmospheric hydrogen/deuterium exchange method for enumeration of labile hydrogens by electrospray ionization mass spectrometry, Anal. Chem. 85 (2013) 5330–5334, https://doi.org/10.1021/ac4006606.
[102] A. Zherebker, I.V. Perminova, Y. Kostyukevich, A. Kononikhin, O. Kharybin, E. Nikolaev, Structural investigation of coal humic substances by selective isotopic exchange and high-resolution mass spectrometry, Faraday Discuss. 218 (2019). https://doi.org/10.1039/c9fd00002j.
[103] T. Solouki, M.A. Freitas, A. Alomary, Gas-phase hydrogen/deuterium exchange reactions of fulvic acids: an electrospray ionization Fourier transform ion cyclotron resonance mass spectral study, Anal. Chem. 71 (1999) 4719–4726, https://doi.org/10.1021/ac990185w.
[104] Y. Kostyukevich, A. Zherebker, A. Kononikhin, I. Popov, I. Perminova, E. Nikolaev, The investigation of the birch tar using ultrahigh resolution Fourier transform ion cyclotron resonance mass spectrometry and hydrogen/deuterium exchange approach, Int. J. Mass Spectrom. 404 (2016) 29–34, https://doi.org/10.1016/j.ijms.2016.03.012.
[105] J. Hertzog, V. Carré, A. Dufour, F. Aubriet, Semi-targeted analysis of complex matrices by ESI FT-ICR MS or how an experimental Bias may be used as an analytical tool, J. Am. Soc. Mass Spectrom. 29 (2018) 543–557, https://doi.org/10.1007/s13361-017-1865-y.
[106] B.P. Koch, T. Dittmar, M. Witt, G. Kattner, Fundamentals of molecular formula assignment to ultrahigh resolution mass data of natural organic matter, Anal. Chem. 79 (2007) 1758–1763, https://doi.org/10.1021/ac061949s.
[107] A. Gaspar, M. Lucio, M. Harir, P. Schmitt-Kopplin, Targeted and non-targeted boron complex formation followed by electrospray Fourier transform ion cyclotron mass spectrometry: a novel approach for identifying boron esters with natural organic matter, Eur. J. Mass Spectrom. 17 (2011) 113–123, https://doi.org/10.1255/ejms.1123.
[108] A.C. Stenson, Fourier transform ion cyclotron resonance mass spectral characterization of metal-humic binding, Rapid Commun. Mass Spectrom. 23 (2009) 465–476, https://doi.org/10.1002/rcm.3889.
[109] H. Waska, A. Koschinsky, M.J. Ruiz Chancho, T. Dittmar, Investigating the potential of solid-phase extraction and Fourier-transform ion cyclotron resonance mass spectrometry

(FT-ICR-MS) for the isolation and identification of dissolved metal-organic complexes from natural waters, Mar. Chem. 173 (2015) 78–92, https://doi.org/10.1016/j.marchem.2014.10.001.

[110] H. Waska, A. Koschinsky, T. Dittmar, Fe- and cu-complex formation with artificial ligands investigated by ultra-high resolution Fourier-transform ion cyclotron resonance mass spectrometry (FT-ICR-MS): implications for natural metal-organic complex studies, Front. Mar. Sci. 3 (2016) 1–19, https://doi.org/10.3389/fmars.2016.00119.

[111] R.M. Boiteau, C.P. Till, T.H. Coale, J.N. Fitzsimmons, K.W. Bruland, D.J. Repeta, Patterns of iron and siderophore distributions across the California current system, Limnol. Oceanogr. 64 (2019) 376–389, https://doi.org/10.1002/lno.11046.

[112] K.D. Duncan, D.A. Volmer, C.G. Gill, E.T. Krogh, Rapid screening of carboxylic acids from waste and surface waters by ESI-MS/MS using barium ion chemistry and on-line membrane sampling, J. Am. Soc. Mass Spectrom. 27 (2016) 443–450, https://doi.org/10.1007/s13361-015-1311-y.

[113] T. Reemtsma, The carbon versus mass diagram to visualize and exploit FTICR-MS data of natural organic matter, J. Mass Spectrom. 45 (2010) 382–390, https://doi.org/10.1002/jms.1722.

[114] S. Kim, R.W. Kramer, P.G. Hatcher, Graphical method for analysis of ultrahigh-resolution broadband mass spectra of natural organic matter, the Van Krevelen diagram, Anal. Chem. 75 (2003) 5336–5344, https://doi.org/10.1021/ac034415p.

[115] A.M. Kellerman, T. Dittmar, D.N. Kothawala, L.J. Tranvik, Chemodiversity of dissolved organic matter in lakes driven by climate and hydrology, Nat. Commun. 5 (2014) 1–8, https://doi.org/10.1038/ncomms4804.

[116] M. Seidel, P.L. Yager, N.D. Ward, E.J. Carpenter, H.R. Gomes, A.V. Krusche, J.E. Richey, T. Dittmar, P.M. Medeiros, Molecular-level changes of dissolved organic matter along the Amazon River-to-ocean continuum, Mar. Chem. 177 (2015) 218–231, https://doi.org/10.1016/j.marchem.2015.06.019.

[117] N. Hertkorn, M. Harir, K.M. Cawley, P. Schmitt-Kopplin, R. Jaffé, Molecular characterization of dissolved organic matter from subtropical wetlands: a comparative study through the analysis of optical properties, NMR and FTICR/MS, Biogeosciences 13 (2016) 2257–2277, https://doi.org/10.5194/bg-13-2257-2016.

[118] P.M. Medeiros, M. Seidel, N.D. Ward, E.J. Carpender, H.R. Gomes, J. Niggemann, A.V. Krusche, J.E. Richey, P.L. Yager, T. Dittmar, Global biogeochemical cycles in the tropical Atlantic Ocean, Global Biogeochem. Cycles (2015) 677–690, https://doi.org/10.1002/2015GB005115.

[119] J. Lv, S. Zhang, S. Wang, L. Luo, D. Cao, P. Christie, Molecular-scale investigation with ESI-FT-ICR-MS on fractionation of dissolved organic matter induced by adsorption on iron oxyhydroxides, Environ. Sci. Technol. 50 (2016) 2328–2336, https://doi.org/10.1021/acs.est.5b04996.

[120] W. Kew, J.W.T. Blackburn, D.J. Clarke, D. Uhrín, Interactive van Krevelen diagrams—advanced visualisation of mass spectrometry data of complex mixtures, Rapid Commun. Mass Spectrom. 31 (2017) 658–662, https://doi.org/10.1002/rcm.7823.

[121] M.P. Barrow, J.V. Headley, K.M. Peru, P.J. Derrick, Data visualization for the characterization of naphthenic acids within petroleum samples, Energy Fuels 23 (2009) 2592–2599, https://doi.org/10.1021/ef800985z.

[122] B.P. Koch, T. Dittmar, Erratum: From mass to structure: an aromaticity index for high-resolution mass data of natural organic matter (Rapid Communications in Mass Spectrometry (2006) 20 (926–932) https://doi.org/10.1002/rcm.2386), Rapid Commun. Mass Spectrom. 30 (2016) 250, https://doi.org/10.1002/rcm.7433.

[123] E. Kendrick, A mass scale based on CH 2 = 14.0000 for high resolution mass spectrometry of organic compounds, Anal. Chem. 35 (1963) 2146–2154, https://doi.org/10.1021/ac60206a048.

[124] C.A. Hughey, R.P. Rodgers, A.G. Marshall, Resolution of 11 000 compositionally distinct components in a single electrospray ionization fourier transform ion cyclotron resonance mass spectrum of crude oil, Anal. Chem. 74 (2002) 4145–4149, https://doi.org/10.1021/ac020146b.

[125] P.J. Roach, J. Laskin, A. Laskin, Higher-order mass defect analysis for mass spectra of complex organic mixtures, Anal. Chem. 83 (2011) 4924–4929, https://doi.org/10.1021/ac200654j.

[126] D. Petras, I. Koester, R. Da Silva, B.M. Stephens, A.F. Haas, C.E. Nelson, L.W. Kelly, L.I. Aluwihare, P.C. Dorrestein, High-resolution liquid chromatography tandem mass spectrometry enables large scale molecular characterization of dissolved organic matter, Front. Mar. Sci. 4 (2017), https://doi.org/10.3389/fmars.2017.00405.

[127] D. Tziotis, N. Hertkorn, P. Schmitt-Kopplin, Kendrick-analogous network visualisation of ion cyclotron resonance Fourier transform mass spectra: Improved options for the assignment of elemental compositions and the classification of organic molecular complexity, Eur. J. Mass Spectrom. 17 (2011) 415–421, https://doi.org/10.1255/ejms.1135.

[128] K. Longnecker, E.B. Kujawinski, Using network analysis to discern compositional patterns in ultrahigh-resolution mass spectrometry data of dissolved organic matter, Rapid Commun. Mass Spectrom. (2016) 2388–2394, https://doi.org/10.1002/rcm.7719.

[129] A. Rivas-Ubach, Y. Liu, T.S. Bianchi, N. Tolić, C. Jansson, L. Paša-Tolić, Moving beyond the van Krevelen diagram: a new stoichiometric approach for compound classification in organisms, Anal. Chem. 90 (2018) 6152–6160, https://doi.org/10.1021/acs.analchem.8b00529.

[130] R. Flerus, O.J. Lechtenfeld, B.P. Koch, S.L. McCallister, P. Schmitt-Kopplin, R. Benner, K. Kaiser, G. Kattner, A molecular perspective on the ageing of marine dissolved organic matter, Biogeosciences 9 (2012) 1935–1955, https://doi.org/10.5194/bg-9-1935-2012.

[131] H. Osterholz, G. Singer, B. Wemheuer, R. Daniel, M. Simon, J. Niggemann, T. Dittmar, Deciphering associations between dissolved organic molecules and bacterial communities in a pelagic marine system, ISME J. 10 (2016) 1717–1730, https://doi.org/10.1038/ismej.2015.231.

[132] R.L. Sleighter, H. Chen, A.S. Wozniak, A.S. Willoughby, P. Caricasole, P.G. Hatcher, Establishing a measure of reproducibility of ultrahigh-resolution mass spectra for complex mixtures of natural organic matter, Anal. Chem. 84 (2012) 9184–9191, https://doi.org/10.1021/ac3018026.

[133] N.J. Nielsen, P. Christensen, C. Stedmon, J.H. Christensen, Examples of unwanted variation when characterising dissolved organic matter using direct injection electrospray mass spectrometry and chemometrics, Anal. Methods 10 (2018) 2636–2646, https://doi.org/10.1039/c8ay00226f.

[134] M.C. Kido Soule, K. Longnecker, S.J. Giovannoni, E.B. Kujawinski, Impact of instrument and experiment parameters on reproducibility of ultrahigh resolution ESI FT-ICR mass spectra of natural organic matter, Org. Geochem. 41 (2010) 725–733, https://doi.org/10.1016/j.orggeochem.2010.05.017.

[135] M. Gonsior, J. Valle, P. Schmitt-Kopplin, N. Hertkorn, D. Bastviken, J. Luek, M. Harir, W. Bastos, A. Enrich-Prast, Chemodiversity of dissolved organic matter in the Amazon Basin, Biogeosciences 13 (2016) 4279–4290, https://doi.org/10.5194/bg-13-4279-2016.

[136] N. Kamjunke, W. von Tümpling, N. Hertkorn, M. Harir, P. Schmitt-Kopplin, H. Norf, M. Weitere, P. Herzsprung, A new approach for evaluating transformations of dissolved

organic matter (DOM) via high-resolution mass spectrometry and relating it to bacterial activity, Water Res. 123 (2017) 513–523, https://doi.org/10.1016/j.watres.2017.07.008.

[137] V.N. Roth, T. Dittmar, R. Gaupp, G. Gleixner, Latitude and pH driven trends in the molecular composition of DOM across a north south transect along the Yenisei River, Geochim. Cosmochim. Acta 123 (2013) 93–105, https://doi.org/10.1016/j.gca.2013.09.002.

[138] A. Mostovaya, J.A. Hawkes, T. Dittmar, L.J. Tranvik, Molecular determinants of dissolved organic matter reactivity in Lake water, Front. Earth Sci. 5 (2017) 1–13, https://doi.org/10.3389/feart.2017.00106.

[139] P.M. Medeiros, M. Seidel, J. Niggemann, R.G.M. Spencer, P.J. Hernes, P.L. Yager, W.L. Miller, T. Dittmar, D.A. Hansell, A novel molecular approach for tracing terrigenous dissolved organic matter into the deep ocean, Global Biogeochem. Cycles 30 (2016) 689–699, https://doi.org/10.1002/2015GB005320.

[140] A.G. Seifert, V.N. Roth, T. Dittmar, G. Gleixner, L. Breuer, T. Houska, J. Marxsen, Comparing molecular composition of dissolved organic matter in soil and stream water: influence of land use and chemical characteristics, Sci. Total Environ. 571 (2016) 142–152, https://doi.org/10.1016/j.scitotenv.2016.07.033.

[141] R.L. Sleighter, Z. Liu, J. Xue, P.G. Hatcher, Multivariate statistical approaches for the characterization of dissolved organic matter analyzed by ultrahigh resolution mass spectrometry, Environ. Sci. Technol. 44 (2010) 7576–7582, https://doi.org/10.1021/es1002204.

[142] A. Mostovaya, J.A. Hawkes, B. Koehler, T. Dittmar, L.J. Tranvik, Emergence of the reactivity continuum of organic matter from kinetics of a multitude of individual molecular constituents, Environ. Sci. Technol. 51 (2017) 11571–11579, https://doi.org/10.1021/acs.est.7b02876.

[143] E.C. Minor, C.J. Steinbring, K. Longnecker, E.B. Kujawinski, Characterization of dissolved organic matter in Lake Superior and its watershed using ultrahigh resolution mass spectrometry, Org. Geochem. 43 (2012) 1–11, https://doi.org/10.1016/j.orggeochem.2011.11.007.

[144] J.J. Mosher, L.A. Kaplan, D.C. Podgorski, A.M. McKenna, A.G. Marshall, Longitudinal shifts in dissolved organic matter chemogeography and chemodiversity within headwater streams: a river continuum reprise, Biogeochemistry 124 (2015) 371–385, https://doi.org/10.1007/s10533-015-0103-6.

[145] P. Herzsprung, W. Von Tümpling, N. Hertkorn, M. Harir, O. Büttner, J. Bravidor, K. Friese, P. Schmitt-Kopplin, Variations of DOM quality in inflows of a drinking water reservoir: linking of van krevelen diagrams with EEMF spectra by rank correlation, Environ. Sci. Technol. 46 (2012) 5511–5518, https://doi.org/10.1021/es300345c.

[146] S. Wagner, R. Jaffé, K. Cawley, T. Dittmar, A. Stubbins, Associations between the molecular and optical properties of dissolved organic matter in the Florida Everglades, a model coastal wetland system, Front. Chem. 3 (2015) 1–14, https://doi.org/10.3389/fchem.2015.00066.

[147] U.J. Wünsch, E. Acar, B.P. Koch, K.R. Murphy, P. Schmitt-Kopplin, C.A. Stedmon, The molecular fingerprint of fluorescent natural organic matter offers insight into biogeochemical sources and Diagenetic state, Anal. Chem. 90 (2018) 14188–14197, https://doi.org/10.1021/acs.analchem.8b02863.

Two-dimensional correlation spectroscopy to assess the dynamics of complex environmental mixtures

5

Mauro Mecozzi
Laboratory of Chemometrics and Environmental Applications, ISPRA, Rome, Italy

Introduction

In analytical chemistry, principal component analysis (PCA), partial least-squares analysis (PLS), and discriminant analysis (LDA) are the most common methods for analysis and pattern recognition of multivariate data samples. These methods were developed as purely statistical techniques and for this reason their use is practically and uniformly diffused in all fields of experimental science [1–3]. Two-dimensional correlation spectroscopy (2DCOS), the subject of this chapter, has a different history. Noda proposed 2DCOS initially in 1988 and 1989 for the spectroscopic interpretation of dynamic structural changes in polymeric systems caused by the action of an external chemical or physical perturbation [4, 5]. For this reason, we can say that 2DCOS has been developed with more targeted aims than those of PCA, PLS, and LDA. This is the first difference between 2DCOS and the previous statistical methods. The second difference is in the types of data because PCA, PLS, and LDA work on both discrete data, reported as tables, and continuous data series of analytical signals like digitized spectra or chromatograms. Conversely, 2DCOS works on digitized analytical signals like spectra and chromatograms only because its applications to discrete data are a nonsense or at least useless.

In the subsequent development of 2DCOS theory, Noda and Ozaki presented the generalized two-dimensional correlation spectroscopy [6, 7]. This further development deepened the concept of dynamic spectra in 2DCOS, elucidating its mathematic supports by means of matrix notation. The generalized 2DCOS also evidenced the natural versatility of 2DCOS extending its application to all the branches of spectroscopy and also chromatography, without excluding "a priori" any typology of analytical signals.

In the last 30 years, we have finally witnessed a vast application of 2DCOS in several fields of polymer and material science, food chemistry, nutrition, biomedical research, and industrial technology, all devoted to the examination of the analytical data set of molecular systems showing dynamic changes caused by an external perturbation [8–12].

Multidimensional Analytical Techniques in Environmental Research. https://doi.org/10.1016/B978-0-12-818896-5.00005-3

Why is 2DCOS so widespread in analytical science? Why is 2DCOS so versatile? Basically, in 2DCOS, the conventional one-dimensional examination of an X matrix, consisting of several spectra or chromatograms, is replaced by the simultaneous and two-dimensional examination of the whole analytical data set. This improves the detection of the molecular characteristics present, often giving information which can be hardly retrieved by means of the conventional mono-dimensional examination which looks at a single spectrum at a time only. The 2DCOS approach has proved to be very efficient, for example, in material science for the study of structural changes in polymers submitted to external chemical and physical agents [6–9, 11, 12].

In environmental science, 2DCOS had in any case a delayed diffusion with respect to other fields of research; in fact, a review of 2DCOS in environmental studies defined it as an uncommon multivariate statistical method [13]. There may be many reasons for the delayed diffusion of 2DCOS with respect to other multivariate techniques, and we can only hypothesize some of them. Surely, as the original aim of 2DCOS was the support for the interpretation of infrared spectra in polymer and material science [4, 5], such a focused task may have narrowed the initial interest of environmental scientists. In addition, as 2DCOS works on digitized spectra instead of discrete data, this could be a further element to explain the delayed diffusion of 2DCOS in environmental studies with respect to other fields of science because 2DCOS requires some experience in signal processing methods.

We wish to demonstrate instead that 2DCOS can work effectively in the study of NOM characteristics and properties. Due to its high complexity, the study of NOM and of all its fractions in environmental matrices such as soils, water, and sediments is a good field for 2DCOS application. In fact, the organic matter, both terrestrial [13, 14] and marine [15–17], has heterogeneous compositions depending on the several vegetal and living organism contributions and on human activities. In NOM of natural waters, seawater, and sediments, we find carbohydrates and polysaccharides, lipids, and proteins and each class of these compounds has a proper heterogeneous composition. Polysaccharides give a significant contribution to marine [15, 18] and soil [19] NOM because it consists of neutral carbohydrates, uronic acids, and aminosugars [15]. Lipids consist of fatty acids and ester fatty acids, hydrocarbons, aliphatic and aromatic alcohols, and waxes [20].

Nitrogen compounds are present as free amino acids, oligopeptides, and proteins [15]. According to this heterogeneous composition, NOM is a mixture of several extremely reactive compounds able to change its characteristics and properties under the effects of environmental conditions such as water and air circulation, solar irradiation, river effluents, rain contributions, and pollutant spills. These interactions cause a long series of dynamic processes such as polymerization reactions, oxidative and degradation reactions, complexation with inorganic elements, and supramolecular interactions among organic constituents [15, 21]. At last under anoxic conditions, Maillard and enzymatic reactions also occur, causing browning development in NOM [22].

For all these reasons, NOM composition and structure evolve according to complex dynamic processes like those described in polymer and material science [7, 23] and then 2DCOS becomes able to investigate the dynamic processes of NOM too.

After a reminder of 2DCOS theory with some simple examples of its application to an FTIR study, in this chapter we describe and review its principal and updated applications in environmental studies.

Theory of 2DCOS

In 2DCOS, the dynamic processes and structural changes in molecular systems are detected by means of the changes of band intensities and positions of the molecular functional groups detected by optical or vibrational spectra [7]. When 2DCOS describes changes in correlations among the wavelengths of the spectra, it is called *wavelength-wavelength* 2DCOS or, alternatively, *variable-variable* 2DCOS. Moreover, Noda and Ozaki have also introduced the *sample-sample* 2DCOS which describes correlations among samples instead, using always the same two-dimensional correlation approach [7].

Over the years, starting from the first presentation and introduction of 2DCOS until today, Noda [24–29] and Ozaki [30] have developed and proposed several new approaches and updates of the 2DCOS theory. The main characteristics of these updates are outside the aim of this chapter; we cite them for the interested readers. Here, let us examine the basic theory of 2DCOS using matrix notation.

Wavelength-wavelength 2DCOS

Each analytical data series consists of n-couples of w_i, x_i data; they are the i-wavelength-variable and the i-signal intensity values, respectively, present in the series. In 2DCOS, the X data matrix of signal intensity values is constructed by sorting the spectral samples of the w, x series according to specific criteria; these criteria could be the sequential order of the time of sample acquisition, the variation concentration of a chemical agent, or any other factor which causes structural changes in the starting conditions of the molecular system under investigation; this is the concept and the meaning of perturbation [4–7]. Alternatively, in some cases of studies encountered in the environmental field, it is also possible to collect spectra without having a precise sampling order; in these studies, the perturbation can be the different sample location factor which causes variations in the sample set [9].

As an example of introduction to 2DCOS, for this chapter let us consider the FTIR spectra of *Callista chione* mollusk shells, sampled in three different sites of the Italian seas with an addition of some fossil samples of the same mollusk shell (Fig. 5.1). We wish to investigate if there are structural differences in this FTIR data set and to investigate some peculiar aspects concerning the interactions among the chemical constituents of mollusk shells. Here, the perturbation required by the 2DCOS theory is the different origins of mollusk shells.

First of all, we wish to understand why 2DCOS applied to FTIR data is more efficient than a traditional examination of an FTIR data set. The visual examination of the spectra in Fig. 5.1, one spectrum at a time, does not improve the detection of the structural differences and interactions existing in the spectral data set. This depends on

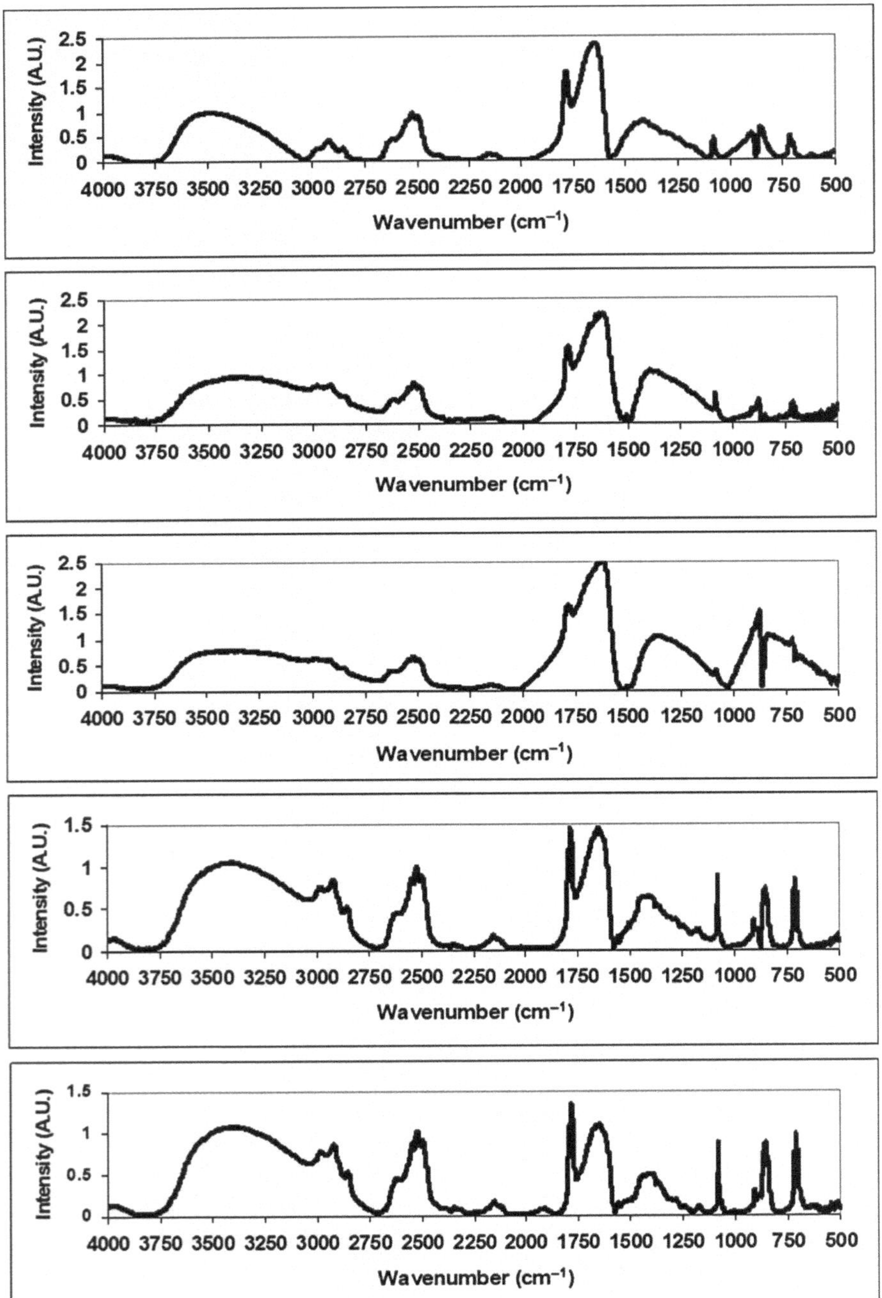

Fig. 5.1 Some FTIR spectra of mollusk shell samples from different sites, used for the construction of the X spectral matrix.

the complex qualitative compositions of mollusk shells where two polymorphic forms of calcium carbonate, aragonite, and calcite, respectively, with, in addition, amorphous calcium carbonate are present. A minority content is due to crystallization water, biogenic silica, and proteins. We observe these chemical components by means of spectra of Fig. 5.1, but we can hardly say more about the structural aspects in mollusk shells.

The comprehension of the structural information present in the spectral data set improves by determining the 2DCOS synchronous (*Synww*), asynchronous (*Asynww*), and disrelation (*Dis*) maps.

Synchronous spectra

Synww spectra describe the structural changes occurring in a molecular system according to linear (i.e., in phase) relationships among the variables. For spectra aligned as column vectors, *Synww* spectra are determined according to

$$Synww = 1/(s-1)X^*X^{\mathrm{T}} \tag{5.1}$$

where X is the spectral data matrix previously described, X^{T} is the corresponding transposed matrix, and s is the number of samples in the matrix.

The *Synww* matrix is a covariance matrix which describes the correlations and the structural interactions among the functional groups present. It is visualized as a counter (i.e., bidimensional) plot called synchronous spectral map (Fig. 5.2). Here we find two peak typologies, autopeaks and crosspeaks. Autopeaks represent the overall correlations among the variables and have two fundamental characteristics: they are placed along the central diagonal axis of the contour plot with the same coordinates (i.e., $w_a = w_b$) on the two axes. Autopeaks are always positive. The autopeaks reported in Fig. 5.2 are assigned to several stretching, bending, and rocking vibrations of the functional groups present. The observed stretching vibrations belong to the —OH group of crystallization water (3400 vs 3400 cm^{-1}), the —C=O group of calcite and aragonite (1780 vs 1780 cm^{-1}), and the —C=O group of calcite and aragonite (1450 vs 1450 cm^{-1}). The observed bending vibrations belong to the —OH group of crystallization water (1650 vs 1650 cm^{-1}). Other minor autopeaks belong to the —C=O group of amorphous carbonate (1085 vs 1085 cm^{-1}) and to the —C=O group of calcite and aragonite (880 vs 880 cm^{-1}). Some minor autopeaks belong to calcite and aragonite (700 vs 700 and 713 vs 713 cm^{-1}) bands having minor relevance. The autopeaks describe signal intensity which changes constantly within the spectral data set (i.e., the overall correlation [7]), and for this reason they have the highest correlation. The diagonal line of the contour plot is referred as to the autocorrelation among signal intensities, which is the mathematical tool for finding repeating patterns in a series of measurements [7].

Crosspeaks are placed outside the diagonal axis of the contour plot with $w_a \neq w_b$ coordinates and in symmetrical position (i.e., w_a, w_b and w_b, w_a) with each other. They represent the linear intercorrelations among different variables and functional groups.

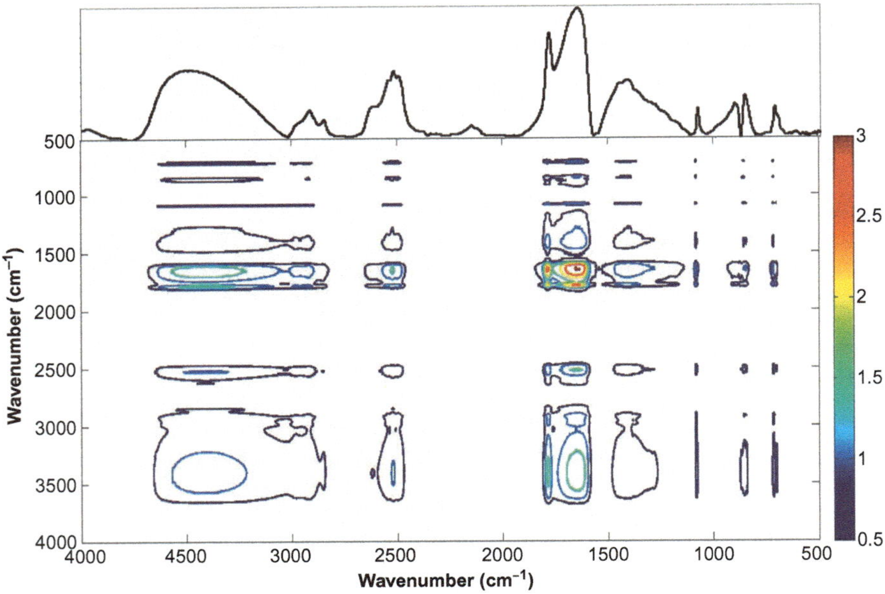

Fig. 5.2 Synchronous 2DCOS map for the *X* matrix. Peaks located in the diagonal line are autopeaks, whereas other peaks are crosspeaks. The symmetrical position with respect to the diagonal line where crosspeaks are placed should be noted. For this case, where the perturbation criteria are the sampling sites, crosspeaks are all positive. Being spectra scaled by means of Pareto's method, the correlation values reported by the *color bar* range between 0 and 3. If data were scaled by autoscaling, the correlation values reported by the *color bar* range between 0 and 1 like usual correlation coefficients.

Crosspeaks can be either positive or negative because they describe both positive and negative correlations among spectral change absorptions. In the 2DCOS synchronous spectra of Fig. 5.2, we observe two intense autopeaks, the first between 3500 and 3400 and the second at 1650 cm^{-1} while other significant crosspeaks are placed at 3450 vs 1650, 1430, 1085, 860, 716, and 700 cm^{-1}. They are assigned to crystallization water, amorphous carbonate, calcite, and aragonite which are the constant characteristics of the mollusk shell set. There is a huge set of crosspeaks which, for the sake of brevity, will be not discussed with details. Looking at wave number/variable coordinates describing each crosspeak, we observe all the functional groups which, according to their specific peak assignment, describe the structural interactions among calcite, aragonite, amorphous carbonate, and crystallization water. This is because the interactions described by crosspeaks depend on all the polar interactions among the above chemical constituents. Analogously, for NOM from soils, water, and sediments, the molecular interactions described by crosspeaks depend on several polar (i.e., intra and inter hydrogen) bonds and nonpolar (van der Walls and supramolecular) interactions [22, 31].

Asynchronous spectra

Asynww spectra describe the nonlinear (out of phase) relationships among the variables of the molecular systems and are determined according to

$$Asynww = 1/(s-1)X^{*}HX^{T} \quad (5.2)$$

where X, X^T, and s have been previously defined already, and H denotes Noda's modification of the Hilbert transform matrix. Like the synchronous matrix, the asynchronous *Asynww* matrix is also visualized by means of a contour plot called asynchronous spectral map (Fig. 5.3). This matrix describes structural differences in the molecular system of mollusk shells, differences related to high order correlations among variables. As asynchronous plots do not describe linear correlations among wavelengths, here we find crosspeaks only. Like crosspeaks in the synchronous plots, they can be positive and negative. Fig. 5.3 reports the 2DCOS asynchronous spectra of the mollusk shell spectra. It is reported in a narrower spectral range, between 600 and 1800 cm^{-1}, because crosspeaks are present in this range only. They are placed in

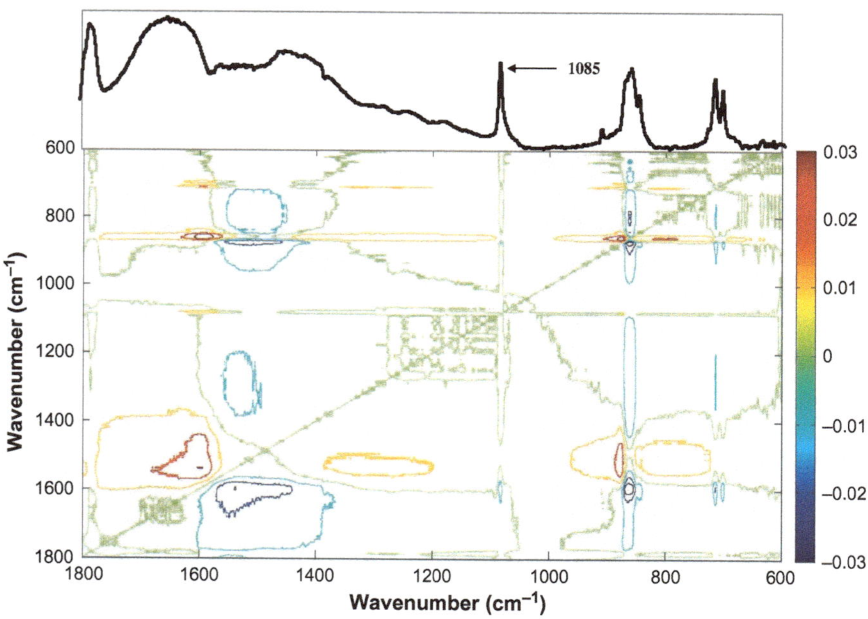

Fig. 5.3 Asynchronous 2DCOS map for the X matrix where only crosspeaks are observed. For the explanation of crosspeaks, see text in "Asynchronous spectra" section. Note the case of the band at 1085 cm^{-1} in the FTIR spectrum assigned to amorphous carbonate, without corresponding crosspeaks in the asynchronous 2DCOS map. This is an evidence of a compound not involved in any structural variation within the data set.

the range (1500 and 1440 cm^{-1}) of the stretching —C=O group of calcite and aragonite, and in the range (900 and 860 cm^{-1}) of the bending —C=O of calcite and aragonite.

These crosspeaks point out some differences in their structural interactions between calcite and aragonite, while the absence of other crosspeaks belonging to amorphous carbonate and crystallization water points out a general homogeneity in the structural characteristics. Let us take the example of the crosspeaks located at 1085 cm^{-1}, assigned to amorphous carbonate. In the synchronous spectra of Fig. 5.2, the band at 1085 cm^{-1} forms crosspeaks with other bands such as 3400, 2500, 1650, 1450, 860, and 715 cm^{-1}. The absence of these crosspeaks in the corresponding asynchronous spectra of Fig. 5.3 means that there are not higher order structural changes interactions involving the band at 1085 cm^{-1} of amorphous carbonate. This also means that the interactions of amorphous carbonate with the other compounds in mollusk shells are comparable for all the samples present in the sample set, without any relevant difference within the spectral set samples. Similar to amorphous calcium carbonate, if crystallization water does not show crosspeaks in Fig. 5.3, then its interactions with calcite and aragonite are comparable within the data set. Conversely, the presence of crosspeaks both in synchronous and asynchronous spectra, related to the band of aragonite, calcite, and crystallization water, is the evidence of differences in the structural characteristics of the samples, depending on calcite and aragonite contents and different structural interactions. Crosspeaks in an asynchronous map like that of Fig. 5.3 are the tool to detect structural differences within a molecular system [7, 23].

Disrelation spectra

Disrelation spectra are the tool to see the whole set of structural differences present in a molecular system. *Dis* spectra are determined according to

$$Dis = \left[Var(X)*Var\left(X^{\mathrm{T}}\right) - Synww^2\right]^{0.5} \tag{5.3}$$

where *Var* is the total variance of the X and X^{T} matrices, and $Synww^2$ is the quadratic *Synww* matrix. *Dis* spectra display both the information occurring with high order correlations (i.e., the case of asynchronous spectra) and in the total absence of correlations [7]. By a mathematic point of view, *Dis* spectra are the visualization of all the variances not explained by linear relationships; if we read Eq. (5.3), we see that *Dis* spectra come from the differences between the total variance and the variance explained by a linear relationship (i.e., the *Synww* matrix). Like the asynchronous spectral maps, *Dis* spectra have crosspeaks only.

The *Dis* spectral map for the mollusk shell set is not reported here because it is almost null, with negligible crosspeaks. For instance, in the study of marine foam structures, the absence of crosspeaks for *Dis* spectra shows that there are not differences within the qualitative compositions of samples [32]. This also means that all the mollusk shells have comparable qualitative compositions and the existing differences, shown by the asynchronous map, depend on structural (i.e., interactions and molecular rearrangements) changes only. In simpler words, the *Dis* map is the tool to test the presence of inhomogeneities in the qualitative composition of samples.

Determination of the sequential events in a dynamic process by 2DCOS

As already mentioned, crosspeaks in synchronous and asynchronous maps show the interrelationships among the variable intensity changes caused by the effect of the perturbation. Now it is important to underline this focal point: if spectra (or chromatograms) are aligned in the X matrix according to the order of variation of the perturbation (increased temperature, increased pH, and so on), 2DCOS also allows to identify the sequential time and dynamic order of the structural changes caused by the perturbation [7]. With this aim, it is necessary to examine the signs in the 2D synchronous and asynchronous crosspeaks according to the following rules of Noda:

1. if the intensities of the bands at w_a, w_b in the data set are changing in the same direction (i.e., simultaneous change), the synchronous 2D crosspeak at w_a, w_b is positive;
2. if the intensities of the bands at w_a, w_b in the data set are changing in the opposite direction (i.e., nonsimultaneous change), the synchronous 2D crosspeak at (w_a, w_b) is negative;
3. if the change at the w_a band precedes the change in the band at w_b, the asynchronous 2D crosspeak at (w_a, w_b) is positive;
4. if the change at w_a takes place after the change in the band at w_b, the asynchronous 2D crosspeak at (w_a, w_b) is negative;
5. if the synchronous 2D crosspeak at (w_a, w_b) is negative, the interpretation of rules 3 and 4 for the asynchronous 2D peak at (w_a, w_b) have to be reversed.

Noda's rules are the highest expression of the concept of "dynamic spectra" [4–7] represented by synchronous and asynchronous 2DCOS spectral maps. The validity of Noda's rules has been the subject of some criticisms, but a recent study concerning their applications to several types of signal functions (both spectroscopic and chromatographic) confirmed their reliability and accuracy definitively [33].

As already mentioned, 2DCOS has also applications in peculiar studies where the sample order of acquisition is unknown like sometimes happens in environmental studies [9]. Obviously, for these studies, we do not need to apply Noda's rules. The book of Noda and Ozaki [7] and other excellent papers report rules, suggestions, and comments for supporting the reliable interpretation of 2DCOS spectral maps and results [34–36].

Comments on the visualization of 2DCOS maps

A final indication concerning the visualization of all the 2DCOS maps is that, in addition to contour plots, they can be reported as three-dimensional mesh plots too. All the rules previously described for the interpretation of a 2DCOS contour plot are preserved for a 2DCOS mesh plot. However, with respect to a contour plot, a mesh plot has major random access memory requirements, longer computational times, and some difficulties in the identification of the correct autopeak and crosspeak positions. Thus, contour plots are largely preferred.

Two-dimensional hetero-correlation spectroscopy

Looking at Eqs. (5.1), (5.2) for the determination of synchronous and asynchronous 2DCOS maps, we see a single analytical typology applied for the X and X^T matrices. However, 2DCOS versatility allows studies without restrictions to the use of a single specific instrumental technique [7–12]. Taking into account Eqs. (5.1), (5.2), we can consider two different spectroscopic techniques, for instance, FTIR spectra for X and FTNIR spectra for X^T. In this case, 2DCOS is termed two-dimensional hetero-correlation spectroscopy (2DHCOS), an extension of 2DCOS having the advantage to exploit the specific information of each spectroscopic technique, using the mutual contribution of each other [7]. Due to the rules of matrix algebra (i.e., the same number of samples to be submitted to the two analytical techniques), we can determine synchronous and asynchronous spectra for 2DHCOS, but it is not possible to determine the disrelation spectra for 2DHCOS due to the different number of variables in the two spectral matrices.

For the same reason, synchronous and asynchronous maps do not have any more symmetrical characteristics present in 2DCOS. In a 2DHCOS counterplot, we will simply observe peaks having the wavelengths of the different spectral techniques as coordinates. Fig. 5.4 reports an example of the synchronous 2DHCOS of mollusk shell samples obtained by means of their FTIR and FTNIR spectra. Here, all the most intense peaks have coordinates 5250 cm^{-1} as abscissa vs 3400, 1780, 1650, and

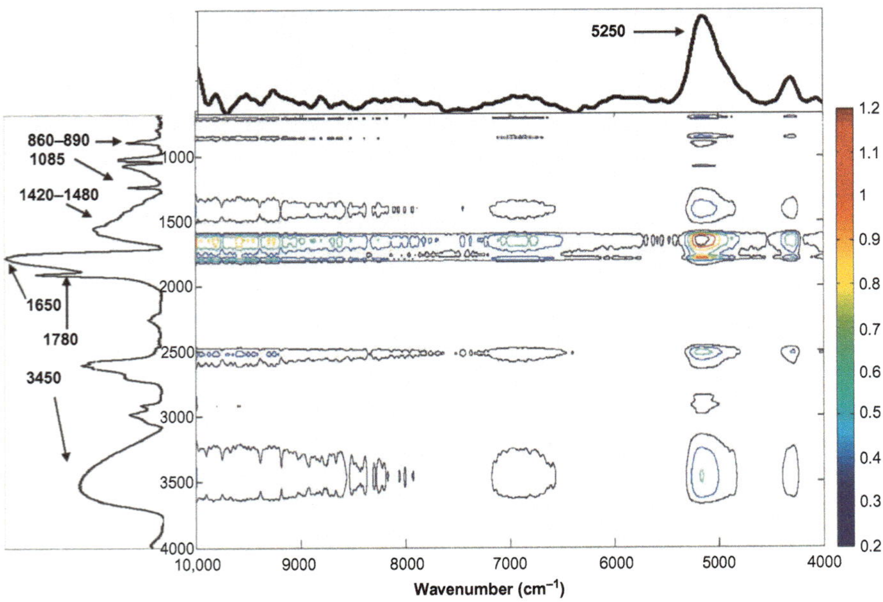

Fig. 5.4 Synchronous FTIR and FTNIR 2HDCOS map. Note the absence of autopeaks depending on the different sizes of FTIR and FTNIR spectral variables. Also note the case of the FTNIR 5250 cm^{-1} band with its interactions with all the FTIR bands of crystallization water, aragonite, calcite, and amorphous carbonate.

1450 cm^{-1} as ordinate. According to the assignment of the FTIR bands previously reported for Fig. 5.2, we can assign and describe the most intense FTNIR band at 5250 cm^{-1} as a combination band among all the polymorphic forms of carbonate with the crystallization water. Then, it should be noted that, in addition to evidence of the structural interactions of crystallization water with all the forms of carbonates present, 2HDCOS is also an efficient tool for the interpretation of analytical signals like those of FTNIR, exploiting the information coming from another technique. The asynchronous 2HDCOR map is not reported because it consisted of noise essentially, here depending on the low number of FTNIR peaks of mollusk shell samples. Noda recommends the application of double two-dimensional correlation spectroscopy (2D2DCOS) in the use of 2DHCOS spectroscopy for its specific ability to generate highly selective bidimensional spectra [26].

Sample to sample 2DCOS

By means of some modifications of the synchronous and asynchronous equations for *ww*-2DCOS, it is also possible to determine the synchronous and asynchronous 2DCOS spectra in the so-called *sample to sample* (*ss*) mode [7, 35, 37].

With respect to the *ww* mode, which describes relationships among variables, the *ss*-mode describes linear and nonlinear relationships among samples. The joint examination of spectral data using both *ww*-2DCOS and *ss*-2DCOS modes gives specific insights into the structure of complex molecular systems [7, 35, 37]. For analytical signals aligned as column vectors, *Synss* spectra are determined according to

$$Synss = 1/(n-1)X^{\mathrm{T}}*X \tag{5.4}$$

where X is the spectral data matrix previously described, X^{T} is its transposed matrix, and n is the number of variables in the spectral or chromatographic data series.

Asynss spectra describe the nonlinear (out of phase) relationships among samples of a molecular system and are determined according to

$$Asynss = 1/(n-1)X^{\mathrm{T}}*HX \tag{5.5}$$

where X, X^{T}, n, and H have the already described meanings. As support to the interpretation of *ss*-2DCOS spectra, we recommend the already cited literature [7, 35, 37].

Software availability for 2DCOS applications

PCA, PLS, and LDA are applied in all the fields of science, then there is a wide availability of software. Conversely, due to the minor diffusion of 2DCOS, there is a minor software availability. In fact, most users of 2DCOS work with in-house routines written in specific language for scientific programming. Bruker and Agilent manufactures make available software packages for 2DCOS in vibrational spectroscopy instrumentations [34]. However, these software packages are close-source software because they are distributed in a manufacture-specific software and format, they are not usable by other manufacture instruments. A free and popular 2DCOS software available for all the analytical instruments is 2D-Shige. It works under Windows operating system and in addition, it works on data files saved in Excel CSV and ASCII format [38].

Other authors report simple and reliable 2DCOS routines in MatLab [39] and R [40] languages. As the readers can see, despite a not widespread use of 2DCOS in environmental studies, software availability is not an insurmountable problem for interested researchers.

Data preprocessing for 2DCOS applications

Like all the multivariate methods, 2DCOS requires appropriate preprocessing treatments necessary for the correct interpretation of analytical signals. In fact, spectral preprocessing (i.e., baseline correction, smoothing, and standardization) of analytical signal series allows to enhance the signal to noise ratio, to improve spectral resolution, and to make independent the signal intensity by the amount of sample taken for the instrumental analysis. Two interesting reviews discuss all the aspects of signal preprocessing necessary to obtain high-quality analytical data in multidimensional and chemometric analysis [41, 42]. Here, let us have a look at some fundamental criteria of spectral preprocessing for 2DCOS, taking into mind that a digitized spectral or chromatographic series consists of $n - w_i$, x_i couples of values, with w_i being the variable and x_i the signal intensity, respectively.

Baseline correction

Baseline correction is necessary to reduce baseline drifts depending on random and unpredictable effects like light scattering and changes in density of samples [7, 41]; for these reasons, a baseline drift depends on the intrinsic characteristics of the samples, and it can remain present in a spectral or in a chromatographic plot even after the background subtraction [41]. A simple and reliable method for baseline correction is the minimum selection points [43]; this is generally available on common spectrophotometers by means of a graphical user interface or by an automatic procedure, so that this operation does not require the exportation of the digitized spectral file in an external software. Other available techniques for baseline correction are the derivative peak subtraction [41], and the polynomial local minimum selection [44]. These methods can be applied using a spreadsheet like Microsoft Excel on files previously exported in ASCII, CSV, or Excel format for Windows.

Smoothing for noise reduction

Smoothing is necessary to reduce the experimental noise and then to improve the definition of spectral peaks, reducing the presence of spectral artifacts, such as ghost peaks. The causes of noise could be the same producing a baseline drift too. In any case, noise can be particularly present in asynchronous spectra because the changes of noise occur out of phase with each other [7]. Perhaps the widest smoothing technique is the Savitzki-Golay method [45] present in most software for spectrophotometer managements. The Savitzki-Golay method works on an odd number of signal points according to

$$x_j = \sum_{i=\frac{m-1}{2}}^{\frac{m-1}{2}} C_j x_j + i \quad \frac{m-1}{2} \leq j \leq n - \frac{m-1}{2} \tag{5.6}$$

where x_j is the smoothed value of the starting x_i value, m is the filter function, C_j is the coefficient of convolution for the x_j value, and n is the number of the couples of w_i, x_i values of the analytical series. Though it may seem complicated, the Savitzki-Golay method is actually quite simple. In fact, the values of the m-filter function and of the convolution C_j coefficients for each m-value are tabulated [43, 45]; so, even in the absence of a smoothing method in the instrumental management software, it can be easily determined using a common spreadsheet like Excel for Windows. The Savitzki-Golay smoothing works using an odd range of x_j-points, and a research group also proposed a new version of this method working on an even number of x_j-points [46]. The performances of the two smoothing versions are highly comparable.

Eilers [47] proposed a new and interesting smoothing method called "the perfect smoother." It has been proposed as an alternative to the Savitzki-Golay method, which can cause peak distortion when an elevated value of the m-filter function is selected for the Eq. (5.6). This smoothing method is based on the approach called penalized least squares which minimizes the sum of squares of differences among the original signal intensities and the smoothed data. To support its use the author reported the full MatLab routine of this method.

A data reduction method like PCA can also be an efficient smoothing and noise filtering method for 2DCOS [7, 48, 49]. Basically, PCA decomposes the X signal intensity matrix according to

$$X = S^* L^{\mathrm{T}} \tag{5.7}$$

where S is the matrix of the uncorrelated and new variables called scores describing the X matrix, L^{T} is the transposed loading matrix, reporting the statistical weights of the w_i spectral variables characterizing the X matrix. Noise in spectral data is a reduced portion of the total variance of the X data set. Very importantly, noise is also uncorrelated with the spectral signal intensities. This means that Eq. (5.7) can be rewritten according to

$$X = S_s{}^* L^{\mathrm{T}}{}_s + E \tag{5.8}$$

where S_s and $L^{\mathrm{T}}{}_s$ are the scores and loading matrices for the factors accounting for a relevant percent ($\approx$90%) of the total variance and E is the noise-error matrix accounting for the remaining and significantly lower portion of the total variance.

The E matrix can be discharged, and we retrieve the original and denoised matrix taking into account the scores and loading of the significant factors according to

$$X_d = S_s * L^T{}_s \tag{5.9}$$

where X_d is the noise reduced spectral matrix. It is obvious that before applying this PCA smoothing method, the X matrix must be previously submitted to the baseline and standardization treatments. However, due to its easy use and diffusion, the Savitzki-Golay smoothing filter still remains the most common smoothing method for analytical signals [41].

Standardization

In many spectroscopic and chromatographic techniques, the signal acquisition is performed starting from solid samples, where it is impossible to use always the same quantity of samples. In this case, in addition to the contribution of molar absorptions of each functional group, the spectral intensities will also depend on the quantity of sample. The inclusion of this contribution will inevitably lead to incorrect 2DCOS results. Standardization, recommended for all the multivariate techniques in chemical analysis [41, 42], is the necessary pretreatment to reduce this undesirable effect. Several standardization methods are available for spectroscopic and chromatographic analysis, and here we briefly describe some of these.

The "maximum signal intensity" standardization method is given by

$$x_i^* = (x_i / x_{max}) \text{ with } i = 1,2,3,\ldots,n \tag{5.10}$$

where x_i, x_{max}, and x_i^* are the original i-value, the maximum absorption in the starting x-series, and the standardized x_i value, respectively. For instance, this standardization is common in infrared studies of biological systems where the absorption of the protein —C=O group is often the highest in the series, and for this selected as the x_{max} value [50].

Other common standardization methods for 2DCOS are mean centered, normalization, autoscaling, and Pareto's method. Let us have a look at these methods. Briefly, in the mean centered method, we obtain

$$x_i^* = (x_i / x_{mean}) \text{ with } i = 1,2,3,\ldots,n \tag{5.11}$$

where x_{mean} is the average value of the x-series and the other terms have the same meaning as the previous equations.

In standardization by normalization, we have

$$x_i^* = (x_i - x_{min}) / (x_{max} - x_{min}) \text{ with } i = 1,2,3,\ldots,n \tag{5.12}$$

where x_{max} and x_{min} are the minimum and maximum values of the x series. Here, the standardized x^* series will have maximum value = 1 and minimum value = 0.

In autoscaling, also called univariate-scaling standardization, we have

$$x_i^* = (x_i - x_{\text{mean}})/\sigma_x \quad \text{with} \quad i = 1,2,3,\ldots,n \tag{5.13}$$

where σ_x is the standard deviation of the x-series. The standardized x^* series has a mean value of 1 and a standard deviation of 0. However, autoscaling has the potential drawback of enhancing noise, especially for a signal having low intensity because of the division by the small values of standard deviation [51]. For this reason, the application of autoscaling needs always requires an accurate preliminary evaluation.

Vilfredo Pareto was an Italian economist and statistician; in the standardization method known by his name, we have

$$x_i^* = x_i/\sqrt{\sigma_x} \quad \text{with} \quad i = 1,2,3,\ldots,n \tag{5.14}$$

where $\sqrt{\sigma_x}$ is the square root of the x-series standard deviation. For the application of 2DCOS, Noda examined and discussed all these standardization methods accurately [27, 51], underlying that Pareto's method could result in the best one because it enhances the fine structure of the spectra without causing noise enhancement, as is sometimes observed by autoscaling.

2DCOS in the dynamic processes of NOM formation, characterization, and reactions

NOM composition and structure are determined by the chemical and physical factors present in the surrounding environment such as pH, temperature, oxic and anoxic conditions, and by the several contributions of living organisms. This chapter reviews the 2DCOS application studies for the structural characterization of all the NOM fractions from aquatic and terrestrial environments, and moreover includes studies on the NOM aggregation process and chemical reactions involved. Other reviewed studies describe the structural characterization of NOM from wastewater and composting treatments. Let us examine these studies.

Characterization of marine organic matter

Humic substances, the major component of NOM in seawater and sediments, consist of three main fractions: fulvic acids—water soluble at any pH, humic acids—water soluble at pH > 2, and humin—insoluble at any pH [15]. The chemical composition and structural characteristics of fulvic acids, humic acids, and humin depend on vegetal and animal contributions, and on the several processes of aggregation and degradation within a complex chemical equilibrium called the "biopolymer degradation and polymerization model" [15]. All the organic fractions take part in this complex process, which also includes the carbon cycle and where nutrients and pollutants interact with NOM.

Studies of fulvic and humic acids extracted from marine sediments performed by FTIR 2DCOS revealed many interesting characteristics involving the roles played by carbohydrates, proteins, and lipids in the mechanisms of NOM aggregation, starting from DOM to particulate organic matter (POM) and to the insoluble humin fraction. In NOM aggregation, several reactions of polymerization and condensation involving carbohydrates and proteins take place; in addition, the interactions of polar (i.e., hydrogen bond) and nonpolar groups produce the supramolecular characteristics of humic and fulvic acids, also confirming the surfactant characteristics of NOM [21, 32, 52].

The formation of incubated macroaggregates of marine organic matter, called mucilages, obtained by the degradation of several algae, was studied by FTIR and UV-Vis spectra supported by 2DCOS and 2HDCOS [31], confirming that the formation of all the types of marine NOM, from DOM and POM until macroaggregates like mucilages, follows the same aggregation and degradation mechanism described by Ishiwatari [15]. In a subsequent study concerning the mucilage formation under anoxic conditions, FTIR and FTNIR 2HDCOR suggested that browning development associated with mucilage formation can simultaneously depend on both nonenzymatic (i.e., Maillard) and enzymatic reactions [22]. Maillard reactions in NOM were also the focal point of a study on the formation of humic-like substances reproduced in laboratory conditions by heating glycine and ribose mixed solutions at 80°C [53]. UV-Vis 2DCOS spectroscopy suggested that the area of the two bands around 280 and 254 nm could be used as a measure of the formation of furfural-like intermediates and humic-like products, respectively.

Abdulla and coworkers [54] applied FTIR and ^{13}C NMR 2HDCOS to investigate variations in the compositional and structural characteristics of high molecular weight DOM samples collected along an estuarine transect submitted to different hydrologic conditions. The 2DCOS maps revealed that heteropolysaccharides, carboxyl-rich compounds, and organic compounds rich in amide and amino functional groups were the three main identified components of fulvic and humic acids. These three main fractions, involved in aggregation and degradation reactions, shared similar backbone structures but with some significant differences depending on the natural abundance of the functional groups present. The same research group applied Fourier transform ion cyclotron resonance mass (FTICR-MS) 2DCOS to study the structure of ultrafiltered DOM sampled in different salinity transects [55]. This study showed that 2DCOS FTICR-MS data of individual peaks were associated with the variation of all the functional groups characterizing DOM. A further study of the same research group applied 2HDCOS with FTIR, ^{1}H NMR, and ^{13}C NMR spectroscopy in DOM under different seawater salinity conditions. The results confirmed the previous results, evidencing in addition that the polysaccharide fraction was always the most relevant [56].

Characterization of NOM from natural water, soils, and other sources

The humic substances of soils and other aquatic systems are very similar to marine NOM, with some little differences only. It consists of fulvic acids water soluble at any pH, humic acids water soluble at pH ≥ 1, and humin, the fraction insoluble at any pH [57]. It is well known that fulvic and humic acids from the aquatic environment are involved into oxidative degradation reactions by microbial activity and solar irradiation. DOM and humic substances are rich in polar groups able to act as ligand sites for metals, heterogeneously distributed in the bulk. A study by fluorescence 2DCOS showed that within the fractions of DOM, fulvic and humic acids and their subfractions, such as tryptophan and tyrosine-like moieties, had a significant ability in binding protons [58].

The variations of rainfall intensity and of pH conditions produced significant changes in the chromophoric structure and composition of DOM in rainwater [59]. This study, performed by fluorescence 2DCOS, evidenced changes in all the chromophoric components of DOM. Moreover, the microbial humic-like fluorophores fraction was always the prevalent one with respect to the other components of tryptophan-like and tyrosine-like substances.

Climate changes have been producing increased temperatures, which affect the chemical composition of DOM, and changing its interactions with pollutants. Fluorescence spectroscopy, supported by parallel factor analysis and 2DCOS, found that the increased temperature induced transitions and variation of DOM in soils and sediments [60]. The sequential order of these changes was protein-like component > fulvic-like component > humic-like component. As far as NOM and pollutant interactions are concerned, this study also showed that the enhanced temperatures weaken the interaction between phenanthrene and DOM. This weakening effect was greater at higher temperature because it depended on specific molecular rearrangements of the hydrophobic groups present in the organic structure.

Humin has a peculiar ability of mediating the electron transfer during the biogeochemical processes of NOM formation. An in situ FTIR spectroscopic technique coupled with 2DCOS was employed for the first time to study the redox properties of humin [61]. FTIR 2DCOS identified that the intensity bands of quinone and phenolic moieties in humin were redox potential dependent, suggesting that they were the most reactive sites in response to the redox variations of soils and sediments. In a specific study performed in a eutrophic lake [62], fluorescence 2DCOS showed that at increased sediment depths, the fluorescent humic-like component was microbially degraded prior to the fulvic-like component. Moreover, the relative abundance of non-fluorescent components decreased with the increased depth of the sediment sampling.

Hur and coworkers [63] investigated the chemical transformation of leaf litter humic and fulvic acids in river waters, submitted to UV irradiation. The results of UV-visible and synchronous fluorescence 2DCOS showed that three types of bands changed intensities sequentially according to the order 290–400 nm → 200–250 nm → 250–290 nm ranges. These findings suggested that aromatic chromophores in amino acids and tannin-like structures were preferentially oxidized, generating non-UV-absorbing compounds. The same research team studied the changes of algal derived organic matter in a microbial fuel cell by FTIR 2DCOS [64]. The application of Noda's rules determined a temporal order of NOM degradation according to sequence proteins → acidic functional groups → polysaccharides → amino acids.

The changes in NOM structure present along the sections of core sediments sampled in the Lake Superior (United States) were studied by FTIR 2DCOS [65].

Synchronous spectra revealed that carbohydrates and aliphatic esters were primarily and significantly degraded with the increasing core depth, also causing a progressive enrichment in biogenic silica and inactive polysaccharides. Asynchronous spectra pointed out that the carbonyl groups of aliphatic esters and the amide groups of proteins were degraded faster than those present in carbohydrate-like aminosugars and aromatic compounds. The study of peat fulvic and humic acids by ^{1}H NMR and ^{13}C NMR spectroscopy supported by 2DCOS indicated that fulvic acids are diagenetically downstream of humic acids because these components were strictly linked among them by the global humification process, known as the biopolymer degradation model [66].

The wet cell walls of the onion *Allium cepa* were submitted to different hydration and mechanically stressed conditions to study the polysaccharide network changes, which were detected by FTIR 2DCOS [67]. The results evidenced that cellulose and xyloglucan associated among each other, while pectin created an independent network exhibiting different reorientation rates. The oscillation stress conditions produced changes first in the pectin chains, and then in the more rigid cellulose chains.

Characterization of NOM in waste managements

The structural variations of NOM during the processes of waste management and disposal are a critical step for the evaluation of the effectiveness of these treatments in sewage. Li and coworkers [68] applied FTIR 2DCOS to investigate the changes of DOM submitted to anaerobic digestion. The results indicated that the sequential order of DOM changes was protein-like groups → structural carbohydrates → carboxylic groups of polysaccharides. The changes in DOM composition involved in wastewater treatment were studied by fluorescence 2DCOS too [69]. The observed order of band changes was 279 nm → 304 nm → 490 nm → 330 nm → 380 nm → 430 nm. According to the band assignments of the molecular compounds present in DOM, the authors established a sequence of decreasing removal and variation order of DOM according to tyrosine-like → tryptophan-like → humic-like → microbial humic-like → fulvic-like components in the wastewater process.

FTIR 2DCOS compared DOM variations in raw sludge submitted to two different treatments: the biodrying digestion and the anaerobic digestion [70]. Biodrying treatment caused a decrease in the content of organic carbon with an associated increase of the molecular weight and aromatic contents. The increased aromatic character also caused the increase of fluorescent intensity of DOM in the sludge matrices. In the anaerobic digestion treatment, DOM sludge was more biostable than DOM submitted to biodrying treatment, then highlighting the treatment with lower performances. Moreover, the FTIR 2DCOS asynchronous map of sludge matrices from the biodrying process showed that the heteropolysaccharide fraction was the first one to present significant structural and compositional changes, followed by the protein-like fraction.

Ozonation is a common procedure for the degradation of NOM in effluent waters, and EES is generally used to evaluate the efficiency of ozonation treatments. However, the large data dimension of both excitation and emission spectral sets makes difficult the interpretation of the results. A combined chemometric approach by parallel factor analysis, necessary to reduce data dimension, and then 2DCOS made possible the simultaneous examination of both excitation and emission spectral data [71]. The results showed that fluorescent protein-like and fulvic-like substances in NOM were preferentially degraded prior to other humic-like substances. The preferential ozonation of protein-like and fulvic-like substances was associated with the oxidations of the amide and phenolic groups. FTIR 2DCOS showed that the ozonation in the tertiary treatment process of domestic wastewater was able to reduce the aromatic structure of DOM, thus minimizing its antiestrogenic activity [72].

The mechanism of DOM degradation from the composting of municipal solid wastes was studied by homo and hetero 2DCOS NMR, FTIR, and synchronous fluorescence spectroscopy [73]. The results suggested that the degradation of DOM followed the sequential order aliphatic substances → proteinaceous compounds → cellulose → hemicellulose and lignin. Moreover, the fluorescence 2DCOS study on removing nitrogen compounds in NOM from domestic and industrial wastewater showed that the protein-like fraction was removed and degraded easier than the other organic fractions [74], thus confirming the results of other studies [71].

Fluorescence 2DCOS described effectively several aspects of changes in DOM from microbial products obtained in batch activated sludge bioreactors and operating under different salinity conditions [75]. The results showed opposite sequences of fluorescent changes between low and high salinity. The spectral variation took place following the order of humic-like → tryptophan-like → tyrosine-like fluorescence at the 1% saline bioreactors. At high salinity, the observed sequence was reversed according to tyrosine-like → tryptophan-like → humic-like fluorescence. The results evidenced that salinity affects the EPS production in the intercellular constituents of microbial organisms.

Graphene and derived graphene compounds are able to eliminate several toxic compounds, such as harmful dye, antibiotics, and nonbiodegradable phenol compounds. The absorption of the EPS component on graphene oxide and on reduced graphene was studied by fluorescence and FTIR spectroscopy supported by 2DCOS [76]. FTIR 2DCOS evidenced that the adsorption of EPS structure related proteins and aromatic components was greater for reduced graphene and lower for graphene oxide. As far as protein structures in EPS are concerned, the adsorption process followed the sequential order α-helix → amide II → β-sheet structures.

Like DOM from municipal solid waste, the knowledge of the mechanism degradations of DOM in the composting of biogas residues is necessary to optimize the conditions of the whole composting process. A study by 2DCOS FTIR spectroscopy determined the order of degradation for the functional groups of DOM, which resulted according to the sequence: 2000–2300 cm^{-1} (C≡C or C=N) → 1288 cm^{-1} (amide III) as the first step; as second step, the order was 2935 cm^{-1} (—CH aliphatic groups) → 1420 cm^{-1} (carboxylic groups) → 3100–3400 cm^{-1} (hydroxyl groups) → 1660 cm^{-1} (aromatic C=C) [77]. These findings suggested that functional C≡C, C=N groups and amide III groups can be degraded prior to aromatic C=C groups, which are characterized by higher chemical stability.

Homo and hetero 2DCOS of FTIR and ^{13}C NMR spectroscopy determined the structural characteristics of the hydrophobic DOM fractions from fine urban air particles collected during different seasons [78]. The results demonstrated that DOM extracted from urban aerosol samples consisted of at least two classes of compounds: one class was rich in both carboxylic and hydroxyl functional groups with a relevant aliphatic character; the second class had a significant component of lignin-derived structures.

Synchronous fluorescence spectra combined with PCA and 2DCOS performed the characterization of DOM obtained from three leachates with different landfill ages and under different hydrophobicity and polarity conditions [79]. The results showed that the bulk DOM consisted of tyrosine, tryptophan, fulvic, and humic subfractions. Tyrosine-like matter, humic-like materials, and fulvic-like substances presented the most relevant compositional changes as a response to the variation of hydrophobicity and polarity conditions.

Biofilms are organized communities of microorganisms produced during the composting of NOM embedded in complex mixtures of EPS. The development and growth of biofilms affected NOM degradation during the composting treatment. 2DHCOS FTIR—^{13}C NMR spectroscopic analysis allowed to observe the trend of NOM degradation in biofilms according to the sequential order heteropolysaccharides > cellulose > proteins [80].

Composition, origin, and dynamic evolutions of DOM in leachate-polluted waters depend essentially on microbial production. Fluorescence spectroscopy supported by several chemometric techniques, including 2DCOS, was applied to characterize the groundwater DOM sampled in two sequential years [81]. In particular, 2DCOS fluorescence spectroscopy showed that the compositional and structural changes of fulvic and humic acids were faster than that of the protein-like substances in the first year of sampling. Conversely, in the second year of sampling, the sequence was reversed. The results depended reasonably on an enhancement of the whole microbial activity.

Biochar is a stable carbon-rich *by*-product synthesized through pyrolysis and carbonization of plant- and animal-based biomass. FTIR 2DCOS spectroscopy was able to describe mechanisms of biochar formation in three different taxonomically biomass productions obtained in the range between 200°C and 650°C [82]. Results indicated that the defragmenting of lignocellulose had three primary reactions consisting of hydrogen cleavage of the network, demethylation, and oxidation involving the conversion of free primary hydroxyl groups to carboxyl groups.

Interaction of NOM and DOM with inorganic elements, minerals, and pollutants

Polysaccharides, proteins, and lipids in NOM are able to develop several polar and nonpolar interactions with minerals, inorganic elements, and pollutants. Perhaps the studies of NOM interactions with organic compounds and inorganic elements are the widest within the applications of 2DCOS in environmental studies because involving the natural cycle and environment fate of oligo-elements and pollutants, and the potential bioremediation ability of NOM to reduce the impact of pollutants.

Interactions of NOM and DOM with a Cu ion

The process of Cu biosorption onto aerobic granular sludge-derived biochar was studied both in the absence and in the presence of NOM, using synchronous fluorescence 2DCOS [83]. The synchronous 2DCOS spectra showed a series of successive fluorescence quenching, observed in various fractions of NOM and depending on the increased Cu concentration. The interpretation of the 2DCOS asynchronous maps according to the Noda's rules showed that fluorescence quenching of NOM in effluents occurred according to the sequential order of 300–400 nm → 250–300 nm. This suggested that the quenching of fulvic-like and humic-like fractions occurred earlier than the quenching of other fluorescence components, such as proteins and lipids.

A study by 2DCOS (both fluorescent and FTIR) spectroscopy and by 2HDCOS (fluorescent vs FTIR) spectroscopy found that carboxyl groups and polysaccharides of fulvic acids were the polar components giving the faster and major responses to Cu binding [84].

A detailed study compared the binding capacity of three DOM typologies coming from municipal solid wastes obtained at the initial temperature, at the high temperature, and at the mature period of the composting process [84]. FTIR, UV-Vis, and fluorescence spectroscopy supported by 2DCOS showed the high heterogeneity of ligands available in the DOM structure. In more detail, the N—H (amide II), phenolic OH, and C—O of alcohols, ethers, and ester functional groups were shown to be the most susceptive and reactive binding sites for Cu in all the three periods of the composting process. Moreover, the results also indicated that protein-like matter had a higher binding capacity than humic-like matter. These results suggest to us that the mechanism of this protein vs Cu interaction could be comparable with the mechanism of the reaction among proteins and Cu applied for the colorimetric determinations of Cu in several environmental matrices [85]. Two studies involving fulvic acids from a eutrophic lake [86] and from hyperthermophilic compost-derived humic acids confirmed the specific Cu-binding capacity of carboxyl and phenolic groups of DOM [87].

The binding capacity of DOM has also relevant importance as a potential bioremediation agent. Fluorescence 2DCOS studied the effect of anthropogenic impoundment on DOM and its Cu-binding affinity [88]. The results showed the different binding capacities of tryptophan and humic-like components in DOM, with the simultaneous and positive effect of reducing Cu bioavailability and toxicity.

A complex study based on ^{29}Si NMR, ^{27}Al NMR, and synchrotron-radiation-based FTIR spectromicroscopy, supported by 2DCOS and 2HDCOS, examined the binding capacity of DOM in soils submitted to different fertilizer treatments [89]. The results showed that the Si—O group of biogenic silicon in DOM could play an important role for binding Cu. However, a study by 2DCOS fluorescence spectroscopy showed that the binding sites of Cu and Hg ions in DOM could present significant variations depending on the origin of the humic substance [90].

The binding sites of Cd and Cu ions in DOM derived from compost and rice straw were investigated by FTIR 2DCOS [91]. The results confirmed those of Guo and coworkers [90], underlying the differences in specific Cd- and Cu-binding features of DOM coming from different sources. UV-Vis 2DCOS investigated DOM extracts from macrophyte and algal dominant sediments, sampled in a eutrophic lake [92]. The aromatic functional groups had a high binding capacity for Zn and Cu, with the latter higher than that observed for Zn in all the types of DOM considered in the study. The results also evidenced the better accuracy of the 2DCOS approach with respect to other chemometric techniques such as parallel factor analysis for the study of Cu-DOM interaction.

Interactions of NOM and DOM with Cr(VI) and Cr(III)

Perhaps Cu(II) is the ion which received the greatest attention due to its interactions with NOM; however, other ions deserve interest. This is the case of Cr(VI). The scientific literature reports two recent 2DCOS applications for the study of interactions between Cr(VI) and NOM. In a study focused on the investigation of binding sites in undissolved humic acids, FTIR 2DCOS and FTIR vs ^{13}C NMR 2HDCOS found that functional groups involving in Cr(VI) binding followed the sequential order carboxyl → phenol → polysaccharide → methyl groups [93]. The same research group also reported another study between Cr(VI) and humic acid interactions, presenting a new and original application of 2DCOS called three-dimensional correlation spectroscopy, where three instrumental techniques, FTIR, ^{13}C NMR, and XPS spectroscopy, were simultaneously applied [94]. The results indicated that free and dissociated carboxylic groups, phenolic groups, and polysaccharide in humic acids, mainly linked to aromatic domains, played a fundamental role for binding Cr(VI).

The binding Cr(VI) capacity of humic acids in soils affects its conversion to Cr(III); 2DCOS analysis for FTIR and ^{13}C-CP/MAS NMR spectroscopy investigated the variations of humic acid functional groups and the related structural changes after reaction with aqueous Cr(VI), under different pH conditions [95]. Based on the changing sequence of functional groups obtained by 2DCOS, the authors identified a complex four-step mechanism for Cr(VI) retention, and then Cr(III) conversion, in DOM. In the first step, the electrostatic adsorption of Cr(VI) to the humic acids surface occurred, followed by the complexation of adsorbed Cr(VI) by the carboxyl and carbonyl groups. Then, the complexed Cr(VI) was reduced to Cr(III) by phenols and polysaccharides before the final complexation of reduced Cr(III) by carboxylic groups.

Interactions of NOM and DOM with the Zn ion

Zn is an essential micronutrient for plant growth. A study based on FTIR 2DCOS with the additional support of XFS spectromicroscopy showed that in soil NOM several sites for binding Zn are available [96]. Zn binding was also studied by means of two different typologies of NOM, the dissolved fraction and the attached fraction, both produced in a eutrophic lake during cyanobacteria blooms [97]. The starting step of the study, performed by ordinary fluorescence spectroscopy and then by fluorescence

2DCOS, assigned the main binding capacity of NOM to the fractions characterized by two main quenching bands at 276 and 232 nm for both dissolved and attached organic matter. Then, the asynchronous 2DCOS maps described a sequential order of quenching, which resulted in 276 nm → 232 nm for dissolved matter and a reversed order, 232 nm → 276 nm, for attached matter.

Interactions of NOM and DOM with Cd

In a study concerning the effects of fertilization treatments on selected soils, FTIR 2DCOS demonstrated that functional groups of soil NOM, which generally determine Cd binding, were modified by the fertilization treatments, thus providing information about the reduced bioavailability of this toxic metal in organic fertilized soils [98].

Interactions of NOM and DOM with Al

Aluminum (Al) salts are widely used as coagulant agents due to the peculiar ability to remove small particles and humic acid moieties from natural waters. Knowledge of DOM and Al interactions is essential to understand the organic carbon storage in soils. FTIR 2DCOS was applied as a tool to explore the organic ligands in DOM soils able to bind with Al as a part of a long-term (21-year) fertilization experiment [99]. This study gave two fundamental results. It described the binding characteristics of both fluorescent and nonfluorescent (i.e., polysaccharides, lipids, and lignin) substances with Al. Meanwhile, FTIR 2DCOS also demonstrated that the sequencing order of organic ligands in binding Al(III) could be also modified by the use of long-term fertilization strategies. A subsequent study by FTIR 2DCOS confirmed that the long-term fertilization strategies were able to modify the binding sites of DOM in soils, providing a clear indication that the –CH group of aliphatic chains was also able to bind Al ions in addition to the well-known ability of polysaccharides [100].

In a study performed in two different acidic conditions, FTIR 2DCOS described the sequences of structural changes in humic acids during coagulation with Al salts [101]. At pH 5, the sequential order of Al(III) binding was according to the sequence COOH → COO^- → NH deformation of amide II → aliphatic hydroxyl C—OH groups. At pH 7, the observed sequential order was changed according to COO^- → aliphatic hydroxyl C—OH.

Interactions of NOM and DOM with other ions

Nakashima and coworkers [102] studied the binding capacity of humic acids with special emphasis on Ca and Pb. Fluorescence 2DCOS showed two major bands at 510 and 435 nm, which gave different constants for describing the binding interactions with the two ions. In the asynchronous spectra, the 510 nm band changed its intensity prior to the 435 nm band; this suggested that the binding interactions with Ca(II) and Pb(II) were faster for the functional groups absorbing at 510 nm and lower for the 435 nm band.

Nitrogen compounds in DOM from leachates are able to give several complexation reactions with metals. FTIR vs fluorescence 2HDCOS described the complexation ability of amines with Mo, Co, Ni, and Cr, while proteinaceous materials had a major affinity with Cr and Ni [103].

In a study involving the specific Pb-binding capacity of humic acids, fluorescence 2DCOS evidenced a structure variation sequence depending on the interaction with Pb according to fulvic-like fraction → humic-like fraction [104]. Moreover, ultraviolet vs fluorescence 2HDCOS suggested that carboxylic and phenolic groups in humic acids were the preferred binding sites of Pb.

One study by UV-Vis vs FTIR 2HDCOS focusing on Pb and DOM interactions showed that the binding sites of DOM depended on the solution chemistry conditions. In fact, the binding sequences involving aryl—C—H, —C=O, —C—OH of polysaccharides and chromophoric groups absorbing at 265 nm showed relevant changes associated with the variations of pH and ionic strength [105].

The role of EPS in the biosorption of Ni onto aerobic and anaerobic granular sludge was studied by fluorescence and FTIR spectroscopy [106]. In more detail, 2DCOS fluorescence spectroscopy showed a contribution of EPS to the biosorption of Ni higher for anaerobic granulated sludge than the contribution in aerobic granulated sludge. Moreover, 2DCOS demonstrated that fluorescence changes occurred following the sequential order humic-like fraction → protein-like fraction in aerobic sludge, whereas in anaerobic sludge, the observed sequence was fulvic-like fraction → protein-like and humic-like fractions.

Interactions of NOM and DOM with nanoparticles

TiO_2 nanoparticles have wide applications for their catalytic, deodorizing, antifouling, and antimicrobial properties, as well as for the role played in H_2 fuel production [107]. However, NOM could affect the morphology of the TiO_2 nanoparticles with a consequent impact on their environmental toxicity and fate. A study for the characterization of the interactions among TiO_2 nanoparticles and NOM found several ligand sites [107]. Carboxylate, amide, quinone, ketone groups, and the C—OH groups of phenol, C—OH in aliphatic chains, and polysaccharide fractions of humic substances have relevant abilities for binding TiO_2 nanoparticles [108]. The study of the interactions between TiO_2 nanoparticles and NOM produced by bacterial cell membranes showed some differences with respect to those of previous studies [109, 110]. In fact, the FTIR 2DCOS asynchronous maps suggested a sequential order of functional groups binding TiO_2 nanoparticles with the order COO^- → aromatic C=C stretching → C=O of ketone → N—H of amide II. These differences depended reasonably on the different origins and organisms producing NOM.

Interactions of NOM and DOM with minerals

The adsorption of EPS onto soil minerals is an important process for understanding the mechanism of bacterial adhesion to mineral surfaces, which affects the environmental cycling of nutrients and contaminants. FTIR 2DCOS investigated these mechanisms involving EPS adhesion onto goethite [111]. Results showed that the sequential order of functional groups in EPS that interacted with goethite was dependent on the electrolyte concentration of the surrounding environment. Molecular compounds characterized by relevant contents of carboxyl and phosphoryl functional groups were the first to be adsorbed at low electrolyte concentration, while amide groups in EPS were the first to be adsorbed at a higher electrolyte concentration.

FTIR 2DCOS also investigated the mechanisms concerning the adhesion of the bacteria species *Shewanella oneidensis* MR-1 onto a goethite surface [112]. The results revealed a complex process according to the following sequential order of adhesion: change in H-bonds of proteins on cell surfaces → formation of monodentate inner-sphere surface complexes → formation of outer-sphere surface complexes → transformation of protein secondary structure on cell surfaces → formation of additional bridging bidentate surface complexes.

Interactions of NOM and DOM with organic compounds and xenobiotics

Ionic liquids have several commercial and industrial applications such as catalysis in organic chemical synthesis and solvent extraction solvents at several temperatures [113]. The large use of ionic liquids increases their diffusion into the environment where they can interact with NOM. A study based on FTIR spectroscopy, synchronous fluorescence spectroscopy, and dynamic light scattering, supported by 2DCOS and 2HDCOS, clarified some fundamental dynamic processes concerning ionic liquids and their interactions with DOM [114]. The cation exchange by means of the carboxylic groups in the humic fraction of DOM was the primary binding agent. With the increasing concentrations of ionic liquids, other functional groups of DOM, such as carbonyl and aromatic groups, interacted with the aliphatic chains of ionic liquids by means of π-π and dipole-dipole mechanisms.

Atrazine is a pesticide widely used in agriculture. The interactions of atrazine with DOM have direct effects on its absorption in soils and on its environmental fate. In a study for the characterization of binding and sorption of atrazine onto DOM coming from black soil, fluorescence 2DCOS showed that the quenching order of the humic-like component occurred earlier than that of the protein humic component. This suggested that the humic-like fraction in DOM was the preferred molecular system for binding atrazine [115]. In a subsequent study, fluorescence 2DCOS confirmed the binding capacity of the humic-like fraction for atrazine, also evidencing a preferential and sequential binding order according to humic-like fraction → protein humic fraction [116].

The presence of microplastics in the aquatic environment is an emerging problem affecting the whole trophic chain [117]. The knowledge of the interactions between DOM and microplastics could contribute to assessing their impact on living organisms and ecosystems. A study performed by FTIR 2DCOS showed that microplastics can be adsorbed on DOM by interacting with its aromatic structures by means of π-π conjugation and via the polar —COOH and C=O groups [118].

Yan and coworkers [119] applied FTIR 2DCOS to study the absorption of the antibacterial agent enrofloxacin onto clay minerals. The electrostatic interaction of the ionic groups in DOM with enrofloxacin was the predominant force to describe its presence in sediments; other significant roles were played by H-donor-acceptor mechanisms and π-π interactions.

Tetracycline antibiotics are extensively used in human medicine and aquaculture industries and due to this wide use, they can enter the aquatic environment where the interactions with DOM become of vital importance to determine their ecotoxicity and environmental fate. A focused study on this topic performed by fluorescence 2DCOS and by FTIR spectroscopy for DOM sampled in eutrophic lakes revealed that several functional groups, including the protein Amide I and Amide II groups, aromatic rings, and aliphatic chains, were the major ones responsible for binding tetracycline [120].

Roxarsone is an organoarsenic additive used to improve feed efficiency and inhibit parasite infections in poultry. Roxarsone present in the secretion products of poultry enters soils where it is fixed by the interactions with NOM. The interactions between roxarsone and NOM were studied by fluorescence 2DCOS [121]. By applying the Noda's rule, this study found that roxarsone binds to NOM according to the sequence humic protein-like fraction ≈ longer wavelength excited humic-like fraction → shorter wavelength excited humic-like fraction. The results of the fluorescence 2DCOS also showed that in the resulting sequence, the aromatic content present in the humic-like fraction played a fundamental role in the total binding capacity.

4-Chlorophenol is widely used as a pesticide in soils and aquaculture wastewaters, and like other xenobiotics, its environmental fate depends on its interactions with DOM [122]. Synchronous fluorescence spectroscopy coupled with 2DCOS found that DOM fractions had different quenching characteristics in binding 4-chlorophenol; the sequential order of fluorescence quenching occurred following the series fulvic-like and humic-like fractions → protein-like fraction. This sequential order was comparable with that observed for atrazine too [115]. Maqbool and Hur [123] investigated the interactions of sodium dodecyl sulfate with three different DOM solutions, including bovine serum albumin (BSA), humic acid (HA), and their mixtures. The study was performed by EEM spectroscopy combined with parallel factor analysis and 2DCOS. Fluorescence 2DCOS demonstrated that the protein-surfactant interactions occurred according to the order tyrosine like → tryptophan-like component for the BSA solution, whereas the sequential order was reversed for the BSA-HA mixtures.

The phytoremediation ability of marine algae depends on the interactions among NOM and pollutants. The phytoremediation ability of the marine macro-alga *Caulerpa racemosa* for reducing the impact of hydrocarbons coming from little spills was tested by UV-Vis and FTIR spectroscopy. In particular, the FTIR disrelation maps showed the significant structural changes in the lipid algal matrix caused by the three different hydrocarbons used for the phytoremediation tests [124].

Essential oils extracted by aromatic plants are potential alternatives to the use of pesticides in agriculture, but the knowledge of their effects on plants needs further studies. The in vivo effects of two types of essential oil mixtures on the secondary protein structure of *Vicia faba* roots were studied by FTIR 2D2DCOS [125]. Asynchronous 2D2DCOS spectra gave a clear evidence of the different structural modifications caused by each type of essential oil mixture on the secondary protein structure of *V. faba* roots.

2DCOS as support to quantitative analysis in environmental studies

Previous sections provided evidences on how 2DCOS always gives great support for the structural studies of NOM and for all its dynamic processes and reactions caused by the interactions with the surrounding environment. However, we believe it is useful to point out some studies where 2DCOS is also a valid support for quantitative determinations in environmental analysis.

Quercetin is a bioactive compound used in traditional Chinese medicine to reduce the effects of toxicological and environmental factors. For the NIR-PLS determination of quercetin in *Ginkgo biloba* leaf extracts, Zhou and coworkers [126] applied 2DCOS as a variable selection tool. This variable selection approach exploited the wavelength ranges where autopeaks of synchronous spectra are placed because here the correlation is maximized [6, 7]. This variable selection method allowed better analytical accuracy for quercetin determination; in addition, with respect to a calibration model using the full spectra, the variable selection method reduced the computational complexity of the multivariate method because a restricted number of variables was used [127]. Analogously, 2DCOS acts as a variable selection method for the fluorescence-PLS determination of anthracene and pyrene in environmental samples [128]. With respect to the conventional PLS fluorescence determination of anthracene and pyrene, the 2DCOS-PLS approach increases the analytical sensitivity in the determination of the two polycyclic aromatic hydrocarbons.

PCR and PLS regression are common methods for multivariate calibration though PLS regression is sometimes preferred due to its ability to take into account nonlinear relationships among concentrations and signal intensities [129]. PCR and PLS were tested for the multivariate determination of bromide, bicarbonate, sulfate, and nitrate in seawater samples by UV spectroscopy [130]. In the range between 260 and 200 nm, PLS regression gave better performances than PCR. The better performances of PLS were explained by the UV asynchronous spectra which revealed crosspeaks in the 260 and 200 nm range depending on nonlinear absorptions in this spectral range. Finally, Barton and coworkers [131] proposed the applications of 2DCOS as a tool to compare the analytical performances of different NIR spectrophotometers.

2DCOS chromatography in environmental studies

The scientific literature reports only a few chromatographic applications of 2DCOS for environmental studies, but they deserve interest. Simon and Felinger [132] proposed the use of 2DCOS in chromatography to verify the reproducibility of HPLC columns for all the types of chemical analysis. This approach has very wide potential applications in intercomparison measurements among laboratories. Recently, the same authors have also proposed the alteration analysis (ALA) in chromatography as an alternative to 2DCOS to study the changes in a series of measurements [133]. In our opinion, the combined use and comparison of 2DCOS and ALA will be an interesting topic in future environmental studies.

Other applications of 2DCOS chromatography are related to the hydrocarbon analysis in environmental samples. The chromatograms of hydrocarbons present in environmental samples are generally characterized by the presence of several peaks. Due to the lack of specific tools and criteria, the qualitative and quantitative comparisons of hydrocarbon distributions in different sampling sites become a difficult task [134]. The so-called two-dimensional correlation coefficient mapping (2DMAP) is an efficient tool for jet fuel classification and for the comparison of hydrocarbon distribution in environmental analysis [135]. Basically, 2DMAP consists of a *sample-sample* 2DCOS analysis of chromatographic samples where a threshold value of the correlation coefficient, generally 0.95, is selected as matching criteria to establish similarity or dissimilarity among samples; then 2DMAP results are reported by a contour map. Samples having a correlation higher than the threshold value are visualized, whereas samples with a correlation lower than the threshold value are hidden and then not reported. According to this simple approach for similarity estimation, 2DMAP becomes a simple and efficient screening method for comparing the hydrocarbon distributions in different sampling sites. The approach of 2DMAP has been further applied for the comparison of hydrocarbon distributions in marine sediments coming from different areas of the Italian coasts [136]. This study showed that several biogenic, anthropogenic, and petrogenic sources of hydrocarbons can produce high qualitative and quantitative dissimilarities even for samples coming from neighboring sites. Finally, in the study of the hydrocarbon distributions in two Antarctic cores, disrelation 2DCOS chromatography showed the differences of anthropogenic, petrogenic, and biogenic sources in the two cores, mainly depending on the location of the sampling sites [137].

2DCOS size exclusion chromatography found different adsorption affinities for TiO_2 and ZnO nanoparticles within the bulk of humic substances, mainly depending on the molecular sizes of humic aggregates [138]. For Cu and DOM interactions, 2DCOS chromatography demonstrated a great potential to describe the metal-binding characteristics of humic substances at high molecular weight resolution. This finding provided a clear picture that metal-humic acid interactions were dependent on the size and molecular weight of DOM [139].

Conclusions

In this chapter, we reviewed the most recent applications of 2DCOS in the study of the dynamic processes involving NOM characteristics and properties. The 2DCOS approach allows the accurate description of the complex mechanisms of NOM formation, degradations, and interactions with organic and inorganic compounds such as toxic metals and xenobiotics; this dynamic behavior can be hardly elucidated by means of PCA, PLS, and LDA techniques, which are multivariate methods for pattern recognition and discriminant analysis, whereas 2DCOS is a family of multivariate methods focused to structural studies. The applications of 2DCOS as a support to spectroscopic and even chromatographic analysis in environmental studies describe dynamic processes such as binding, complexation, redox reactions, and supramolecular interactions between NOM and chemicals. Moreover, if 2DCOS was originally developed for the spectroscopic studies of polymer materials, its applications to chromatography give an excellent example of its high versatility because no analytical technique limits 2DCOS applications. Obviously, this study does not want to point out that 2DCOS has higher performances than PCA, PLS, and LDA, but vice versa; 2DCOS has different targets often complementary to these last methods as shown by some studies [6, 60, 71, 140]. The final purpose of this chapter is to contribute to a future wider diffusion of 2DCOS in environmental studies.

Abbreviations

2D2DCOS	double two-dimensional correlation spectroscopy
2DCOS	two-dimensional correlation spectroscopy
2DHCOS	two-dimensional hetero correlation spectroscopy
2DMAP	two-dimensional correlation coefficient mapping
DOM	dissolved organic matter
EEM	synchronous excitation emission matrix
EES	excitation emission spectroscopy
EPS	extracellular polymeric substance
FTIR	Fourier transform infrared spectroscopy
FTNIR	Fourier transform near infrared spectroscopy
HPLC	high-performance liquid chromatography
LDA	linear discriminant analysis
NOM	natural organic matter
PCA	principal component analysis
PCR	principal component regression
PLS	partial least square
POM	particulate organic matter
***ss* 2DCOS**	*sample-sample* two-dimensional correlation spectroscopy
UV-Vis	ultraviolet and visible spectroscopy
***ww* 2DCOS**	*wavelength-wavelength* two-dimensional correlation spectroscopy
XFS	X-ray fluorescence spectroscopy

References

[1] V. Zitko, Principal component analysis in the evaluation of environmental data, Mar. Pollut. Bull. 28 (12) (1994) 718–722.

[2] S. Masa, A. de Juan, R. Tauler, A.C. Olivieri, G.M. Escandar, Application of chemometric methods to environmental analysis of organic pollutants: a review, Talanta 80 (3) (2010) 1052–1067.

[3] I.S. Helland, Some theoretical aspects of partial least squares regression, Chemom. Intell. Lab. Syst. 58 (2) (2001) 97–107.

[4] I. Noda, Two-dimensional infrared (2D IR) spectroscopy. A new tool for interpreting infrared spectra, Microchim. Acta 94 (1–6) (1988) 101–103.

[5] I. Noda, Two-dimensional infrared spectroscopy, J. Am. Chem. Soc. 111 (1) (1989) 8116–8118.

[6] I. Noda, Y. Ozaki, Generalized two-dimensional correlation spectroscopy, Appl. Spectrosc. 54 (7) (2000) 236A–248A.
[7] I. Noda, Y. Ozaki, Two-Dimensional Correlation Spectroscopy, Application in Vibrational and Optical Spectroscopy, John Wiley & Sons, Chichester, 2004.
[8] I. Noda, 2DCOS and I, three decades of two-dimensional correlation spectroscopy, J. Mol. Struct. 1124 (1) (2016) 3–7.
[9] I. Noda, Two-dimensional correlation analysis of spectra collected without knowing sampling order, J. Mol. Struct. 1156 (1) (2018) 418–423.
[10] Y. Park, S. Jin, I. Noda, Y. Mee Jung, Recent progress in two-dimensional correlation spectroscopy, J. Mol. Struct. 1168 (1) (2018) 1–21.
[11] I. Noda, Frontiers of Two-Dimensional Correlation Spectroscopy. Part 1. New concepts and noteworthy developments, J. Mol. Struct. 1069 (1) (2014) 3–22.
[12] I. Noda, Frontiers of Two-Dimensional Correlation Spectroscopy. Part 2. Perturbation methods, fields of applications, and types of analytical probes, J. Mol. Struct. 1069 (1) (2014) 23–49.
[13] M. Mecozzi, M.B. Tudino, M.G. Finoia, M.E. Conti, Uncommon multivariate statistical methods in environmental studies: a review, Trends Environ. Anal. Chem. 6–7 (2015) 31–38.
[14] J. Neuman, Soil organic matter maintenance in no-till and crop rotation management systems, in: Reference Module in Earth Systems and Environmental Sciences, Elsevier, 2017. https:/www.sciencedirect.com/science/article/pii/B9780124095489106530.
[15] R.R. Ishiwatari, Macromolecular materials (humic substance) in the water column and sediments, Mar. Chem. 39 (1–3) (1992) 151–166.
[16] C.J. Schubert, S.E. Calvert, Nitrogen and carbon isotopic composition of marine and terrestrial organic matter in Arctic Ocean sediments: implications for nutrient utilization and organic matter composition, Deep-Sea Res. I 48 (3) (2001) 789–810.
[17] J.A. Baldock, C.A. Masiello, Y. Gélinas, J.I. Hedges, Cycling and composition of organic matter in terrestrial and marine ecosystems, Mar. Chem. 92 (1–4) (2004) 39–64.
[18] M. Mecozzi, P. Dragone, M. Amici, E. Pietrantonio, Ultrasound assisted extraction and determination of the carbohydrate fraction in marine sediments, Org. Geochem. 31 (12) (2000) 1797–1803.
[19] J. Six, R.T. Conant, E.A. Paul, K. Paustian, Stabilization mechanisms of soil organic matter: implications for C-saturation of soils, Plant Soil 241 (2) (2002) 155–176.
[20] M. Mecozzi, M. Amici, G. Romanelli, E. Pietrantonio, A. Deluca, Ultrasound extraction and thin layer chromatography–flame ionization detection analysis of the lipid fraction in marine mucilage samples, J. Chromatogr. A 963 (1–2) (2002) 363–373.
[21] A. Piccolo, The supramolecular structure of humic substances, Soil Sci. 166 (11) (2001) 810–832.
[22] M. Mecozzi, R. Acquistucci, L. Nisini, M.E. Conti, Mechanisms of browning development in aggregates of marine organic matter formed under anoxic conditions: a study by mid-infrared and near-infrared spectroscopy, Infrared Phys. Technol. 63 (2014) 74–83.
[23] Z. Chen, J.N. Hay, M.J. Jenkins, FTIR spectroscopic analysis of poly(ethylene terephthalate) on crystallization, Eur. Polym. J. 48 (9) (2012) 1586–1610.
[24] I. Noda, Kernel analysis for two-dimensional (2D) correlation spectroscopy, J. Mol. Struct. 799 (1–3) (2006) 34–40.
[25] I. Noda, Projection two-dimensional correlation analysis, J. Mol. Struct. 974 (1–3) (2010) 116–126.
[26] I. Noda, Double two-dimensional correlation analysis—2D correlation of 2D spectra, J. Mol. Struct. 974 (1–3) (2010) 108–115.

[27] I. Noda, Techniques useful in two-dimensional correlation and co-distribution spectroscopy (2DCOS and 2DCDS) analyses, J. Mol. Struct. 1124 (2016) 29–41.
[28] I. Noda, Quadrature two-dimensional correlation spectroscopy (Q-2DCOS), J. Mol. Struct. 1124 (2016) 42–52.
[29] I. Noda, Two-dimensional correlation spectroscopy (2DCOS) analysis of polynomials, J. Mol. Struct. 1124 (2016) 53–60.
[30] S. Morita, Y. Ozaki, Moving-window two-dimensional correlation spectroscopy and perturbation-correlation moving-window two-dimensional correlation spectroscopy, Chemom. Intell. Lab. Syst. 168 (2017) 114–120.
[31] M. Mecozzi, M. Pietroletti, V. Gallo, M.E. Conti, Formation of incubated marine mucilages investigated by FTIR and UV–VIS spectroscopy and supported by two-dimensional correlation analysis, Mar. Chem. 116 (1–4) (2009) 18–35.
[32] M. Mecozzi, M. Pietroletti, Chemical composition and surfactant characteristics of marine foams investigated by means of UV-vis, FTIR and FTNIR spectroscopy, Environ. Sci. Pollut. Res. 23 (22) (2016) 22418–22432.
[33] T. Zhou, L. Peng, Y. Liu, Y. Zhan, F. Liu, A. Zhang, An insight into the sequential order in 2D correlation spectroscopy using polymer transitions: Boltzmann sigmoid, Gaussian cumulative, Lorentz cumulative, and asymmetric sigmoid. Findings in experiments and simulations, Vib. Spectrosc. 70 (2014) 137–161.
[34] P. Lasch, I. Noda, Two-dimensional correlation spectroscopy (2D-COS) for analysis of spatially resolved vibrational spectra, Appl. Spectrosc. 73 (4) (2019) 359–379.
[35] S. Šašic, A. Muszynki, Y. Ozaki, A new possibility of the generalized two-dimensional correlation spectroscopy 1. Sample-sample correlation spectroscopy, J. Phys. Chem. A 104 (27) (2000) 6380–6387.
[36] M.A. Czarnecki, Determination of relative rate of spectral events by novel modification of two-dimensional correlation spectroscopy, Anal. Chim. Acta 702 (1) (2011) 72–80.
[37] S. Šašic, J. Jiang, Y. Ozaki, Potential of variable-variable and sample-sample, generalized and statistical, two-dimensional correlation spectroscopy in investigation of chemical reactions, Chemom. Intell. Lab. Syst. 65 (1) (2003) 1–15.
[38] S. Morita, Y. Ozaki, https:/sites.google.com/view/shigemorita/home/2dshige.
[39] T. Pazderka, V. Kopecký, http:/dsp.vscht.cz/konference_matlab/MATLAB08/prispevky/081_pazderka.pdf.
[40] R. Geitner, R. Fritzsch, J. Popp, T.W. Bocklitz, Corr2D: implementation of two-dimensional correlation analysis in R, J. Stat. Softw. 90 (3) (2019) 1–9.
[41] J.H. Christensen, G. Tomasi, Practical aspects of chemometrics for oil spill fingerprinting, J. Chromatogr. A 1169 (1–2) (2007) 1–18.
[42] J. Engel, J. Gerretzen, E. Szymanska, J.J. Jansen, G. Downey, L. Blanchet, L.M. C. Buydens, Breaking with trends in pre-processing, Trends Anal. Chem. 50 (2013) 96–106.
[43] J. Zupan, Algorithms for Chemists, John Wiley & Sons, New York, 1989.
[44] M. Mecozzi, Polynomial curve fitting method for baseline drift correction in the chromatographic analysis of hydrocarbons in environmental samples, APCBEE Procedia 10 (2014) 2–6.
[45] A. Savitzky, M.J.E. Golay, Smoothing and differentiation of data by simplified least squares procedures, Anal. Chem. 36 (8) (1964) 1627–1639.
[46] J. Luo, K. Ying, J. Bai, Savitzky–Golay smoothing and differentiation filter for even number data, Signal Process. 85 (7) (2005) 1429–1434.
[47] P.H.C. Eliers, A perfect smoother, Anal. Chem. 75 (14) (2003) 3631–3636.

[48] Y. Mee Jung, Principal component analysis based two-dimensional correlation spectroscopy for noise filtering effect, Vib. Spectrosc. 36 (2) (2004) 267–270.
[49] T. Pazderka, V. Kopecky Jr., Two-dimensional correlation analysis of Raman optical activity. Basic rules and data treatment, Vib. Spectrosc. 60 (2012) 193–199.
[50] C. Yu, J. Irudayaraj, Spectroscopic characterization of microorganisms by Fourier transform infrared microspectroscopy, Biopolymers 77 (6) (2005) 368–377.
[51] I. Noda, Scaling techniques to enhance two-dimensional correlation spectra, J. Mol. Struct. 883–884 (2008) 216–227.
[52] M. Mecozzi, E. Pietrantonio, M. Pietroletti, The roles of carbohydrates, proteins and lipids in the process of aggregation of natural marine organic matter investigated by means of 2D correlation spectroscopy applied to infrared spectra, Spectrochim. Acta A 71 (5) (2009) 1877–1884.
[53] Y. Nakaya, S. Nakashima, M. Morizumu, Nondestructive spectroscopic tracing of simulated formation processes of humic-like substances based on the Maillard reaction, Appl. Spectrosc. 72 (8) (2018) 1189–1198.
[54] A.H.N. Abdulla, E.C. Minor, P.G. Hatcher, Using two-dimensional correlations of ^{13}CNMR and FTIR spectroscopy to investigate changes in the chemical composition of dissolved organic matter along an estuarine transect, Environ. Sci. Technol. 44 (21) (2010) 8044–8049.
[55] A.H.N. Abdulla, R.L. Sleighter, P.G. Hatcher, Two dimensional correlation analysis of Fourier transform on cyclotron mass spectra of dissolved organic matter: a new graphical analysis of trends, Anal. Chem. 85 (8) (2013) 3895–3902.
[56] H.A.N. Abdulla, P.G. Hatcher, Dynamics of dissolved organic matter: a view from two dimensional correlation spectroscopy techniques, J. Mol. Struct. 1069 (1–3) (2014) 313–317.
[57] C. Steelink, Investigating humic acids in soils, Anal. Chem. 74 (11) (2002) 327A–333A.
[58] F. Song, F. Wu, B. Xing, T. Li, W. Feng, J.P. Giesy, W. Guo, H. Wang, S. Liu, Y. Bai, Protonation-dependent heterogeneity in fluorescent binding sites in sub-fractions of fulvic acid using principle component analysis and two-dimensional correlation spectroscopy, Sci. Total Environ. 616–617 (2018) 1279–1287.
[59] Y. Zhou, X. Yao, Y. Zhang, K. Shi, Y. Zhang, E. Jeppesen, G. Gao, G. Zhu, B. Qin, Potential rainfall-intensity and pH-driven shifts in the apparent fluorescent composition of dissolved organic matter in rainwater, Environ. Pollut. 224 (2017) 638–648.
[60] W. Chen, X.Y. Liu, H.Q. Yu, Temperature-dependent conformational variation of chromophoric dissolved organic matter and its consequent interaction with phenanthrene, Environ. Pollut. 222 (2017) 23–31.
[61] Y. Yuan, X. Cai, B. Tan, S. Zhou, B. Xing, Molecular insights into reversible redox sites in solid-phase humic substances as examined by electrochemical in situ FTIR and two-dimensional correlation spectroscopy, Chem. Geol. 494 (2018) 136–143.
[62] H. Xu, L. Guo, H. Jiang, Depth-dependent variations of sedimentary dissolved organic matter composition in a eutrophic lake: implications for lake restoration, Chemosphere 146 (2016) 551–559.
[63] J. Hur, K. Young Jung, Y. Mee Young, Characterization of spectral responses of humic substances upon UV irradiation using two-dimensional correlation spectroscopy, Water Res. 45 (9) (2011) 2965–2974.
[64] J. Hur, B.M. Lee, K.S. Choi, B. Min, Tracking the spectroscopic and chromatographic changes of algal derived organic matter in a microbial fuel cell, Environ. Sci. Pollut. Res. 21 (3) (2014) 2230–2239.

[65] H. Li, E.C. Minor, P.K. Zigah, Diagenetic changes in Lake Superior sediments as seen from FTIR and 2D correlation spectroscopy, Org. Geochem. 58 (2013) 125–136.
[66] N. Hertkorn, A. Permin, I. Perminova, D. Kovalevskii, M. Yudov, V. Petrosyan, A. Kettrup, Comparative analysis of partial structures of a peat humic and fulvic acid using one- and two-dimensional nuclear magnetic resonance spectroscopy, J. Environ. Qual. 31 (2) (2002) 375–387.
[67] R.H. Wilson, A.C. Smith, M. Kačuráková, P.K. Saunders, N. Wellner, K.W. Waldron, The mechanical properties and molecular dynamics of plant cell wall polysaccharides studied by Fourier-transform infrared spectroscopy, Plant Physiol. 124 (1) (2000) 397–405.
[68] X. Li, X. Dai, J. Takahashi, N. Li, J. Jin, L. Dai, B. Dong, New insight into chemical changes of dissolved organic matter during anaerobic digestion of dewatered sewage sludge using EEM-PARAFAC and two-dimensional FTIR correlation spectroscopy, Bioresour. Technol. 159 (2014) 412–420.
[69] H. Pan, H. Yu, Y. Wang, R. Liu, H. Lei, Investigating variations of fluorescent dissolved organic matter in wastewater treatment using synchronous fluorescent spectroscopy combined with principal component analysis and two-dimensional correlation, Environ. Technol. 39 (19) (2018) 2495–2502.
[70] X. Li, X. Dai, L. Dai, Z. Liu, Two-dimensional FTIR correlation spectroscopy reveals chemical changes in dissolved organic matter during the biodrying process of raw sludge and anaerobically digested sludge, RSC Adv. 5 (100) (2015) 82087–82096.
[71] H. Yu, F. Qu, X. Zhang, S. Shao, H. Rong, H. Liang, L. Bai, J. Ma, Development of correlation spectroscopy (COS) method for analyzing fluorescence excitation emission matrix (EEM): a case study of effluent organic matter (EfOM) ozonation, Chemosphere 228 (2019) 35–43.
[72] X. Tang, Q.Y. Wu, X. Zhao, Y. Du, H. Huang, X.L. Shi, H.Y. Hu, Transformation of anti-estrogenic-activity related dissolved organic matter in secondary effluents during ozonation, Water Res. 48 (2014) 605–612.
[73] X.S. He, B.D. Xi, Z.Y. Zhang, R.T. Gao, W.B. Tan, D.Y. Cui, Insight into the evolution, redox, and metal binding properties of dissolved organic matter from municipal solid wastes using two-dimensional correlation spectroscopy, Chemosphere 117 (2014) 701–707.
[74] J. Li, J. Wei, H.H. Ngo, W. Guo, H. Liu, B. Du, Q. Wei, D. Wei, Characterization of soluble microbial products in a partial nitrification sequencing batch biofilm reactor treating high ammonia nitrogen wastewater, Bioresour. Technol. 249 (2018) 241–246.
[75] T. Maqbool, J. Cho, J. Hur, Changes in spectroscopic signatures in soluble microbial products of activated sludge under different osmotic stress conditions, Bioresour. Technol. 255 (2018) 29–38.
[76] B.M. Lee, J. Hur, Adsorption behavior of extracellular polymeric substances on graphene materials explored by fluorescence spectroscopy and two-dimensional Fourier transform infrared correlation spectroscopy, Environ. Sci. Technol. 50 (14) (2016) 7364–7372.
[77] X. Guo, C. Li, Q. Zhu, T. Huang, Y. Cai, N. Li, J. Liu, X. Tan, Characterization of dissolved organic matter from biogas residue composting using spectroscopic techniques, Waste Manag. 78 (2018) 301–309.
[78] R.M.B.O. Duarte, S.M.S.C. Freire, A.C. Duarte, Investigating the water-soluble organic functionality of urban aerosols using two-dimensional correlation of solid-state 13C NMR and FTIR spectral data, Atmos. Environ. 116 (2015) 245–252.
[79] B.S. Su, Z. Qu, X.-S. He, Y.H. Song, L.M. Jia, Characterizing the compositional variation of dissolved organic matter over hydrophobicity and polarity using fluorescence spectra

combined with principal component analysis and two-dimensional correlation technique, Environ. Sci. Pollut. Res. 23 (9) (2016) 9237–9244.
[80] G.H. Yu, Z. Tang, Y.C. Xu, Q.R. Shen, Multiple fluorescence labelling and two dimensional FTIR–^{13}C NMR heterospectral correlation spectroscopy to characterize extracellular polymeric substances in biofilms produced during composting, Environ. Sci. Technol. 45 (21) (2011) 9224–9231.
[81] X.S. He, B.D. Xi, R.T. Gao, L. Wang, Y. Ma, D.Y. Cui, W.B. Tan, Using fluorescence spectroscopy coupled with chemometric analysis to investigate the origin, composition, and dynamics of dissolved organic matter in leachate-polluted groundwater, Environ. Sci. Pollut. Res. 22 (11) (2015) 8499–8506.
[82] O.R. Harvey, B.E. Herbert, L. Kuo, P. Louchouarn, Generalized two-dimensional perturbation correlation infrared spectroscopy reveals mechanisms for the development of surface charge and recalcitrance in plant-derived biochars, Environ. Sci. Technol. 46 (19) (2012) 10641–10650.
[83] D. Wei, H.H. Ngo, W. Guo, W. Xu, B. Du, M.S. Khan, Q. Wei, Biosorption performance evaluation of heavy metal onto aerobic granular sludge-derived biochar in the presence of effluent organic matter via batch and fluorescence approaches, Bioresour. Technol. 249 (2018) 410–416.
[84] W. Chen, N. Habibul, X.Y. Liu, G.P. Sheng, H.Q. Yu, FTIR and synchronous fluorescence heterospectral two-dimensional correlation analyses on the binding characteristics of copper onto dissolved organic matter, Environ. Sci. Technol. 49 (4) (2015) 2052–2058.
[85] O.H. Lowry, N.J. Rosebrough, A.L. Farr, R.J. Randall, Protein measurement with the Folin phenol reagent, J. Biol. Chem. 193 (1) (1951) 265–275.
[86] W. Li, F. Zhang, Q. Ye, D. Wu, L. Wang, Y. Yu, B. Deng, J. Du, Composition and copper binding properties of aquatic fulvic acids in eutrophic Taihu Lake, China, Chemosphere 172 (2017) 496–504.
[87] J. Tang, L. Zhuang, Z. Yu, X. Liu, Y. Wang, P. Wen, S. Zhou, Insight into complexation of Cu(II) to hyperthermophilic compost-derived humic acids by EEM PARAFAC combined with heterospectral two dimensional correlation analyses, Sci. Total Environ. 656 (2019) 29–38.
[88] B. Hu, P. Wang, C. Wang, J. Qian, J. Hou, X. Cui, N. Zhang, The effect of anthropogenic impoundment on dissolved organic matter characteristics and copper binding affinity: insights from fluorescence spectroscopy, Chemosphere 188 (2017) 424–433.
[89] F. Sun, Y. Li, X. Wang, Z. Chi, G. Yu, Using new hetero-spectral two-dimensional correlation analyses and synchrotron-radiation-based spectromicroscopy to characterize binding of cu to soil dissolved organic matter, Environ. Pollut. 223 (2017) 457–465.
[90] X.J. Guo, Y.Z. Li, Y.H. Feng, D.H. Yuan, Using fluorescence quenching combined with two-dimensional correlation fluorescence spectroscopy to characterise the binding-site heterogeneity of dissolved organic matter with copper and mercury in lake sediments, Environ. Chem. 14 (2) (2017) 91–98.
[91] M. Huang, Z. Li, B. Huang, N. Luo, Q. Zhang, X. Zhai, G. Zeng, Investigating binding characteristics of cadmium and copper to DOM derived from compost and rice straw using EEM-PARAFAC combined with two-dimensional FTIR correlation analyses, J. Hazard. Mater. 344 (2018) 539–548.
[92] H. Xu, G. Yu, L. Yang, H. Jiang, Combination of two-dimensional correlation spectroscopy and parallel factor analysis to characterize the binding of heavy metals with DOM in lake sediments, J. Hazard. Mater. 263 (2013) 412–421.

[93] J. Zhang, H. Yin, L. Chen, F. Liu, H. Chen, The role of different functional groups in a novel adsorption complexation-reduction multi-step kinetic model for hexavalent chromium retention by undissolved humic acid, Environ. Pollut. 237 (2018) 740–746.
[94] J. Zhang, H. Yin, B. Samuel, F. Liu, H. Chen, A novel method of three-dimensional heterospectral correlation analysis for the fingerprint identification of humic acid functional groups for hexavalent chromium retention, RSC Adv. 8 (2018) 3522–3529.
[95] J. Zhang, L. Chen, H. Yin, S. Jin, F. Liu, H. Chen, Mechanism study of humic acid functional groups for Cr(VI) retention: two-dimensional FTIR and ^{13}C-CP/MAS NMR correlation spectroscopic analysis, Environ. Pollut. 225 (2017) 86–92.
[96] F.S. Sun, G.H. Yu, M.L. Polizzotto, W. Rana, Q.R. Shen, Toward understanding the binding of Zn in soils by two-dimensional correlation spectroscopy and synchrotron-radiation-based spectromicroscopies, Geoderma 337 (2019) 238–245.
[97] X. Liu, H. Lv, H.C. Xu, Two-dimension fluorescence correlation spectroscopy to characterize the binding of organic ligands with zinc in eutrophic lake, Chin. Chem. Lett. 26 (2) (2015) 205–209.
[98] F. Sun, M.L. Polizzotto, D. Guanc, J. Wu, Q. Shen, W. Ran, B. Wang, G. Yua, Exploring the interactions and binding sites between Cd and functional groups in soil using two-dimensional correlation spectroscopy and synchrotron radiation based spectromicroscopies, J. Hazard. Mater. 326 (2017) 18–25.
[99] G.H. Yu, M.J. Wu, G.R. Wei, Y.H. Luo, W. Ran, B.R. Wang, J.C. Zhang, Q.R. Shen, Binding of organic ligands with Al(III) in dissolved organic matter from soil: implications for soil organic carbon storage, Environ. Sci. Technol. 46 (11) (2012) 6102–6109.
[100] Y. Wen, H. Li, J. Xiao, C. Wang, Q. Shen, W. Ran, X. He, Q. Zhou, G. Yu, Insights into complexation of dissolved organic matter and Al(III) and nanominerals formation in soils under contrasting fertilizations using two-dimensional correlation spectroscopy and high resolution-transmission electron microscopy techniques, Chemosphere 111 (2014) 441–449.
[101] P. Jin, J. Song, X.C. Wang, X. Jin, Two-dimensional correlation spectroscopic analysis on the interaction between humic acids and aluminium coagulant, J. Environ. Sci. 64 (2018) 181–199.
[102] K. Nakashima, S. Xing, Y. Gong, T. Miyajima, Characterization of humic acids by two-dimensional correlation fluorescence spectroscopy, J. Mol. Struct. 883–884 (2008) 155–159.
[103] X.S. He, B.D. Xi, Z.Y. Zhang, R.T. Gao, W.B. Tan, D.Y. Cui, Y. Yuan, Composition, removal, redox, and metal complexation properties of dissolved organic nitrogen in composting leachates, J. Hazard. Mater. 283 (2015) 227–233.
[104] N. Habibul, W. Chen, Structural response of humic acids upon binding with lead: a spectroscopic insight, Sci. Total Environ. 643 (2018) 479–485.
[105] H. Xu, M. Yan, W. Li, H. Jiang, L. Guo, Dissolved organic matter binding with Pb(II) as characterized by differential spectra and 2D UV-FTIR heterospectral correlation analysis, Water Res. 144 (2018) 435–443.
[106] N. Li, D. Wei, S. Wang, L. Hu, W. Xu, B. Du, Q. Wei, Comparative study of the role of extracellular polymeric substances in biosorption of Ni(II) onto aerobic/anaerobic granular sludge, J. Colloid Interface Sci. 490 (2017) 754–761.
[107] S.M. Gupta, M. Tripathi, A review of TiO_2 nanoparticles, Chin. Sci. Bull. 56 (16) (2011) 1639–1657.
[108] W. Chen, C. Qian, X.Y. Liu, H.Q. Yu, Two-dimensional correlation spectroscopic analysis on the interaction between humic acids and TiO_2 nanoparticles, Environ. Sci. Technol. 48 (19) (2014) 11119–11126.

[109] G. Huang, T.W. Ng, T. An, G. Li, B. Wang, D. Wu, H.Y. Yip, H. Zhao, P.K. Wong, Interaction between bacterial cell membranes and nano-TiO_2 revealed by two-dimensional FTIR correlation spectroscopy using bacterial ghost as a model cell envelope, Water Res. 118 (2017) 104–113.
[110] G. Huang, T.W. Ng, T. An, G. Li, B. Wang, D. Wu, H.Y. Yip, H. Zhao, P.K. Wong, Corrigendum to "Interaction between bacterial cell membranes and nano-TiO_2 revealed by two-dimensional FTIR correlation spectroscopy using bacterial ghost as a model cell envelope" [Water Res. 118 (2017) 104-113], Water Res. 129 (2018) 522.
[111] P. Caia, D. Lin, C.L. Peacockc, W. Peng, Q. Huang, EPS adsorption to goethite: molecular level adsorption mechanisms using 2D correlation spectroscopy, Chem. Geol. 494 (2018) 127–135.
[112] W. Yan, H. Wang, C. Jing, Adhesion of *Shewanella oneidensis* MR-1 to Goethite: a two-dimensional correlation spectroscopic study, Environ. Sci. Technol. 50 (8) (2016) 4343–4349.
[113] K. Ghandi, A review of ionic liquids, their limits and applications, Green Sustain. Chem. 4 (2014) 44–53.
[114] X.Y. Liu, W. Chen, C. Qian, H.Q. Yu, Interaction between dissolved organic matter and long-chain ionic liquids: a microstructural and spectroscopic correlation study, Environ. Sci. Technol. 51 (9) (2017) 4812–4820.
[115] Y. Wang, X. Zhang, X. Zhang, Q. Meng, F. Gao, Y. Zhang, Characterization of spectral responses of dissolved organic matter (DOM) for atrazine binding during the sorption process onto black soil, Chemosphere 180 (2017) 531–539.
[116] L.J. Zhu, Y. Zhao, Y.N. Chen, H.Y. Cui, Y.Q. Wei, H.L. Liu, X.M. Chen, Z.M. Wei, Characterization of atrazine binding to dissolved organic matter of soil under different types of land use, Ecotoxicol. Environ. Saf. 147 (2018) 1065–1072.
[117] A.A. Koelmans, N.H. Mohamed Nor, E. Hermsen, M. Kooi, S.M. Mintenig, J. De France, Microplastics in freshwaters and drinking water: critical review and assessment of data quality, Water Res. 155 (2019) 410–422.
[118] W. Chen, Z.Y. Ouyang, C. Qian, H.Q. Yu, Induced structural changes of humic acid by exposure of polystyrene microplastics: a spectroscopic insight, Environ. Pollut. 233 (2018) 1–7.
[119] W. Yan, J. Zhang, C. Jing, Adsorption of Enrofloxacin on montmorillonite: two-dimensional correlation ATR/FTIR spectroscopy study, J. Colloid Interface Sci. 390 (1) (2013) 196–203.
[120] L. Bai, Z. Zhao, C. Wang, C. Wang, X. Liu, H. Jiang, Multi-spectroscopic investigation on the complexation of tetracycline with dissolved organic matter derived from algae and macrophyte, Chemosphere 187 (2017) 421–429.
[121] Q.L. Fu, J.Z. He, L. Blaney, D.M. Zhou, Roxarsone binding to soil-derived dissolved organic matter: insights from multi-spectroscopic techniques, Chemosphere 155 (2016) 225–233.
[122] L. Huang, M. Li, H.H. Ngo, W. Guo, W. Xu, B. Du, Q. Wei, D. Wei, Spectroscopic characteristics of dissolved organic matter from aquaculture wastewater and its interaction mechanism to chlorinated phenol compounds, J. Mol. Liq. 263 (2018) 422–427.
[123] T. Maqbool, J. Hur, Changes in fluorescent dissolved organic matter upon interaction with anionic surfactant as revealed by EEM-PARAFAC and two dimensional correlation spectroscopy, Chemosphere 161 (2016) 190–199.
[124] M. Pietroletti, A. Capobianco, E. Ragosta, M. Mecozzi, Preliminary evaluation of hydrocarbon removal power of *Caulerpa racemosa* in seawater by means of infrared and visible spectroscopic measurements, Spectrochim. Acta A 77 (3) (2010) 673–679.

[125] M. Mecozzi, E. Sturchio, Effects of essential oil treatments on the secondary protein structure of *Vicia faba*: a mid-infrared spectroscopic study supported by two-dimensional correlation analysis, Spectrochim. Acta A 137 (2015) 90–98.
[126] X. Zhou, B. Xiang, M. Zhang, Novel spectral selection method based on synchronous two-dimensional correlation spectroscopy, Anal. Lett. 46 (2) (2013) 340–348.
[127] C.M. Andersen, R. Bro, Variable selection in regression—a tutorial, J. Chemom. 24 (11 – 12) (2010) 728–737.
[128] R. Yang, G. Dong, X. Sun, Y. Yang, Y. Yu, H. Liu, W. Zhang, Feasibility of the simultaneous determination of polycyclic aromatic hydrocarbons based on two-dimensional fluorescence correlation spectroscopy, Spectrochim. Acta A 190 (2018) 342–346.
[129] R.G. Brereton, Introduction to multivariate calibration in analytical chemistry, Analyst 125 (11) (2000) 215–2154.
[130] M. Mecozzi, Y.B. Monakhova, Application of multivariate methods in the monitoring of marine environment: simultaneous determination of bromide, bicarbonate, nitrate and sulphide in seawater by ultraviolet spectroscopy, Int. J. Environ. Health 6 (3) (2013) 235–251.
[131] F.E. Barton II, J.A. de Haseth, D.S. Himmelsbach, The use of two-dimensional correlation spectroscopy to characterize instrumental differences, J. Mol. Struct. 799 (1–3) (2006) 221–225.
[132] J. Simon, A. Felinger, Two-dimensional correlation analysis of the reproducibility of high-performance liquid chromatography columns, J. Chromatogr. A 1384 (2015) 115–123.
[133] J. Simon, A. Felinger, Exploring the changes in a series of measurements—the comparison of the two-dimensional correlation analysis and the alteration analysis, Chemom. Intell. Lab. Syst. 168 (2017) 28–37.
[134] G.S. Fraser, J. Ellis, L. Hussain, An international comparison of governmental disclosure of hydrocarbon spills from offshore oil and gas installations, Mar. Pollut. Bull. 56 (1) (2008) 9–13.
[135] G. Wang, J. Karnes, C. Bunker, E.M. Lei Geng, Two-dimensional correlation coefficient mapping in gas chromatography: jet fuel classification for environmental analysis, J. Mol. Struct. 799 (1–3) (2006) 247–252.
[136] M. Pietroletti, S. Mattiello, F. Moscato, F. Oteri, M. Mecozzi, One step ultrasound extraction and purification method for the gas chromatographic analysis of hydrocarbons from marine sediments: application to the monitoring of Italian coasts, Chromatographia 75 (17–18) (2012) 961–971.
[137] M. Mecozzi, M. Pietroletti, G. Trifirò, L. Nisini, Two-dimensional correlation analysis and multivariate time series analysis as innovative tools for the study and the comparison of the hydrocarbon distributions in marine Antarctic cores, Mar. Chem. 159 (2014) 9–18.
[138] D.D. Phong, J. Hur, Using two-dimensional correlation size exclusion chromatography (2D-CoSEC) and EEM-PARAFAC to explore the heterogeneous adsorption behavior of humic substances on nanoparticles with respect to molecular sizes, Environ. Sci. Technol. 52 (2) (2018) 427–435.
[139] Y.K. Lee, J. Hur, Using two-dimensional correlation size exclusion chromatography (2D-CoSEC) to explore the size-dependent heterogeneity of humic substances for copper binding, Environ. Pollut. 277 (2017) 490–497.
[140] M. Mecozzi, L. Nisini, The differentiation of biodegradable and non-biodegradable polyethylene terephthalate (PET) samples by FTIR spectroscopy: a potential support for the structural differentiation of PET in environmental analysis, Infrared Phys. Technol. 101 (2019) 119–126.

Excitation-emission fluorescence mapping and multiway techniques for profiling natural organic matter

6

Zhidan Wen[a], *Yingxin Shang*[a,b], *Ying Zhao*[a], *Kaishan Song*[a]
[a]Northeast Institute of Geography and Agroecology, Chinese Academy of Sciences, Changchun, China, [b]University of Chinese Academy of Sciences, Beijing, China

Introduction

In aquatic ecosystems, natural organic matter is a key source of energy driving ecosystem functions. As a major fraction of carbon in inland waters, dissolved organic carbon (DOC) plays a crucial role in carbon cycling on a global scale. It has important influence on light penetration, secondary production, and nitrogen dynamics processes in aquatic systems [1, 2]. DOC concentration and composition in aquatic ecosystems are related to many global change processes [3]. In the recent decades, DOC concentrations have increased in many inland aquatic ecosystems in North America and Europe, the future DOC concentrations in boreal lakes are predicted to continue to rise, with unpredicted consequences for the global carbon cycle [4–7].

Chromophoric dissolved organic matter (CDOM), the colored fraction of dissolved organic matter (DOM) in aquatic environments, is a photolytic reactive photochemical intermediate of both natural DOM and trace organic pollutants. It can be allochthonous substances, which originate from terrestrial substances, and autochthonous substances which are from the decay of algae by microorganisms. Because the fluorescent fraction of CDOM (i.e., FDOM) can emit fluorescence at longer wavelength after absorbing light at certain excitation wavelength, fluorescence spectroscopy has been applied to trace detailed information about FDOM components due to its high selectivity and sensitivity. In the recent years, the excitation-emission matrix (EEM) fluorescence spectroscopy has been widely used to characterize FDOM in marine and coastal waters, fresh and saline waters, river waters, urban waters, and wastewaters. There are three methods that can be conducted to study EEM spectroscopy based on the previous reports: the traditional "peak-picking" method, EEM coupled with parallel factor analysis (EEM-PARAFAC), and EEM coupled with fluorescence regional integration (EEM-FRI). The traditional peak-picking method makes it difficult to assess FDOM dynamics in waters due to the unreliability of fluorescence peaks. Traditional fluorescence techniques, including fluorescence emission spectrometry and synchronous fluorescence scanning, applied to examine CDOM components have the drawback that the output was restricted to a linear scan [8]. In the recent years, EEM-PARAFAC and EEM-FRI techniques can provide detailed information on environmental FDOM

Multidimensional Analytical Techniques in Environmental Research. https://doi.org/10.1016/B978-0-12-818896-5.00006-5

dynamics in aquatic ecosystems. EEM-PARAFAC can decompose EEM spectra of FDOM into individual fluorescent components to investigate the correlations between the EEM-PARAFAC-extracted fluorescent components and water quality parameters. However, the EEM-PARAFAC technique can provide only several paired fluorescence intensity data with limited FDOM components information. Compared to the EEM-PARAFAC method, EEM-FRI as a quantitative technique is associated with all the wavelength-dependent paired fluorescence intensity data in an EEM. EEM-FRI has been proven to be an effective method to integrate the volumes beneath defined EEM regions, and the integrated fluorescence intensities represent the FDOM components. EEM could be separated into five regions by FRI, which were related to different FDOM components: R1, tyrosine-like component; R2, tryptophan-like component; R3, fulvic-like component; R4, microbial protein-like component; and R5, humic-like component, which the excitation and emission wavelengths characteristic of each region were shown in Table 6.1.

There is increasing interest in understanding the role of organic carbon (OC) derived from multiple sources in inland water systems in response to global change [10–12]. If we can identify the source and composition of OC in inland waters, this will improve the understanding of the status of inland waters in terrestrial carbon cycle.

Profiling natural organic matter in rivers

Rivers play an important role in connecting the land with the oceans and atmosphere. River carbon flux can reflect biogeochemical processes (i.e., plant production, weathering, sediment, land use/land cover, and pollution within catchment areas), and it also provides important information on the content and source of carbon substances imported into lakes, reservoirs, seas, and oceans [13–15]. The total quantity of river carbon plays a significant role in the global carbon cycle [16–18]. The total OC transported to the oceans by rivers across the world ranges from 0.4 to 0.9 Pg year^{-1}, about 55% of which in the form of DOC [19]. Riverine DOC concentrations for different rivers are mainly impacted by river discharge, precipitation, pollution, climate, land use/land cover, and soil organic carbon [20, 21]. In addition, the CDOM

Table 6.1 Excitation and emission wavelength ranges of the five integrated regions identified by fluoresence regional integration (FRI).

FRI region	Excitation (nm)	Emission (nm)	Source
R1	200–250	250–330	Tyrosine-like protein
R2	200–250	330–350	Tryptophan-like protein
R3	200–250	350–500	Fulvic acid-like
R4	250–280	250–380	Microbial-like
R5	280–400	380–500	Humic-like

Five regions identified by FRI method according to Ref. [9].

fluorescent regions in river waters still have site-specific optical properties in relation to pollution sources/levels and land use/cover under similar hydrological and climate conditions. Therefore, the EEM-FRI method has the advantage of investigating the dynamics of CDOM fluorescent regions for river waters in detail.

Characterization of DOM in rivers

FRI-divided FDOM components

It is known that FRI can be applied to large continental-scale examinations of CDOM sources in inland waters. Using FRI methods, five FDOM components have been identified, and these components may vary with sources and also the hydrologic conditions, geographical settings, climatic zones, and regional landscape characteristics with complex conditions of inland waters.

The EEM spectrum varied with the intensities of the five marked fluorescence fractions for various types of river water samples. Four types of river samples were selected to illustrate FDOM characteristics (Fig. 6.1 and Table 6.1). Type I water was primarily from terrestrial sources with marked fulvic-like (R3) and humic-like (R5) components. Type II water derived from polluted water input with marked tryptophan-like (R2) component. Type III water was typically from the Qinghai-Tibetan Plateau with much higher fluorescence intensities of the tyrosine-like (R1) and microbial protein-like (R4) component. Type IV derived from algal decay with distinct fluorescence intensities of the humic-like (R5) component. The excitation-emission area volumes Φ_i and P_i ($i=1$, 2, 3, 4, and 5) were proportional to the total fluorescence intensities (F_{SUM}) and the relative contributions of five different components to the total fluorescence intensities, respectively (Fig. 6.1E and F). F_{SUM} and P_i differed from type to type in river water samples. F_{SUM} decreased from 18,000 to 4000 nm for river water of types I–IV. P_i also differed from type to type, for example, P_5 is about 65% in type I, 56% in type II, and 44% in type III. The fluorescence intensities Φ_4 (F_{R4}) and Φ_5 (F_{R5}) were predominant in type III.

Spatial distribution of FRI fluorescent components

With various hydrologic, geographical, and climatic characteristics, different processes can control the FDOM components in different river basins. The river water samples collected from different river basins in China were pooled together to study the spatial variation of the five fluorescent components [22]. The average F_{Ri} ($i=1, 2,$ 3, 4, and 5) in the seven major river basins had spatial variation (Table 6.2). For each river basin, the fluorescence intensities of the river waters were all dominated by the fulvic-like R3 and humic-like substance R5. The highest F_{R3} and highest F_{R5} were both present in the Songhua River (SHR) Basin among these different river basins in China, attributed to much higher vascular plant and black soil characteristics, which could provide a high background value of the allochthonous substances (i.e., the fulvic-like R3 and the humic-like R5 components) in Northeast China. The lowest F_{R3} and F_{R5} were presented in the Hai (HR) River Basin, and Pearl River (PR) Basin,

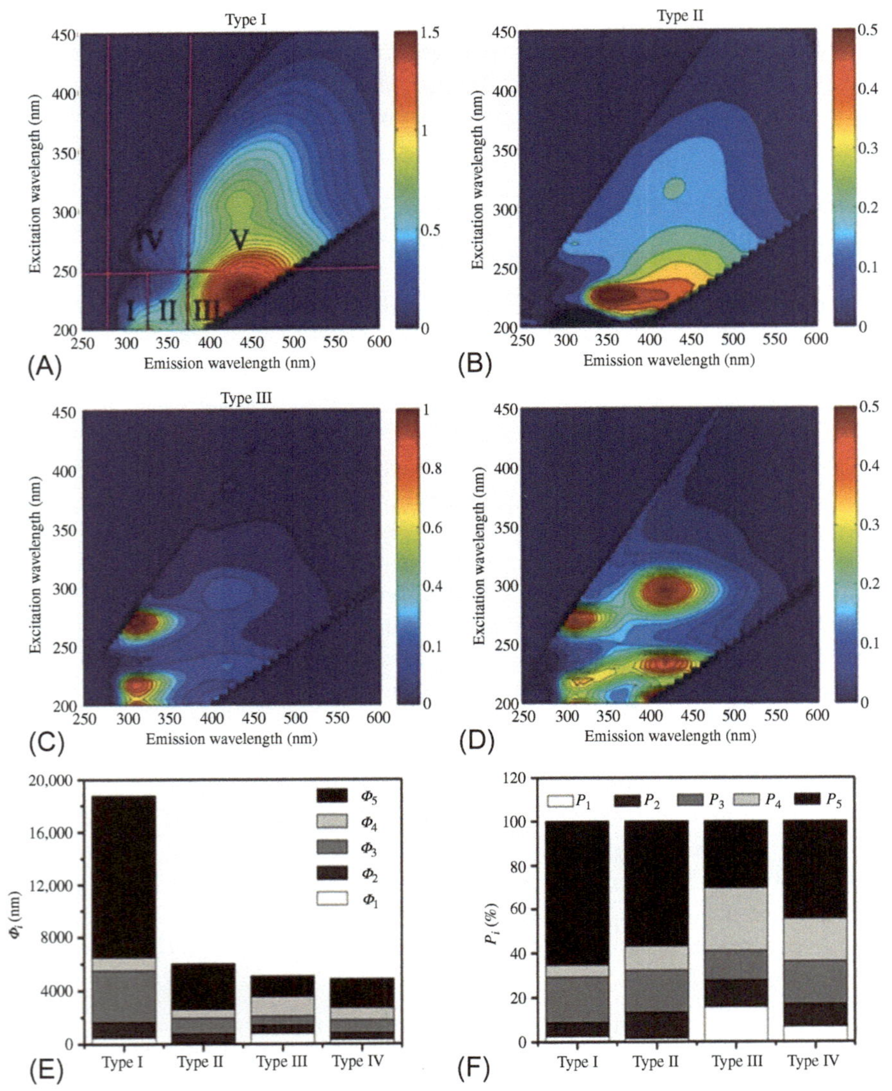

Fig. 6.1 Four typical EEM fluorescence spectra. (A) Type I, primarily from terrestrial sources; (B) type II, derived from polluted water input; (C) type III, typically from the Qinghai-Tibetan Plateau with much higher fluorescence intensities of the tyrosine-like and microbial protein-like components; and (D) type IV, derived from algal decay with distinct fluorescence intensities of the humic-like component. (E) Distributions of FRI-extracted FDOM components in types I–IV and (F) distributions of percentages of FRI-extracted FDOM components in types I–IV.

respectively. Furthermore, FDOM in river waters exhibited a relative high fluorescence intensity of the tryptophan-like component when impacted by wastewater [23]. In this study, the highest F_{R2} for the tryptophan-like R2 and highest F_{R4} for the microbial protein like were all present in the highly polluted Liao River (LR)

Table 6.2 Average total fluorescence intensity F_{SUM}, and the humification index (HIX) for the river waters collected from seven river basins in China.

River basins[a]	F_{SUM} (nm)	HIX
SHR	$(12.84 \pm 5.77) \times 10^3$	5.85 ± 1.39
LR	$(9.99 \pm 4.42) \times 10^3$	4.29 ± 1.41
HR	$(4.36 \pm 2.82) \times 10^3$	7.03 ± 3.56
YHR	$(6.60 \pm 3.25) \times 10^3$	3.30 ± 1.52
YR	$(4.65 \pm 3.02) \times 10^3$	4.14 ± 1.84
PR	$(4.39 \pm 3.53) \times 10^3$	2.75 ± 0.87
Inflow	$(4.67 \pm 2.67) \times 10^3$	4.32 ± 2.43

[a] *HR*, Hai River; *Inflow*, Inflow River in China; *LR*, Liao River; *PR*, Pearl River; *SHR*, Songhua River; *YHR*, Yellow River and Huai River; *YR*, Yangtze River.

Basin. The average total fluorescence intensities F_{SUM} in the seven major river basins in China differed under spatial variation (Table 6.2). The highest average value of F_{SUM} in the SHR Basin demonstrated that the percentage of the fluorescence substances in CDOM was much higher.

DOM in the highly polluted tributary

The whole SHR, located in Jilin and Heilongjiang provinces of Northeast China, and composed by the upper and lower SHR, plays an important role in drinking water, freshwater supply for industry and agriculture, shipping business, fishery, and recreation. With rapid development of urban, industry, and agriculture, the whole SHR has been heavily polluted by organic contamination due to household sewage and industrial-agricultural wastewater, particularly for the Yinma River (YMR) and Yitong River (YTR), which is the important tributary of upper SHR. Therefore, the composition of CDOM in the SHR watershed dominated by allochthonous humic-like substances has been greatly changed. EEM-FRI was, therefore, applied to characterize CDOM fluorescent regions in the SHR and in its highly polluted tributary in Northeast China.

FRI divided CDOM regions

The fluorescence intensities of the five marked fluorescence regions R1-R5 of FRI-EEM varied with the site-specific optical properties of river waters affected by various sources of organic pollution (i.e., urban sewage, industrial, and agricultural sources) and land use/cover [22, 24, 25], respectively. It was noted that the highest fluorescence intensities of the tryptophan-like region (R2) were present in the highly polluted tributary. The fluorescence intensities of the tyrosine-like (R1), fulvic-like (R3), and microbial by-product-like (R4) regions almost all increased simultaneously with the fluorescence intensities of the tryptophan-like (R2) region in the highly polluted tributary. The highest fluorescence intensities of R3 and R5 region were present in the river influenced by forest, wetlands, grassland, and blank soil type, which means that the fulvic- and humic-like regions can be used to distinguish between the highly polluted waters originated from allochthonous anthropogenic organic matter and the waters dominated by terrestrial substances influenced by land cover.

Principal component analysis (PCA) was further conducted using the relative scores of the five FRI fluorescent regions and the HIX, to assess the relative distributions of all water samples. The first two PCA axes explained 90.40% of the total variance in the data set with factors 1 and 2 accounting for 63.84% and 26.56%, respectively. The five FRI fluorescent regions all showed positive factor 1 loadings. The fulvic-like (R3), humic-like (R5) regions, and the HIX showed positive factor 2 loadings. FRI-PCA could separate fluorescent characteristics of the allochthonous fulvic-like (R3) and humic-like (R5) regions and the autochthonous tyrosine-like (R1), tryptophan-like (R2), and microbial by-product-like (R4) regions. The water samples from the highly polluted tributary (YMR and YTR) were scattered with factor 1 scores in the range of −1 to 3.5 and factor 2 scores ranging from −2 to 0.5, falling within range of the water samples with mild pollution (SHR and second SHR). In particular, the water samples from the highly polluted YTR were scattered with much higher factor 1 scores, which can be attributed to severe pollution resulting in much higher autochthonous substances (R1, R2, and R4) and anthropogenic organic matter. In contrast, most of the water samples with mild pollution clustered with much higher factor 2 scores and much lower factor 1 scores. These results indicated that the fulvic-like (R3) and humic-like (R5) regions in the SHR (TJ) were highly dominated in CDOM derived from wetlands, forest, and peatlands as a consequence of much higher HIX compared with that of other river waters [26].

Correlations between CDOM fluorescence and DOC

The linkage between FRI regions and DOC is very complex due to fluorescent and free fluorescent CDOM fractions. Significant strong relationships were observed between DOC and either R3 ($r_p = 0.850$; $P < .01$) or R5 ($r_p = 0.815$; $P < .01$), showing that the R3 and R5 fluorescent regions were dominated in DOC. On the other hand, there were weak relationships between DOC and the autochthonous substances (R1, R2, and R4). However, there were significant strong positive relationships between R2 and R1 ($r_p = 0.887$; $P < .01$) and between R2 and R4 ($r_p = 0.857$; $P < .01$), indicating they were probably originated from similar sources. A strong significant relationship between the fulvic-like (R3) and humic-like (R5) regions was also observed for all water samples suggesting they may derive from a common terrestrially derived source. The correlation coefficient (R^2) between R3 and R5 was improved to 0.86 and the slope increased to 2.05 when excluding the water samples from the highly polluted tributary. However, there was a strong significant natural logarithm relationship ($R^2 = 0.97$; $P < .01$) between R3 and R5 exclusively for the data from the highly polluted tributary.

The fulvic-like R3 and humic-like R5 were all dominated in DOC for river waters, and they can be used to distinguish between the highly polluted waters and the waters derived from terrestrial sources in this section. Therefore, the relationship between the fulvic-like R3 and humic-like R5 was also investigated. There was a significant strong relationship between R3 and R5 for the water samples from the mainstream of SHR. The strong significant natural logarithm relationship between R3 and R5 for the water samples from the highly polluted tributary further indicated that the fluorescence intensities of the fulvic-like R3 increased fast than that of R5 for the highly polluted waters.

Excitation-emission fluorescence application in rivers

Total fluorescence intensities of DOC

EEM-FRI was applied to obtain five fluorescence intensity components and to calculate the carbon-specific fluorescence intensities (*F*/DOC) of a total of 301 water samples from seven river basins (Table 6.2) between 2013 and 2015, which cover a wide geographical, hydrologic, climatic, and landscape range across China. A relationship was established between the total fluorescence intensities (F_{SUM}) and DOC concentrations based on the threshold values (8000 L nm mg^{-1}) for the total carbon-specific fluorescence intensities F_{SUM}/DOC [27]. This may be useful for retrieving DOC concentrations based on the FDOM properties in river waters across China. The riverine DOC concentration for all water samples ranged from 0 to 48.15 mg/L (Fig. 6.2A). There were significant differences in DOC concentrations between the SHR Basin and six other river basins (*t*-test, $P < .001$), attributed to higher DOC background values resulting from humic-dominated substances in Northeast China [28]. The F_{SUM} of five defined regions for all river water samples varied from 267.04 nm in the Inflow River Basin to 28,351.25 nm in the SHR Basin with a mean value of 5875.80 $\pm$ 4561.79 nm (Fig. 6.2B). The lowest average F_{SUM} was observed in the Inflow River Basin in Northwest China and the highest average F_{SUM} in the SHR Basin, corresponding to much higher average DOC concentrations due to humic-dominated substances in the SHR Basin in Northeast China [28].

PCA was conducted using the relative scores of the carbon-specific FRI fluorescent regions (i.e., R1/DOC, R2/DOC, R3/DOC, R4/DOC, R5/DOC, F_{SUM}/DOC), DOC concentrations, and HIX to assess the relative distributions of all water samples. The carbon-specific fulvic-like R3/DOC, humic-like R5/DOC regions, total fluorescence intensities F_{SUM}/DOC, DOC concentrations, and HIX showed positive factor 2 loadings. The DOC concentrations were close to factor 1 loadings and close to zero for factor 2 loadings. The HIX was close to the factor 2 loadings and close to zero for the factor 1 loadings. Therefore, the FRI-PCA in this section could separate carbon-specific fluorescent characteristics of the allochthonous substances (R3/DOC and R5/DOC) from the autochthonous substances (R1/DOC, R2/DOC, and R4/DOC regions).

F_{SUM} vs DOC based on F_{SUM}/DOC grouping

When all river water samples were pooled together, there was a weak correlation between the total fluorescence intensities F_{SUM} and DOC ($r_p = 0.636$; $P < .01$). Further studies on grouping of water samples with much higher correlation coefficient R^2 between F_{SUM}

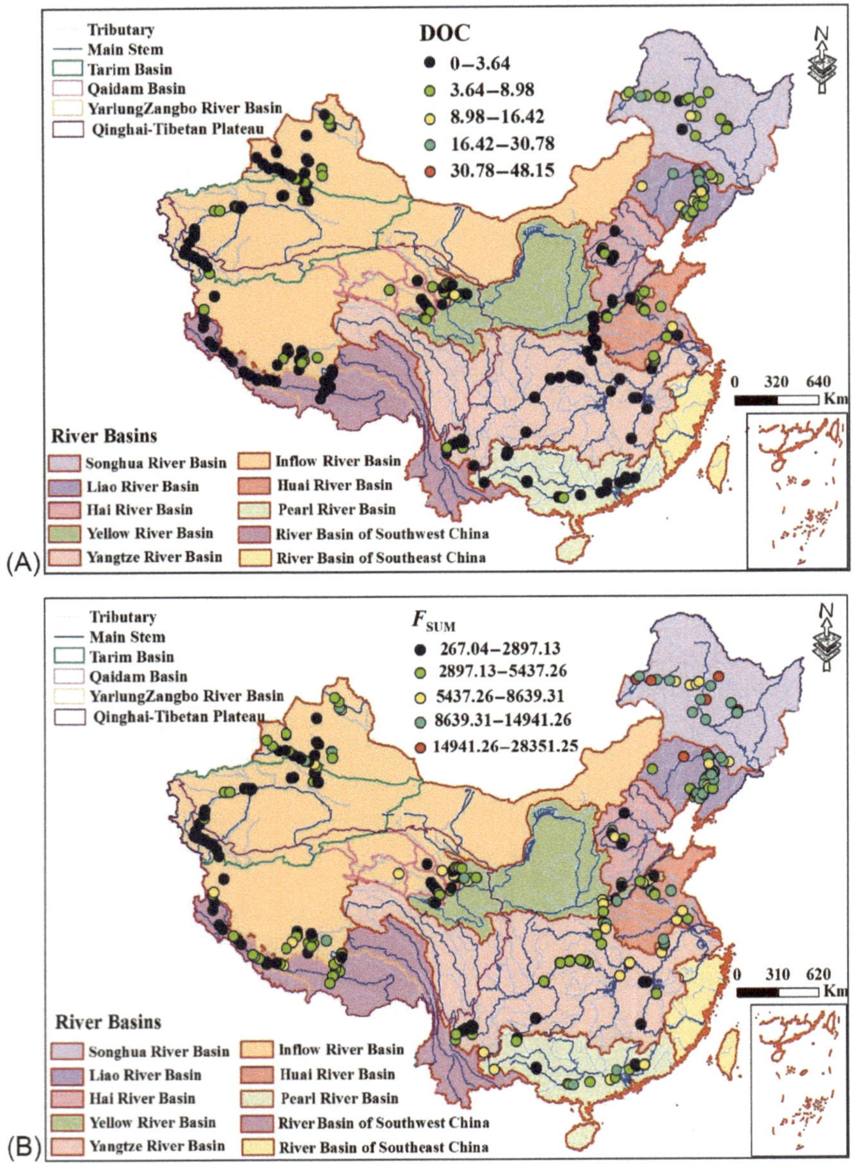

Fig. 6.2 Map of (A) DOC concentrations (mg/L) and (B) the total fluorescence intensities F_{SUM} (nm) from eight river basins across China. Inset shows the South China Sea. The Qinghai-Tibetan Plateau is overlapped with the Yellow River Basin, the Yangtze River Basin, the Inflow River Basin, and the River Basin of Southwest China.

and DOC for each group should be investigated. The total carbon-specific fluorescence intensities F_{SUM}/DOC for 90% or more of all water samples were lower than 3000 L nm mg^{-1}. Based on the minimum numbers of groups with correlation coefficient R^2 higher than 0.8 for each group, 90% or more of all water samples with F_{SUM}/DOC at an interval of 1000 L nm mg^{-1} were divided into three groups. The remaining water samples with F_{SUM}/DOC higher than 3000 L nm mg^{-1} but lower than 8000 L nm mg^{-1} were defined as the fourth groups. Therefore, based on the threshold value for F_{SUM}/DOC, all 301 water samples excluding five outliers with F_{SUM}/DOC higher than 8000 L nm mg^{-1} were grouped into four subsets of F_{SUM} and DOC: group 1, $N=77$, F_{SUM}/DOC < 1000 L nm mg^{-1}; group 2, $N=127$, 1000 L nm mg^{-1} < F_{SUM}/DOC < 2000 L nm mg^{-1}; group 3, $N=70$, 2000 L nm mg^{-1} < F_{SUM}/DOC < 3000 L nm mg^{-1}; and group 4, $N=22$, 3000 L nm mg^{-1} < F_{SUM}/DOC < 8000 L nm mg^{-1}. There were strong positive correlations between the total fluorescence intensities F_{SUM} and DOC concentrations in each group, respectively (group 1, $R^2=0.843$, $P<.01$; group 2, $R^2=0.928$, $P<.01$; group 3, $R^2=0.964$, $P<.01$; and group 4, $R^2=0.953$, $P<.01$). This indicated that the total carbon-specific fluorescence intensities F_{SUM}/DOC varied with water samples even in a specific river basin, which were affected by different hydrologic, geographical, climatic and landscape characteristics, physical/chemical properties, land use/cover, soil type, and pollution. The majority of the F_{SUM}/DOC values for all water samples were present in the range of 1000–2000 L nm mg^{-1} in group 2 ($N=127$). For each river basin, most of the water samples with the lowest total carbon-specific fluorescence intensities F_{SUM}/DOC from the LR Basin were in group 1 attributed to microbial-dominated substances and anthropogenic organic matters exhibiting higher DOC concentrations than F_{SUM} (i.e., decreased carbon-specific fluorescence intensities F_{SUM}/DOC). Furthermore, most of the water samples from the Inflow River Basin were in group 2. For the fourth group, the F_{SUM}/DOC values of the water samples with the least number ($N=22$) were much higher than that of three other groups due to allochthonous-dominated substances exhibiting higher F_{SUM} than DOC concentrations (i.e., increased carbon-specific fluorescence intensities F_{SUM}/DOC). Therefore, it has been demonstrated that the DOC concentrations in river waters across China can be retrieved by the fluorescence intensities F_{SUM} according to the subsets of the F_{SUM}/DOC groups.

Summary

1. Absorption and fluorescence can be used to characterize CDOM source and quality in river waters. Five fluorescent regions $F_{R(1-5)}$ were obtained by EEM-FRI method, and weak correlations between fluorescence intensities $F_{R(1-5)}$, F_{SUM}, and DOC concentrations for all river water samples were observed, respectively.
2. The DOC concentrations and the fluorescent regions have site-specific properties for river water samples due to the effect of various sources of organic pollution (i.e., urban sewage, industrial, and agricultural sources) and land use/cover.
3. The average fluorescence intensities of the five fluorescent components and F_{SUM} differed under spatial variation among the different river basins: the fulvic-like F_{R3} and the humic-like F_{R5} showed a strong positive linear relationship, indicating that the two components R3 and R5 may originate from similar sources.
4. River water samples can be divided into different groups based on the threshold values for F_{SUM}/DOC and strong positive correlations between FSUM and DOC were observed in each group. The riverine DOC concentrations on large scale across China can be estimated directly from FDOM properties.

Profiling natural organic matter in lakes

DOM is considered the largest pool of organic matter in natural waters, accounting for >90% of the total organic matter [29, 30]. Lake is the most important part of inland waters. As an optically active substance, CDOM absorption properties are significantly affected by several factors, including DOM concentration and its chemical composition [31], photoinduced and microbial processes in aquatic environments, seasonal conditions [32], and salinity [33]. Studies are needed for characterizing CDOM absorption and fluorescence characteristics in inland saline and fresh lakes [34, 35]. EEMs fluorescence techniques can be applied to large continental-scale examinations of CDOM sources in different types of inland lakes (freshwater and saline). This is crucial to quantify the role of inland saline waters and fresh lakes to the global carbon cycle [30, 34].

Evaluation of DOM sources in lakes

Characterization of CDOM by EEM-PARAFAC

A total of 127 water samples were collected from 19 lakes across Northeast China in September 2012 and September 2015. The four fluorescent components (C1–C4) (Table 6.3) and the corresponding excitation and emission loadings of lake CDOM were validated by EEM-PARAFAC. The first fluorescent component C1 was similar to the terrestrial humic-like fluorescence peak A [37]. The second fluorescent component (C2) was in accord with the tryptophan-like peak T. The C3 component was a blue-shifted humic-like component similar to that of C1. The C4 component is similar to that of the tyrosine-like component B. In particular, the fluorescence peak with a red-shifted excitation wavelength (Ex/Em = 290/306 nm) was close to the first-order Rayleigh scatter peak because of the strong scatter in some colloidal water samples from brackish water lakes. Therefore, the component C4 was not included in subsequent analyses.

The fluorescence indices FI_{370} (<1.4) and FI_{310} (<0.8) in most of the tested water bodies revealed that CDOM sources were derived from terrestrial humic-like substances rather than autochthonous substances. There were no significant differences in the two fluorescence indices FI_{370} and FI_{310} between freshwater and brackish water lakes (*t*-test, $P > .05$). We can speculate there may be no difference in CDOM sources between freshwater and brackish water lakes. The average maximum fluorescence intensity (F_{max}) for the three fluorescent components C1–C3 in CDOM was different from lake to lake. The average fluorescence intensities were highly dominated by the humic-like component C1 for most of the lakes in this study. The great difference in distribution of the three fluorescent components may result from large variations in water quality among the lakes. There was a positive linear correlation between the components C1 and C3 ($R^2 = 0.68$, $P < .01$). This result indicated that the two humic-like components were derived from a common source [38]. A positive linear relationship was also found between FI_{370} and FI_{310} ($R^2 = 0.67$, $P < .01$), showing that the FI_{370} and FI_{310} represented similar indication in terms of CDOM sources. The positive linear correlations between the fluorescent component C1

Table 6.3 Positions of fluorescence maximum peaks of the four components identified by PARAFAC modeling in this study compared to those previously identified.

Components	Ex_{max} (nm)	Em_{max} (nm)	Description and source	Label 1	Label 2
C1	230 (300)	425	Marine humic-like (phytoplankton degradation)	M	6
C2	255 (350)	460	Terrestrial humic-like	A and C	1 and 4
C3	225 (290)	360	Autochthonous tryptophan-like	T	
C4	220 (275)	320	Autochthonous tyrosine-like	B	8

Secondary excitation maxima are given in brackets.
Label 1—fluorescence peaks named by Coble et al. [36, 37] and Zhang et al. [38, 39]; Label 2—fluorescence peaks named by Stedmon and Markager [40].

and total nitrogen (TN) ($R^2=0.76$ and 0.81, $P<.01$) were also found. These linear relationships indicated that nitrogen in CDOM of lakes was dominated by organic nitrogen, which was derived from the terrestrial humic-like substances.

The EEM spectra of water samples from these lakes revealed that the fluorescence intensity of the tyrosine-like component C4 could be obscured by first-order Rayleigh scatter. This can be attributed to fine colloidal particles (<220 nm) that remained in the water samples after filtering through 0.22 mm millipore cellulose filters. The colloidal particles, observed in transmission electron microscopy (TEM) images, likely result from saturating amounts of small humic-like molecules that are present in the water bodies, which are composed of macromolecular structures that form via self-assembly and interaction with various cations [41, 42]. In high ionic strength environments (i.e., water bodies with high salt concentrations), the humic-like substances tend to form globular aggregates and ring-like structures [42, 43]. These macromolecular structures of humic-like substances can affect the solubility of toxic elements and organic pollutants in waters. The strong Rayleigh scatter in EEM spectra can be removed after a secondary filtration of the brackish water samples through a 0.22 mm millipore cellulose filter.

Seasonal characterization of DOM in lakes

EEM characterization of CDOM

The seasonal characteristics of fluorescent components in CDOM for lakes in the semiarid region of Northeast China were examined by EEM-PARAFAC. Based on the EEM "peak-picking" technique, the key fluorescence peaks can be observed in these lake water samples: two humic-like and two protein-like substances [37, 44] (Table 6.3). The humic-like components consist of a mixture of aromatic and aliphatic compounds with a terrestrial and phytoplankton origin. With respect to the protein-like components (tyrosine- and tryptophan-like substances), they mainly consist of dissolved amino acids. The first fluorescent component (C1) result from biological degradation of humic-like component, comparable to humic-like peaks (M and N) in marine and phytoplankton degradation experiments from inland waters [37, 45]. Component 2 (C2) is consistent with the humic-like peaks (A and C) defined by Coble [37]. Component 3 (C3) resembles the tryptophan-like (T) component as found by Baker et al. [46] and Hudson et al. [8]. Component 4 (C4) is likely related to the tyrosine-like component (B) [8]. C3 and C4 represent autochthonous semilabile CDOM associated with bacteria activity and phytoplankton degradation [44, 47]. In particular, there is a shoulder (i.e., deflection but not peak) at the excitation wavelength 310–330 nm in C4 and 330–340 nm in C3, which may be due to the residual Raman peaks in some water samples. In this study, not all of the four components were present in all lake samples.

Temporal distribution of PARAFAC components

These lakes in Northeast China are endowed with similar geological, hydrological, and climatic settings; thus, it is presumed that similar processes may control the CDOM components. When a model II-ANOVA using season and lake as random effect factors was performed, it shows that the seasonal variability ($F > F_{crit}$, $P < .05$) is higher than between-lake variability. Therefore, the water samples from different lakes for every season were pooled together in order to study the seasonal variation of the fluorescent components. The average fluorescence intensity of the four components had seasonal variation. The total fluorescence intensity also differed under seasonal variation: it was highest in June, decreased in August, and then increased in February and again reduced to the lowest intensity of $1.57 \pm 0.55\,nm^{-1}$ in April 2014. The intensities of the four fluorescent components from the samples collected in June exhibited similar trends to that of the pooled data set. These values were higher than the seasonal average, except for C2 ($0.32 \pm 0.06\,nm^{-1}$). This can be explained by enhanced activities from plant degradation and microbial activities, but fewer terrestrial substances were imported to the waterbodies in June; therefore, the fluorescence intensity of C2 was lower than the seasonal average. Compared to the fluorescence intensity in June, the three fluorescence intensities from the samples collected in August were reduced. Particularly, the fluorescence intensities of two protein-like components (C3 and C4) showed an obvious difference. This can be attributed to substantially increased precipitation up to 180 mm from June to August 2013; therefore, floods occurred when rainfall continued to increase in August. Gradually, DOM contained in terrestrial CDOM was flushed by rainfall to the lakes; therefore, the C2 ($0.42 \pm 0.05\,nm^{-1}$) fluorescence intensity became higher. In accordance to Cheng and coworkers, the rainwater CDOM for this study was largely characterized by protein-like components [48]. The fluorescence intensity of the rainwater CDOM was very weak, and the rainwater CDOM also had a much lower humic-like concentration. The intensities of the other three components decreased because of dilution resulting from heavy rain and relatively weak microbial decomposition of plants. The ice CDOM samples collected from the under-ice surface water in February 2014 were characterized by the tyrosine-like component (C4). However, the underlying lake CDOM included both humic-like (C1 and C2) and protein-like (C3 or C4) components. When the ice in the lakes melted during warming weather and biological degradation took place, and human activity was weak, the lake CDOM was diluted by the ice meltwater, leading to a decrease in the fluorescence intensity, reaching its lowest value in early spring.

DOM in saline and freshwater lakes

FRI-based CDOM fluorescent components

Inland waters account for a significant portion of the global freshwater storage [49], and in China saline lakes comprise a large share of all inland waters [50]. FRI technique was used to identify the FDOM components in 234 lakes with different salinity and elevation across 5 lake regions in China. Based on the broad regional variations of landforms and climate characteristics, the lakes in China have been grouped into five regions: Northeastern lake region (NLR), Inner Mongolia-Xinjiang lake region (MXR), Tibetan-Qinghai Plateau lake region (TQR), Yungui Plateau lake region (YGR), and Eastern lake region (ELR) [50].

Among the five lake regions of China, DOC concentration ranged from 5.25 to 25 $mg\,C\,L^{-1}$, with a mean DOC concentration of 27.4 and 6.68 $mg\,C\,L^{-1}$ for saline and freshwater lakes, respectively. The DOC concentration in NLR, MXR, and TQR region lakes were higher than that of lakes from the ELR and YGR regions. The $a_{CDOM}(254)$ of saline lakes were significantly different from that of fresh lakes ($P < .01$). When all five lake regions were considered together, the $a_{CDOM}(254)$ in saline waters were much higher than that of the fresh waters in the NLR, MXR, and TQR. Another interesting result of this study is that, for the saline lakes in the NLR, MXR, and TQR region, the $a_{CDOM}(254)$ decreased significantly with increased lakes elevation. The diversity of $a_{CDOM}(254)$ may be due to the following factors. First, the CDOM absorption is dependent on the total DOM concentration in the lake waters. The regional hydrogeological conditions and the diversity of climatic situations additionally affect the variability of DOM in lake waters. Because of the prevailing dry climate in the areas surrounding most of the saline lakes, high rates of evaporation are likely to occur and that may have contributed to the high DOM concentrations measured in these lakes. Second, the saline waters are generally terminal lakes in semiarid or arid regions, and consequently DOM from terrestrial sources accumulates in these lakes (and is not exported downstream). EEM fluorescence spectra from Lake Seli Co (TQR), Lake Aibi (MXR), and Lake Khanka (NLR) were selected as representative of lakes in these regions. The total fluorescence intensity (F_T) ranged from 3.78×10^{10} to 4.55×10^{12} nm for all water samples, and the F_T in saline waters (4.81×10^{11}) was higher than that of the fresh waters in lakes (3.94×10^{11}).

In both saline and freshwater lakes in the different regions, the F_5 (humic-like) and F_3 (fulvic-like) components were predominant in FDOM. Furthermore, the percentage contribution of F_5 to F_T in saline lakes was higher than that of freshwater lakes in the NLR, MXR, and TQR regions (in fresh water, 64% and in saline water, 68%). The opposite situation was observed for the F_3 fluorescent component (in fresh water, 25% and in saline water, 21%). These results indicated that the fluorescence intensity of the five components and their relative contribution to the total fluorescence intensity differed between saline and freshwater lakes across the five lake regions of China.

However, the average humification index (HIX) for freshwater lakes (8.45) was nearly twofold higher than that for saline waters (4.39). The PCA results for saline waters indicated that the fulvic-like (F_3) and humic-like (F_5) components were predominant in FDOM for saline waters. With increased lake altitude, the contribution of autochthonous substances increases in saline waters. The PCA score of the fresh waters indicated that the fluorescent CDOM varied with different hydrogeological conditions within individual lake regions, and that the allochthonous substances (F_3 and F_5) were predominant in the terminal lakes of NLR and MXR regions, while the more autochthonous substances (F_1, F_2, and F_4) were produced in lakes of the YGR and TQR regions.

CDOM sources for saline and fresh waters

In both saline and freshwater lakes, correlations among the five fluorescent components and with other optical parameters (the absorption coefficient at 254 nm as $a_{CDOM}(254)$, HIX and humic-like fluorescence intensity indicated as $F_n(355)$) showed strong positive linear relationships. For instance, there is a positive linear correlation between F_1 and F_4 (in fresh water, slope = 3.50, $R^2 = 0.93$ and in saline water, slope = 4.29, $R^2 = 0.71$), suggesting that parts of the tyrosine-like component F_1 and microbial protein-like component F_4 are likely from the same autochthonous sources. Similar results were also found between F_3 and F_5 (in fresh water, slope = 2.04, $R^2 = 0.89$ and in saline water, slope = 2.96, $R^2 = 0.85$), indicating that the fulvic-like components F_3 and the humic-like components F_5 may originate from similar allochthonous sources. Consequently, organic materials (autochthonous and allochthonous) progressively accumulate in these lakes along with salinity, which may have contributed to the highest concentrations of CDOM and FDOM measured in the saline lakes [51, 52]. Strong positive correlations between the tryptophan-like F_2 and fulvic-like F_3 components were also observed in saline waters (slope = 0.03, $R^2 = 0.67$) and freshwater lakes (slope = 0.09, $R^2 = 0.54$), indicating that part of the tryptophan-like fluorophore may have originated from the fulvic-like components, and that the ratio of tryptophan-like component to fulvic-like component was higher in fresh waters than in saline waters. Moreover, a positive correlation between $a_{CDOM}(254)$ and $F_n(355)$ was also found for saline waters (slope = 0.27, $R^2 = 0.54$) and fresh waters (slope = 0.18, $R^2 = 0.61$) individually. These results suggest that the CDOM was dominated with allochthonous substances in both saline waters and fresh waters. Long periods of exposure to sunlight have likely increased the extent of photochemical oxidation processes in saline waters. The photoproduction process significantly increases with salinity in natural waters [53, 54]. The results obtained in this study indicate that the linkage between the FDOM components and CDOM absorption coefficients in saline waters and fresh waters can be very complex because of various hydro-geographical and climatic factors, including precipitation, salinity, wastewater discharge, and trophic status of lakes.

As shown in Fig. 6.3, a good correlation was also found between elevation and $a_{CDOM}(254)$ ($R^2 = 0.51$) in saline lakes (Fig. 6.3A). These trends indicate that the CDOM absorption was lower in high-altitude than in lower-altitude lakes, and that the molecular weight and aromaticity of CDOM decrease with increasing lakes elevation. There was also a strong correlation between the humic-like F_5 and elevation

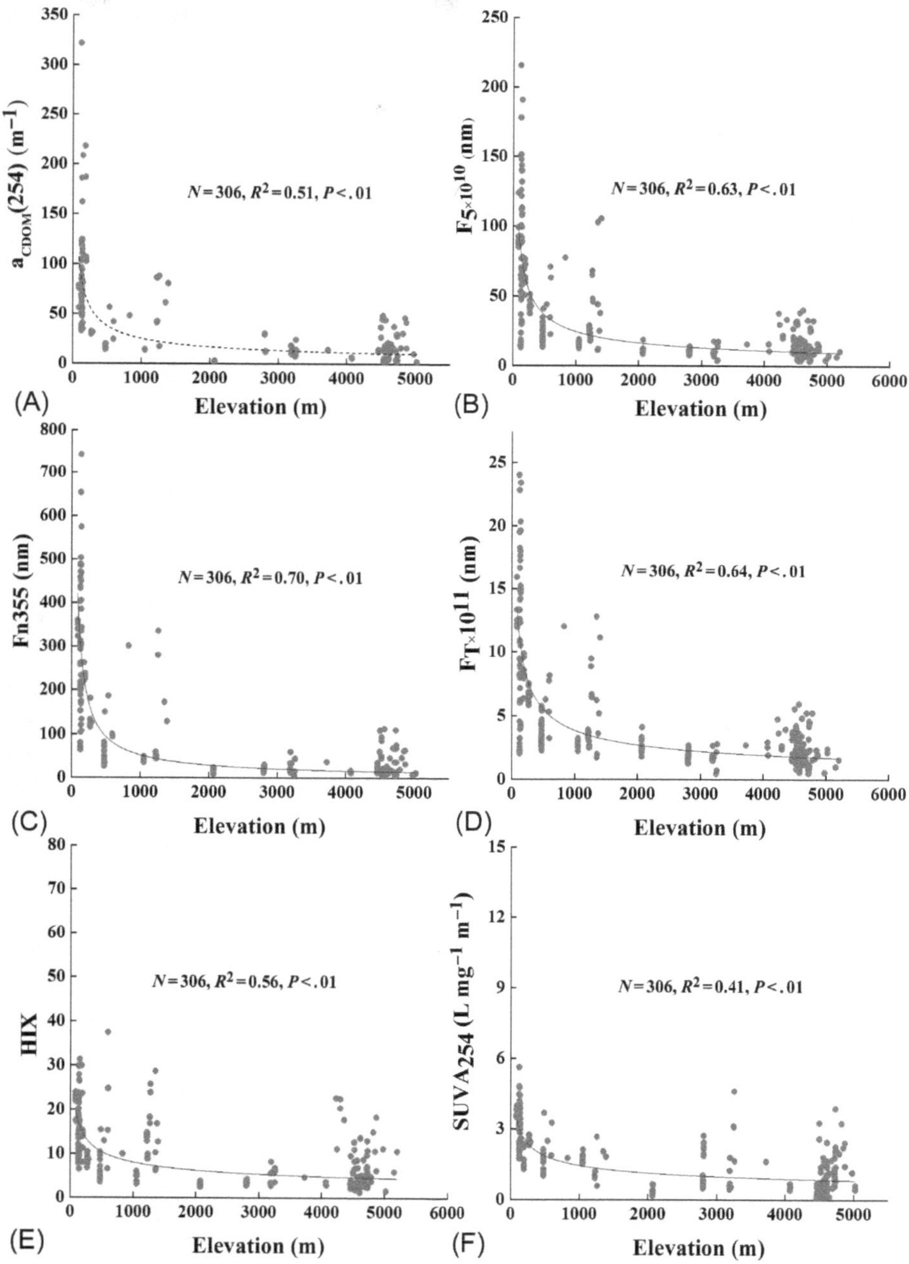

Fig. 6.3 Correlation between elevation and CDOM absorption and FRI-EEM extracted indices for saline waters, (A) $a_{CDOM}(254)$ vs elevation, (B) F_5 vs elevation, (C) $F_n(355)$ vs elevation, (D) F_T vs elevation, (E) HIX and elevation, and (F) $SUVA_{254}$ vs elevation.

($R^2=0.63$) (Fig. 6.3B). Meanwhile, a similar strong correlation was found between the F_T and elevation ($R^2=0.64$) (Fig. 6.3D), suggesting that humic-like materials were predominant in the FDOM in saline waters and decrease with increasing altitude. This interpretation would be consistent with the moderate negative correlation between elevation and HIX ($R^2=0.56$) (Fig. 6.3E), and the strong correlation between elevation and $F_n(355)$ ($R^2=0.70$) (Fig. 6.3C), indicating that the concentration of humic-like substances and the contribution of allochthonous sources of CDOM seem to decline in the high-altitude lakes due to possible degradation processes.

Summary

1. The CDOM components in lakes were affected by spatial variation in land cover and pollution sources in hypereutrophic brackish water lakes. The spatial variation in land cover may affect terrestrially derived humic-like acids in CDOM.
2. There was a difference in F_{max} distribution for the fluorescent components C1–C3 from agricultural sources and from industrial or domestic wastewater in eutrophic brackish water lakes.
3. The average fluorescence intensity of the four components C1–C4 in CDOM differed between seasons from June to August 2013 and from February to April 2014.
4. The EEM-PARAFAC technique can be used to evaluate the seasonal dynamics of CDOM fluorescent components for inland waters in the semiarid regions of Northeast China, and to quantify CDOM components for other waters with similar environmental conditions.
5. The low molecular weight CDOM fractions were relatively more abundant in saline than in freshwater lakes, and the photooxidation process is much stronger in high-altitude lakes. The EEM-FRI is useful to differentiate between the CDOM sources and composition in saline and fresh waters.

Profiling natural organic matter in reservoirs and urban waters

Reservoirs of inland water bodies play an important role in transportation, transformation, and storage of large amounts of carbon from terrestrial-derived substances [34]. On the other hand, these reservoirs are disturbed by anthropogenic activities (i.e., agricultural, industrial, and household organic pollutants), which also disturb the carbon balance [55]. An increasing number of reservoirs and dams are being constructed worldwide; suggesting that the influence of reservoirs on CDOM absorption cannot be neglected. Urban water bodies and reservoirs positively affect people welfare in urban communities. However, residential/commercial wastewater, industrial discharge, automobiles, garbage, and polluted storm water runoff from urban landscapes are discharged into urban water bodies due to a rapid urbanization and industrial development. As such, the urban surface water quality is being severely deteriorated, resulting in variations in CDOM concentrations and composition. The EEM fluorescence fingerprint has been used to distinguish the source of DOM in reservoirs, being also applied to retrieve information on CDOM pollution sources in urban water bodies [56]. This EEM tool will help to understand the general distribution of CDOM in China, which is beneficial for further studying the carbon cycle in aquatic systems in China and evaluating the amount of dissolved carbon storage for inland waters.

Characterization of DOM in reservoirs

CDOM absorption and trophic states

China is the greatest dam-building country in the world and has more than 87,000 reservoirs with a total water storage capacity of 706 billion m^3 [57]. Nationwide, reservoirs account for a significant amount of the total urban drinking water sources. Water flowing into urban water bodies is mainly discharged underground, as tap water, rainfall, and sewage effluents [56]. In many parts of the country, however, the water supply is being threatened by eutrophication. In addition to nutrients availability and primary productivity in aquatic systems, the trophic state can be affected by environmental pollution from anthropogenic activity. The FDOM has been widely used to trace CDOM sources, distribution, and composition [58]. Particularly, anthropogenic contamination may result in a high contribution of tryptophan-like fractions [59].

There was a wide range of variation in DOC concentrations of 109 reservoirs, which indicated the complex aquatic systems in various regions of China [60]. The average DOC concentration ranged from 2.61 to 7.20 mg/L, and the reservoirs with highest and lowest DOC contents were in NLR and YGR, respectively. The DOC concentration was linked to the climate and regional landscape characteristics [61]. The average CDOM absorbance of samples in NLR, MXR, YGR, ELR, and TQR (from highest to lowest) at 355 nm was 4.07 ± 2.31, 2.68 ± 3.04, 2.41 ± 1.41, 2.22 ± 1.51, and $0.79 \pm 0.67\,m^{-1}$, respectively. In terms of reservoirs, the CDOM absorbance varied in the order: Bajiahu reservoir in MXR ($12.11\,m^{-1}$), Kulipao reservoir in NLR ($10.34\,m^{-1}$), and Yamdrok Lake ($0.31\,m^{-1}$). Li et al. [62] reported that the $a_{CDOM}(355)$ at STKM reservoir in the NLR ranged from 5.32 to $11.66\,m^{-1}$. These results are in agreement and comparable to those reported in the previous studies [45, 63, 64]. The reason for these variations in CDOM absorption among five lake regions is accountable to the highly complex factors that affect the concentration of CDOM, such as the geographical, environmental, and human-induced activities, as well as climate conditions.

With increasing trophic status, CDOM absorption increased significantly ($P < .05$), with the average $a_{CDOM}(355)$ ranging from 1.47 to $2.75\,nm^{-1}$, and further to $4.21\,nm^{-1}$ for oligotrophic, mesotrophic, and eutrophic reservoirs, respectively. With increasing eutrophication, the influence of autochthonous sources also increases [38].

Meanwhile, additional environmental influences would affect the terrestrial input and the activities of phytoplankton.

EEM characterization of CDOM

Examples of EEM fluorescence spectra from Yamdrok Lake (oligotrophic), Xiaolangdi Reservoir (mesotrophic), Hamatong Reservoir (eutrophic), and Xiashan Reservoir (hyper-eutrophic) were selected as representative of reservoirs in various trophic gradients. The total fluorescence intensity (F_T) ranged from 5.48×10^{10} to 4.51×10^{12} nm for all water samples, and the average F_T for oligotrophic, mesotrophic, and eutrophic status was 1.32×10^{11} nm, 2.64×10^{11} nm, and 6.04×10^{11} nm, respectively (Fig. 6.4A).

The relative contribution of individual components to total fluorescence intensity differed among water reservoirs with different trophic status (Fig. 6.4B). It was found that F_5 (humic-like) and F_3 (fulvic-like) components were predominant in FDOM for all types of reservoirs. With increased trophic status, the percentage of F_1 and F_4 decreased, while the percentage of F_3 and F_5 increased significantly ($P < .01$). The average HIX from oligotrophic, mesotrophic, and eutrophic reservoirs was in the order of 3.26, 8.08, and 8.76, respectively. Using a comprehensive data set from several lakes in China, Zhang et al. [65] noted a parallel increase between CDOM absorption and trophic state. The percentages distribution of the individual fluorescent components and the increasing proportion of allochthonous fluorescent components (F_3 and F_5) in reservoirs indicated that the CDOM fluorescence from reservoirs was dominated by the humic-like components, and allochthonous input from the catchment formed an important contribution to total fluorescence with the increasing of trophic states. With the potential establishment of the distribution of CDOM fluorescence with handheld spectrometers, and land-observing satellite sensors [66, 67], it would be possible to estimate the long-term continuous changes of trophic states of reservoirs over the world.

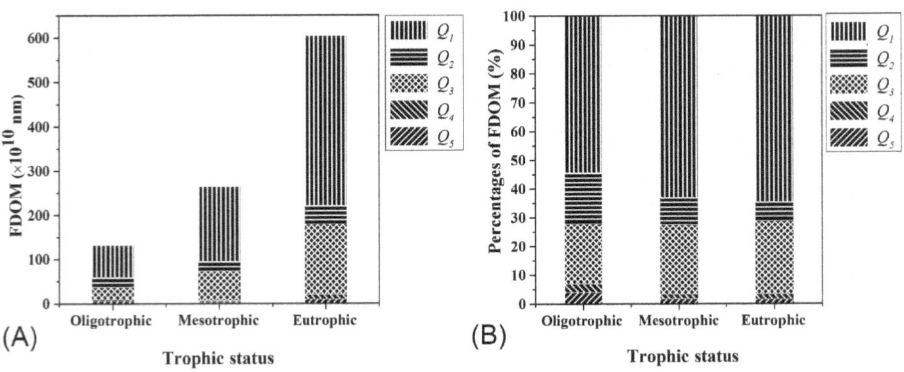

Fig. 6.4 (A) Fluorescence of CDOM for reservoirs with different trophic states and (B) the percentages of individual components.

Characterization of DOM in urban waters

Characterization of CDOM EEM fluorescence spectra

It has been shown that the distribution of DOC concentrations and CDOM absorption in urban water samples from Chinese cities varied from one water body to another as well as geographically [24]. Looking at three provincial capital cities (Shenyang, Changchun, and Harbin) in Northeastern China and two municipalities (Beijing and Tianjin) in the Northern China, it was found that the humic-like peaks (A and C), tryptophan-like peak (T), and tyrosine-like peak (B) were detected in 84 urban water samples [37]. Fluorescence intensity depends on the concentrations of the main fluorophores dissolved in the waters. Five types of urban water bodies were recorded based on the intensities and locations of the marked fluorescence peaks (Fig. 6.5). The distinctly high protein-like intensity in wastewater and sewage is explained by the difference in CDOM quality between natural and sewage water, which is consistent with the findings reported by Ma et al. [68] and Baker [23]. The first fluorescent component (C1) identified by PARAFAC analysis was a humic-like substance with two excitation maxima (230 and 330 nm) at a single emission wavelength (442 nm), which was consistent with the humic-like peaks (A and C) defined by Coble [37]. Component C2 exhibits two excitation maxima (220 and 285 nm) and one emission maximum (348 nm), which is similar to tryptophan-like substances from anthropogenic sewage CDOM inputs. The humic-like C1 and tryptophan-like C2 components are similar to those derived from PARAFAC modeling of EEM spectra of different sewage waters [69]. Component C3 is likely a tyrosine-like substance, which was characterized by maxima excitation at 270 nm and emission at 300 nm. Components C2 and C3 represent autochthonous semilabile CDOM associated with bacterial activities or phytoplankton decomposition [44].

Variations in PARAFAC components of CDOM

The spatial changes and physicochemical properties of urban waters have significant effects on CDOM dynamics [70]. The DOC concentrations in urban waters within a given area are associated with relative pollution levels [71]. Water samples with three DOC concentration gradients (type I, very low DOC concentration with the mean value of 5.79 mg/L; type II, intermediate DOC concentration with the mean value of 13.08 mg/L; and type III, very high DOC concentration with the mean value of 21.46 mg/L) were selected from each city to represent typical lightly polluted to severely polluted waters and sewage water. Only type I and II samples were selected in Tianjin because of the absence of a sewage sample (type III). The intensities of fluorescent components C1 and C2 of these urban water samples increased simultaneously with increasing DOC concentrations from types I to III. These results indicate that the fluorescence intensities of the three CDOM components in water samples from the same city differed due to different pollution levels. Fluorescence intensity was attributed to component C2 when the water was severely polluted, as reported by Baker and Inverarity [72], which can obscure humic-like C1 and tyrosine-like C3 components due to the distinctively high tryptophan-like C2 component. Notably, the fluorescence

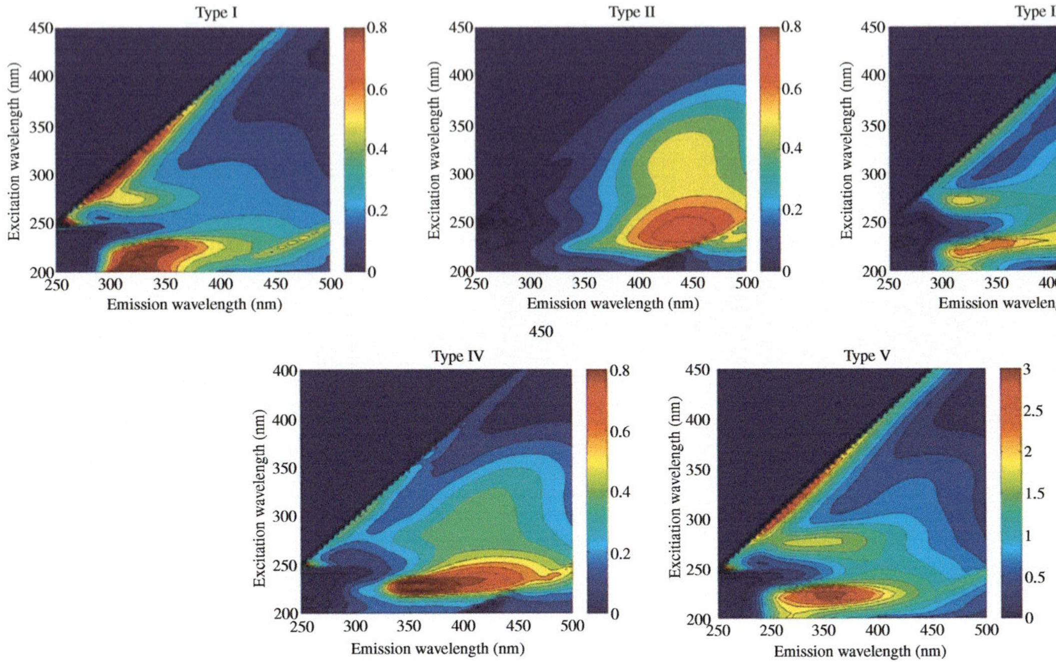

Fig. 6.5 Five fluorescence spectra types in urban waters sampled from cities of Northern and Northeastern China (Raman: nm^{-1}). Type I, mainly from underground and tap waters; type II, derived from terrestrial sources or plant decay; type III, freshly supplied underground water and polluted sewage water from industrial and household sources; type IV, lakes in parks containing treated polluted or untreated wastewater effluent; and type V, typically derived from substances in sewage water.

intensity of C1 (i.e., humic-like component) for the selected samples increased gradually from Tianjin to Changchun and to Harbin, regardless of whether the samples were type I or II. This can be explained by the gradual accumulation of CDOM in water from allochthonous sources originated from landscapes with gradients of soil organic enrichment from Northern to Northeastern China.

Positive correlations between F_{max} (C2) and TN ($R^2=0.78$) were also observed. The tryptophan-like C2 intensity was moderately correlated with NH_4-N ($R^2=0.68$) and COD_{Mn} ($R^2=0.52$), despite the unavailability of data for 22 water samples collected at Harbin. The positive correlations between the fluorescence intensities F_{max} (C2) and traditional water quality parameters indicated that the protein-like C2 can be used as a quick-and-easy detection method to evaluate urban water pollution levels across China rather than relying on traditional water testing approaches, which are time and labor consuming.

Summary

1. The significant increase in humic-like (F_5) and fulvic-like (F_3) fluorescent components follows the increase of trophic state in reservoirs, based on EEM-FRI analyses.
2. Average fluorescence intensities of the three CDOM components varied spatially and according to the pollution level in cities of Northern and Northeastern China.
3. EEM-PARAFAC could be used to characterize the different fluorescent components in urban waters, being also useful to identify the origin and to evaluate the dynamics of CDOM fluorescent components in urban waters.

References

[1] J.J. Cole, Y.T. Prairie, N.F. Caraco, W.H. McDowell, L.J. Tranvik, R.G. Striegl, C. M. Duarte, P. Kortelainen, J.A. Downing, J.J. Middelburg, J. Melack, Plumbing the global carbon cycle: integrating inland waters into the terrestrial carbon budget, Ecosystems 10 (1) (2007) 171–184.

[2] K. Song, Z. Wen, Y. Xu, H. Yang, L. Lyu, Y. Zhao, C. Fang, Y. Shang, J. Du, Dissolved carbon in a large variety of lakes across five limnetic regions in China, J. Hydrol. 563 (2018) 143–154.

[3] T.B. Parr, S.P. Inamdar, M.J. Miller, Overlapping anthropogenic effects on hydrologic and seasonal trends in DOC in a surface water dependent water utility, Water Res. 148 (2019) 407–415.

[4] S. Larsen, T. Andersen, D.O. Hessen, Climate change predicted to cause severe increase of organic carbon in lakes, Glob. Chang. Biol. 17 (2) (2011) 1186–1192.

[5] D.T. Monteith, J.L. Stoddard, C.D. Evans, H.A. de Wit, M. Forsius, T. Hogasen, A. Wilander, B.L. Skjelkvale, D.S. Jeffries, J. Vuorenmaa, B. Keller, J. Kopacek, J. Vesely, Dissolved organic carbon trends resulting from changes in atmospheric deposition chemistry, Nature 450 (7169) (2007) 537–U9.

[6] N. Roulet, T.R. Moore, Browning the waters, Nature 444 (7117) (2006) 283–284.

[7] F. Worrall, T. Burt, R. Shedden, Long term records of riverine dissolved organic matter, Biogeochemistry 64 (2) (2003) 165–178.

[8] N. Hudson, A. Baker, D. Reynolds, Fluorescence analysis of dissolved organic matter in natural, waste and polluted waters—a review, River Res. Appl. 23 (6) (2007) 631–649.

[9] W. Chen, P. Westerhoff, J.A. Leenheer, K. Booksh, Fluorescence excitation-emission matrix regional integration to quantify spectra for dissolved organic matter, Environ. Sci. Technol. 37 (2003) 5701–5710.
[10] L. Guo, R.W. Macdonald, Source and transport of terrigenous organic matter in the upper Yukon River: evidence from isotope (delta C-13, delta C-14, and delta N-15) composition of dissolved, colloidal, and particulate phases, Glob. Biogeochem. Cycles 20 (2) (2006) GB2011.
[11] E. Mayorga, A.K. Aufdenkampe, C.A. Masiello, A.V. Krusche, J.I. Hedges, P.D. Quay, J. E. Richey, T.A. Brown, Young organic matter as a source of carbon dioxide outgassing from Amazonian rivers, Nature 436 (7050) (2005) 538–541.
[12] K. Watanabe, T. Kuwae, How organic carbon derived from multiple sources contributes to carbon sequestration processes in a shallow coastal system? Glob. Chang. Biol. 21 (7) (2015) 2612–2623.
[13] Z. Xu, Y.J. Xu, Dissolved carbon transport in a river-lake continuum: a case study in a subtropical watershed, USA, Sci. Total Environ. 643 (2018) 640–650.
[14] S.M. Mitrovic, D.S. Baldwin, Allochthonous dissolved organic carbon in river, lake and coastal systems: transport, function and ecological role, Mar. Freshw. Res. 67 (9) (2016) I–IV.
[15] J.H. Gao, J. Jia, A.J. Kettner, F. Xing, Y.P. Wang, X.N. Xu, Y. Yang, X.Q. Zou, S. Gao, S. Qi, F. Liao, Changes in water and sediment exchange between the Changjiang River and Poyang Lake under natural and anthropogenic conditions, China, Sci. Total Environ. 481 (2014) 542–553.
[16] T.J. Battin, S. Luyssaert, L.A. Kaplan, A.K. Aufdenkampe, A. Richter, L.J. Tranvik, The boundless carbon cycle, Nat. Geosci. 2 (9) (2009) 598–600.
[17] L. Ran, X.X. Lu, H. Sun, J. Han, R. Li, J. Zhang, Spatial and seasonal variability of organic carbon transport in the Yellow River, China, J. Hydrol. 498 (2013) 76–88.
[18] P.J. Mulholland, J.A. Watts, Transport of organic-carbon to the oceans by rivers of North America—a synthesis of existing data, Tellus 34 (2) (1982) 176–186.
[19] W. Ludwig, J.L. Probst, S. Kempe, Predicting the oceanic input of organic carbon by continental erosion, Glob. Biogeochem. Cycles 10 (1) (1996) 23–41.
[20] Z. Tao, Q. Gao, W. Guo, Z. Wang, Y. Zhang, C. Xie, X. Huang, H. Zhong, Temporal and spatial variability of dissolved organic carbon concentration in the Xijiang River, South China, J. Mt. Sci. 8 (5) (2011) 694–703.
[21] R.G.M. Spencer, B.A. Pellerin, B.A. Bergamaschi, B.D. Downing, T.E.C. Kraus, D. R. Smart, R.A. Dahgren, P.J. Hernes, Diurnal variability in riverine dissolved organic matter composition determined by in situ optical measurement in the San Joaquin River (California, USA), Hydrol. Process. 21 (23) (2007) 3181–3189.
[22] Y. Zhao, K. Song, Y. Shang, T. Shao, Z. Wen, L. Lv, Characterization of CDOM of river waters in China using fluorescence excitation-emission matrix and regional integration techniques, J. Geophys. Res. Biogeosci. 122 (8) (2017) 1940–1953.
[23] A. Baker, Fluorescence excitation-emission matrix characterization of some sewage-impacted rivers, Environ. Sci. Technol. 35 (5) (2001) 948–953.
[24] Y. Zhao, K. Song, S. Li, J. Ma, Z. Wen, Characterization of CDOM from urban waters in Northern-Northeastern China using excitation-emission matrix fluorescence and parallel factor analysis, Environ. Sci. Pollut. Res. 23 (15) (2016) 15381–15394.
[25] H. Sun, J. Han, D. Li, X. Lu, H. Zhang, W. Zhao, Organic carbon transport in the Songhua River, NE China: influence of land use, Hydrol. Process. 31 (11) (2017) 2062–2075.
[26] J. Shi, H. Cui, L. Jia, L. Qiu, Y. Zhao, Z. Wei, J. Wu, X. Wen, Bioavailability of riverine dissolved organic carbon and nitrogen in the Heilongjiang watershed of northeastern China, Environ. Monit. Assess. 188 (2) (2016) 113.

[27] Y. Zhao, K. Song, Relationships between DOC and CDOM based on the total carbon-specific fluorescence intensities for river waters across China, J. Geophys. Res. Biogeosci. 123 (8) (2018) 2353–2361.

[28] Y. Zhao, K. Song, Z. Wen, C. Fang, Y. Shang, L. Lv, Evaluation of CDOM sources and their links with water quality in the lakes of Northeast China using fluorescence spectroscopy, J. Hydrol. 550 (2017) 80–91.

[29] J. Para, P.G. Coble, B. Charrière, M. Tedetti, C. Fontana, R. Sempéré, Fluorescence and absorption properties of chromophoric dissolved organic matter (CDOM) in coastal surface waters of the northwestern Mediterranean Sea, influence of the Rhône River, Biogeosciences 7 (12) (2010) 4083–4103.

[30] P.G. Coble, Marine optical biogeochemistry: the chemistry of ocean color, ChemInform 38 (20) (2007) 402–418.

[31] C. Minero, S. Chiron, G. Falletti, V. Maurino, E. Pelizzetti, R. Ajassa, M.E. Carlotti, D. Vione, Photochemincal processes involving nitrite in surface water samples, Aquat. Sci. 69 (1) (2007) 71–85.

[32] G.C. Shank, R. Lee, A. Vähätalo, R.G. Zepp, E. Bartels, Production of chromophoric dissolved organic matter from mangrove leaf litter and floating Sargassum colonies, Mar. Chem. 119 (1) (2010) 172–181.

[33] S. Singh, E.J. D'Sa, E.M. Swenson, Chromophoric dissolved organic matter (CDOM) variability in Barataria Basin using excitation-emission matrix (EEM) fluorescence and parallel factor analysis (PARAFAC), Sci. Total Environ. 408 (16) (2010) 3211–3222.

[34] L.J. Tranvik, J.A. Downing, J.B. Cotner, S.A. Loiselle, R.G. Striegl, T.J. Ballatore, P. Dillon, K. Finlay, K. Fortino, L.B. Knoll, P.L. Kortelainen, T. Kutser, S. Larsen, I. Laurion, D.M. Leech, S.L. McCallister, D.M. McKnight, J.M. Melack, E. Overholt, J.A. Porter, Y. Prairie, W.H. Renwick, F. Roland, B.S. Sherman, D.W. Schindler, S. Sobek, A. Tremblay, M.J. Vanni, A.M. Verschoor, E. von Wachenfeldt, G. A. Weyhenmeyer, Lakes and reservoirs as regulators of carbon cycling and climate, Limnol. Oceanogr. 54 (6) (2009) 2298–2314.

[35] C.M. Duarte, Y.T. Prairie, C. Montes, J.J. Cole, R. Striegl, J. Melack, J.A. Downing, CO2 emissions from saline lakes: a global estimate of a surprisingly large flux, J. Geophys. Res. Biogeosci. 113 (G4) (2008) G04041.

[36] P.G. Coble, C.E. Del Castillo, B. Avril, Distribution and optical properties of CDOM in the Arabian Sea during the 1995 Southwest Monsoon, Deep-Sea Res. II Top. Stud. Oceanogr. 45 (10–11) (1998) 2195–2223.

[37] P.G. Coble, Characterization of marine and terrestrial DOM in seawater using excitation-emission matrix spectroscopy, Mar. Chem. 51 (4) (1996) 325–346.

[38] Y. Zhang, E. Zhang, Y. Yin, M.A. van Dijk, L. Feng, Z. Shi, M. Liu, B. Qin, Characteristics and sources of chromophoric dissolved organic matter in lakes of the Yungui Plateau, China, differing in trophic state and altitude, Limnol. Oceanogr. 55 (6) (2010) 2645–2659.

[39] Y. Zhang, Y. Yin, X. Liu, Z. Shi, L. Feng, M. Liu, G. Zhu, Z. Gong, B. Qin, Spatial-seasonal dynamics of chromophoric dissolved organic matter in Lake Taihu, a large eutrophic, shallow lake in China, Org. Geochem. 42 (5) (2011) 510–519.

[40] C.A. Stedmon, S. Markager, Tracing the production and degradation of autochthonous fractions of dissolved organic matter by fluorescence analysis, Limnol. Oceanogr. 50 (5) (2005) 1415–1426.

[41] S.C.B. Myneni, J.T. Brown, G.A. Martinez, W. Meyer-Ilse, Imaging of humic substance macromolecular structures in water and soils, Science 286 (5443) (1999) 1335–1337.

[42] W.C. Chin, M.V. Orellana, P. Verdugo, Spontaneous assembly of marine dissolved organic matter into polymer gels, Nature 391 (6667) (1998) 568–572.

[43] C. Chen, X. Wang, H. Jiang, W. Hu, Direct observation of macromolecular structures of humic acid by AFM and SEM, Colloids Surf. A Physicochem. Eng. Asp. 302 (1–3) (2007) 121–125.
[44] C.A. Stedmon, S. Markager, R. Bro, Tracing dissolved organic matter in aquatic environments using a new approach to fluorescence spectroscopy, Mar. Chem. 82 (3–4) (2003) 239–254.
[45] Y. Zhang, M.A. van Dijk, M. Liu, G. Zhu, B. Qin, The contribution of phytoplankton degradation to chromophoric dissolved organic matter (CDOM) in eutrophic shallow lakes: field and experimental evidence, Water Res. 43 (18) (2009) 4685–4697.
[46] A. Baker, D. Ward, S.H. Lieten, R. Periera, E.C. Simpson, M. Slater, Measurement of protein-like fluorescence in river and waste water using a handheld spectrophotometer, Water Res. 38 (12) (2004) 2934–2938.
[47] M. Borisover, Y. Laor, A. Parparov, N. Bukhanovsky, M. Lado, Spatial and seasonal patterns of fluorescent organic matter in Lake Kinneret (Sea of Galilee) and its catchment basin, Water Res. 43 (12) (2009) 3104–3116.
[48] Y.-y. Cheng, W.-d. Guo, A.-m. Long, S.-y. Chen, Study on optical characteristics of chromophoric dissolved organic matter (CDOM) in rainwater by fluorescence excitation-emission matrix and absorbance spectroscopy, Spectrosc. Spectr. Anal. 30 (9) (2010) 2413–2416.
[49] T. Oki, S. Kanae, Global hydrological cycles and world water resources, Science 313 (5790) (2006) 1068–1072.
[50] R. Ma, G. Yang, H. Duan, J. Jiang, S. Wang, X. Feng, A. Li, F. Kong, B. Xue, J. Wu, S. Li, China's lakes at present: number, area and spatial distribution, Sci. China Earth Sci. 54 (2) (2011) 283–289.
[51] K. Song, Z. Wen, Y. Shang, H. Yang, L. Lyu, G. Liu, C. Fang, J. Du, Y. Zhao, Quantification of dissolved organic carbon (DOC) storage in lakes and reservoirs of mainland China, J. Environ. Manag. 217 (2018) 391–402.
[52] N. Catalan, R. Marce, D.N. Kothawala, L.J. Tranvik, Organic carbon decomposition rates controlled by water retention time across inland waters, Nat. Geosci. 9 (7) (2016) 501.
[53] C.L. Osburn, D.W. O'Sullivan, T.J. Boyd, Increases in the longwave photobleaching of chromophoric dissolved organic matter in coastal waters, Limnol. Oceanogr. 54 (1) (2009) 145–159.
[54] M. Nieto-Cid, X.A. Álvarez-Salgado, F.F. Pérez, Microbial and photochemical reactivity of fluorescent dissolved organic matter in a coastal upwelling system, Limnol. Oceanogr. 51 (3) (2006) 1391–1400.
[55] Z.-g. Wang, W.-q. Liu, N.-j. Zhao, H.-b. Li, Y.-j. Zhang, W.-c. Si-Ma, J.-g. Liu, Composition analysis of colored dissolved organic matter in Taihu Lake based on three dimension excitation-emission fluorescence matrix and PARAFAC model, and the potential application in water quality monitoring, J. Environ. Sci. 19 (7) (2007) 787–791.
[56] J. Wu, M.N. Pons, O. Potier, Wastewater fingerprinting by UV-visible and synchronous fluorescence spectroscopy, Water Sci. Technol. 53 (4–5) (2006) 449–456.
[57] J. Yang, X. Yu, L. Liu, W. Zhang, P. Guo, Algae community and trophic state of subtropical reservoirs in southeast Fujian, China, Environ. Sci. Pollut. Res. Int. 19 (5) (2012) 1432–1442.
[58] K. Song, Y. Shang, Z. Wen, P.-A. Jacinthe, G. Liu, L. Lyu, C. Fang, Characterization of CDOM in saline and freshwater lakes across China using spectroscopic analysis, Water Res. 150 (2019) 403–417.
[59] J. Bridgeman, A. Baker, C. Carliell-Marquet, E. Carstea, Determination of changes in wastewater quality through a treatment works using fluorescence spectroscopy, Environ. Technol. 34 (23) (2013) 3069–3077.

[60] Y. Shang, K. Song, Z. Wen, L. Lyu, Y. Zhao, C. Fang, B. Zhang, Characterization of CDOM absorption of reservoirs with its linkage of regions and ages across China, Environ. Sci. Pollut. Res. Int. 25 (16) (2018) 16009–16023.
[61] R. Jiang, R. Hatano, Y. Zhao, K. Kuramochi, A. Hayakawa, K.P. Woli, M. Shimizu, Factors controlling nitrogen and dissolved organic carbon exports across timescales in two watersheds with different land uses, Hydrol. Process. 28 (19) (2014) 5105–5121.
[62] S. Li, J. Zhang, G. Mu, S. Ha, C. Sun, H. Ju, F. Zhang, Y. Chen, Q. Ma, Optical properties of chromophoric dissolved organic matter in the Yinma River watershed and drinking water resource of Northeast China, Pol. J. Environ. Stud. 25 (3) (2016) 1061–1073.
[63] Z.D. Wen, K.S. Song, Y. Zhao, J. Du, J.H. Ma, Influence of environmental factors on spectral characteristics of chromophoric dissolved organic matter (CDOM) in Inner Mongolia Plateau, China, Hydrol. Earth Syst. Sci. 20 (2) (2016) 787–801.
[64] C. Nima, B. Hamre, O. Frette, S.R. Erga, Y.-C. Chen, L. Zhao, K. Sorensen, M. Norli, D.-R. Lu, Q.-G. Xing, Y.-M. Ma, N. Gelsor, J.J. Stamnes, Impact of particulate and dissolved material on light absorption properties in a High-Altitude Lake in Tibet, China, Hydrobiologia 768 (1) (2016) 63–79.
[65] Y. Zhang, Y. Zhou, K. Shi, B. Qin, X. Yao, Y. Zhang, Optical properties and composition changes in chromophoric dissolved organic matter along trophic gradients: implications for monitoring and assessing lake eutrophication, Water Res. 131 (2018) 255–263.
[66] W. Zhu, Q. Yu, Y.Q. Tian, R.F. Chen, G.B. Gardner, Estimation of chromophoric dissolved organic matter in the Mississippi and Atchafalaya river plume regions using above-surface hyperspectral remote sensing, J. Geophys. Res. 116 (C2) (2011) C02011.
[67] T. Kutser, The possibility of using the Landsat image archive for monitoring long time trends in coloured dissolved organic matter concentration in lake waters, Remote Sens. Environ. 123 (2012) 334–338.
[68] H.Z. Ma, H.E. Allen, Y.J. Yin, Characterization of isolated fractions of dissolved organic matter from natural waters and a wastewater effluent, Water Res. 35 (4) (2001) 985–996.
[69] J.L. Vasel, E. Praet, On the use of fluorescence measurements to characterize wastewater, Water Sci. Technol. 45 (4–5) (2002) 109–116.
[70] R. Jaffe, D. McKnight, N. Maie, R. Cory, W.H. McDowell, J.L. Campbell, Spatial and temporal variations in DOM composition in ecosystems: the importance of long-term monitoring of optical properties, J. Geophys. Res. Biogeosci. 113 (G4) (2008) G04032.
[71] D. Florescu, A.M. Iordache, D. Costinel, E. Horj, R.E. Ionete, M. Culea, Validation procedure for assessing the total organic carbon in water samples, Rom. J. Physiol. 58 (1–2) (2013) 211–219.
[72] A. Baker, R. Inverarity, Protein-like fluorescence intensity as a possible tool for determining river water quality, Hydrol. Process. 18 (15) (2004) 2927–2945.

Multidimensional liquid chromatography and capillary electrophoresis coupled to high-resolution detectors applied to complex environmental samples

7

Pedro F. Brandão, Antoine S. Almeida, Armando C. Duarte, Regina M.B.O. Duarte
Department of Chemistry & CESAM, University of Aveiro, Aveiro, Portugal

Introduction

Since the beginning of Analytical Chemistry as a major field of science, the development of increasingly more efficient separation techniques has been one of its main goals. Separation techniques such as one-dimensional liquid chromatography and capillary electrophoresis, 1D-LC, and CE (Fig. 7.1), respectively, when coupled to different detection methods, can give us significant information on the chemical identity of analytes. However, even though these methods possess several advantages, they are still unable to fully resolve all the different coeluting compounds in more complex samples, such as those of environmental matrices. In order to solve this issue, alternative separation methods with higher peak capacity based on more than one separation mechanism have been shaped to resolve as many compounds as possible. Of these, those based on two-dimensional chromatography (2D-LC) have been widely applied, with direct impact on the overall separation efficiency (Fig. 7.1). Still, there are no strict limitations on increasing dimensionality for three or more dimensions, excluding the sequential cumulative dilution of the sample. As depicted in Fig. 7.1, and irrespective of the employed separation method, the chemical knowledge extracted from the analytical process is always strictly reliant on the type of detectors used.

Multidimensional chromatography may appear a recent development in separation technology, however, it was first described in 1944 [1]. More than three decades later, the first comprehensive two-dimensional liquid chromatography (LC $\times$ LC) method was reported by Erni and Frei [2]. The first applications of multidimensional liquid chromatography (MDLC) for environmental research were performed at the end of the 20th century for what was known as column switching mode, widely used for the separation and analysis of pesticides [3–8]. Later, research on MDLC took a slightly different path, where a MDLC system combining a separation and an online reduction column was employed for the determination of nitropolycyclic aromatic

Multidimensional Analytical Techniques in Environmental Research. https://doi.org/10.1016/B978-0-12-818896-5.00007-7

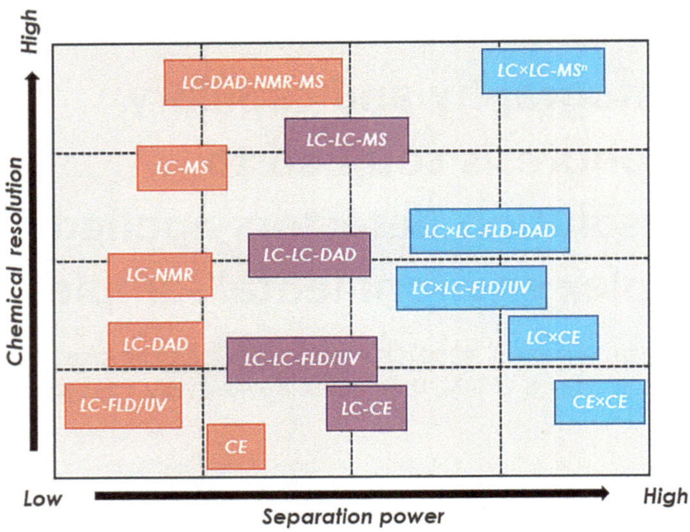

Fig. 7.1 Schematic representation of the range of LC- and CE-based techniques currently employed in environmental and natural products research as a function of their separation power and completeness of chemical information achieved. Acronyms: *CE × CE*, two-dimensional comprehensive capillary electrophoresis; *DAD*, diode array detector; *FLD*, fluorescence detector; *LC*, one-dimensional liquid chromatography; *LC × CE*, comprehensive liquid chromatography coupled to capillary electrophoresis; *LC × LC*, two-dimensional comprehensive LC; *LC-CE*, heart-cutting liquid chromatography coupled to capillary electrophoresis; *LC-LC*, two-dimensional heart-cutting LC; MS^n, mass spectrometry detector; *NMR*, nuclear magnetic resonance spectroscopy; *UV*, ultraviolet detector. Due to limited space, coupling of different types of detector to CE, and their resolution capabilities are not represented.

hydrocarbons [9–12]. Likewise, restricted-access medium (RAM) columns, combined with other separation columns, were used in MDLC for online sample cleanup and concentration [13–15]. However, to the best of our knowledge, the first reports on the use of a comprehensive two-dimensional liquid chromatography system, namely LC × LC, for the analysis of environmental samples were only published in 2006 by Pól and coworkers [16]. In this work, acidic compounds present in atmospheric aerosols were separated using a multidimensional chromatographic system for the first time, proving the capabilities of LC × LC systems to effectively separate complex mixtures that would be too difficult to separate properly using 1D-LC. Also, other analytical techniques, like capillary electrophoresis (CE) have been successfully coupled to an LC system in a multidimensional configuration (LC-CE). These hybrid multidimensional separations exhibit some unique advantages that will be further discussed. Although vastly underreported, two-dimensional electrophoresis (CE-CE) will also be briefly discussed.

When compared to conventional 1D-LC, the addition of an additional separation dimension poses extra complexity that lead to new challenges, including the selection of appropriate chromatographic stationary phases (i.e., separation mechanisms), compatibility of separation modes and mobile phases, suitability and sensitivity of detection techniques, as well as optimization of separation conditions, and data analysis (i.e., application of algorithms for processing 2D chromatograms). The wide research using 2D-LC has also given rise to a range of different concepts, which justified the proposition of new, proper and unambiguous nomenclature and symbols to enable communication between analysts [17]. Undeniably, since the pivotal work of Erni and Frei [2], 2D-LC has emerged as a frontline tool for analysis of samples of wide-ranging complexity in different research fields. In this regard, readers interested in specific application areas are recommended to check the recent reviews and references therein, on the use of 2D-LC in Traditional Chinese Medicines [18], biopharmaceutical analysis and lipidomics [19–21], proteomics [22], food analysis [23], and polymer analysis [24]. In most of the existing research and review works, a great effort has been made to streamline the applicability of 2D-LC to efficiently and successfully resolve the most challenging and complex samples. Recently, two excellent review works were published where a great deal of consideration has been given to the basic principles [21], as well as the latest technical progress, method development, and optimization strategies [19] in 2D-LC, which are essential keys for the design of efficient 2D-LC separation approaches for the targeted and untargeted analysis of complex samples. Also, a review article strictly focusing on LC × LC separation of environmental samples and natural products has been published recently by our research group [25]. Furthermore, some excellent reviews on multidimensional capillary electrophoresis have been published and readers should be referred to them for a more detailed overview on the subject [26–29]. Under this scenario, to advance the frontiers of knowledge within the research field of complex samples separation, it is mandatory to provide the scientific community with an up-to-date and critical assessment on the use of two-dimensional separation strategies, including also two-dimensional electrophoretic separations, coupled to high-resolution detectors for the separation and targeted or untargeted profiling of complex environmental samples. This chapter's main goal is to present the scientific community with a different perspective on the benefits of using these multidimensional strategies for gaining new insights into the nature of those complex matrices.

Heart-cutting and comprehensive 2D-LC strategies: Setting up the scene

It is of common knowledge to chromatographers in many application areas that there are two main modes of operation in 2D-LC: heart-cutting and comprehensive; plus two intermediate modes: multiple heart-cutting and selective comprehensive [19, 21] (Fig. 7.2).

Heart-cutting 2D-LC, denoted in the literature as LC-LC, is a multidimensional methodology comprising two or more chromatographic columns connected by a switching valve, which ensures the selective and online transfer of specific fractions (e.g., a single peak, a specific time segment, and a portion of a peak) from one to

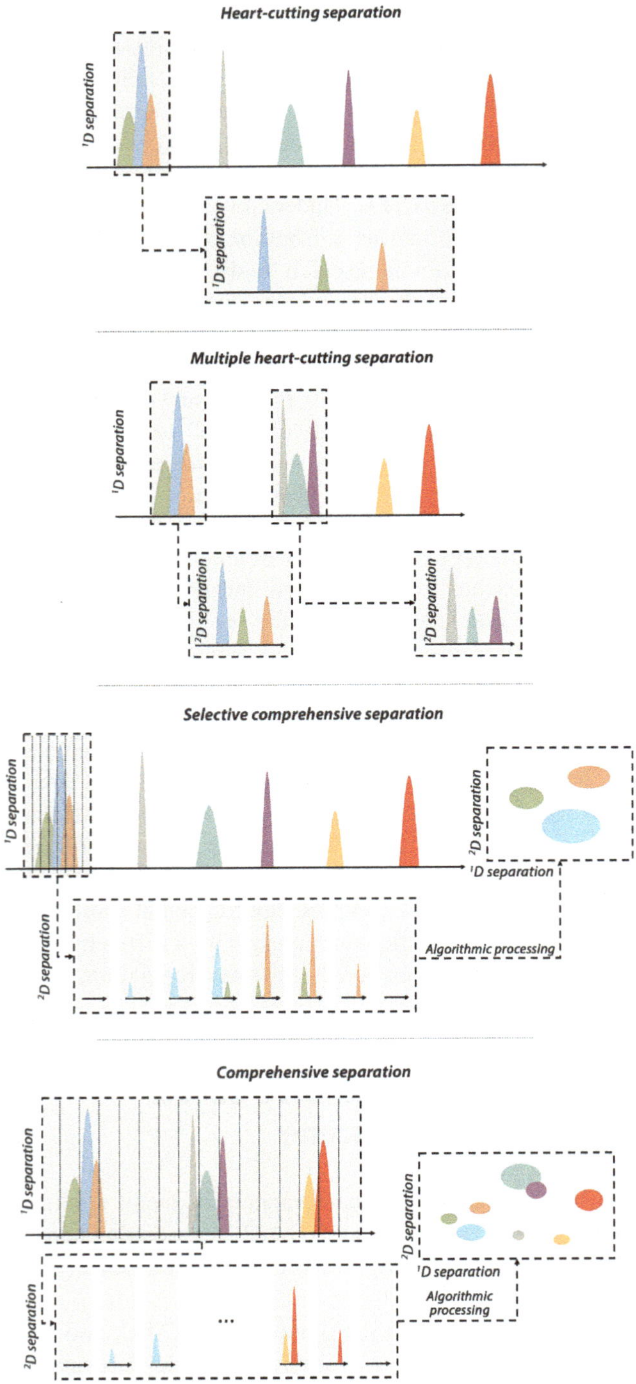

Fig. 7.2 Schematic representation of the two main modes of 2D-LC separations, LC × LC and LC-LC, and of the two intermediate separation modes, selective comprehensive and multiple heart-cutting.

another column. The LC-LC methodology is especially suited for the separation of a limited number of target analytes since it requires the definition of the collection time of the fractions, according to the elution times of the analytes of interest. Thus, the main advantage of this method is that after knowing the elution time of the analytes under study in the different columns, it becomes possible to optimize a procedure sufficiently selective to separate a given analyte independently of the complexity of the environmental matrix in which it is embedded. However, this advantage turns out to be its major limitation: when either the analytes are unknown, or the standards required for the optimization process are not available, then the application of this method becomes impracticable. Consequently, this advantage/limitation makes this technique ideal for target analysis of a limited number of compounds instead of a global characterization of the sample, regardless of the detection system. This fact helps to understand the reason why this technique is associated to a low degree of completeness of chemical information when compared to 1D-LC-based techniques, as schematically shown in Fig. 7.1. Besides, the selectivity of LC-LC limits the overall information that can be obtained from other constituents in the sample. This technique has been used in the last decades for the analysis of compounds present in biological and environmental samples at low concentrations, such as trace-level quantification of low-molecular mass carbonyl compounds in air [30], acidic pesticides in soils [14], endocrine disrupting compounds in water [31], and estrogens in sediments [32].

On the other hand, when the main objective is to carry out a nontargeted screening of a given sample, comprehensive MDLC, namely LC × LC, is a more adequate option. In order to attain a true comprehensive 2D separation, a few set of conditions must be fulfilled: (1) the whole sample must be subjected to two independent separation mechanisms within the same run; (2) the whole sample constituents passes through the detection system or at least in equal percentages that guarantees that the obtained chromatogram is representative of the entire sample; and (3) the resolution attained in the first dimension (^{1}D) should be kept (as much as possible) in the second dimension [33–35]. These three main criteria were defined by Giddings and are generally accepted for LC × LC [33–35].

Selective comprehensive and multiple heart-cutting can be seen as the middle of "spectra" between proper comprehensive and heart-cutting separations. In a multiple heart-cutting separation, more than one fraction is further separated in the second dimension. The selective comprehensive separations are similar to typical comprehensive separation methods but are only applied to a specific section of the chromatogram [19].

Since some detectors provide information that can be considered "multidimensional," as is the case of multichannel Mass Spectrometry (MS^n) or Diode Array detectors (DAD), few researchers may consider the detection step as an additional dimension. Nonetheless, in MDLC, when one refers to "dimensions," usually it refers to "separation dimensions" (e.g., separation columns). Although high degrees of orthogonality can be achieved in LC × LC by using suitable column selectivity in the two dimensions (i.e., different retention mechanisms), the use of an additional dimension represented by the detection step may pose difficulties to retrieve useful information from the 2D chromatograms. Understandably, the use of an extra dimension leads to a more complex visual representation of the acquired chromatographic data since these are typically represented as 2D contour plots or three-dimensional surfaces. This 2D chromatographic data typically contains a vast amount of information that needs to be further processed through different algorithms in order to be readily accessible to the analyst. In this field, three main approaches have been proposed to deal with LC × LC data structures [36]. In the first approach, data from the LC × LC is viewed as a set of consecutive one-dimensional chromatograms. These chromatograms are then treated individually, taking advantage of all the knowledge and large amount of software already available for data treatment in 1D chromatography. The second approach consists in dealing directly with the data matrix, which requires knowledge and expertise on complex chemometric algorithms. This approach is especially suited to deal with three or higher order data structures, such as data derived from 2D chromatographic systems coupled to multichannel detectors (e.g., MS^n and DAD). Finally, the third approach, converts the 2D matrix data into an image and uses the high diversity of image processing algorithms and tools for data handling and treatment [36]. These different data processing methods have been discussed in the literature in detail and we suggest the works of Matos et al. [36] and Pierce et al. [37] for the interested readers. It should also be mentioned that LC × LC coupled to high-resolution detectors is an analytical tool much more entwined with qualitative rather than quantitative analysis and, consequently, there are relatively few LC × LC studies in which quantitative analysis is discussed [38]. Although the use of LC × LC systems seem to be difficult by nonexperts, the introduction of a commercial version of this equipment and the development of more "user-friendly" data processing software are very likely to boost a growing interest for including these LC × LC techniques in laboratories in order to deal with the complexity of environmental samples.

It is also important to note that LC × LC can be carried out either in online or offline modes. In offline mode, sample is injected in a single conventional 1D-LC system, and fractions of the effluent are manually collected and injected, at a later time, into a 1D-LC system with a different separation column. As highlighted by Stoll and Carr [21], a couple of interesting features can be assigned to offline LC × LC approach: (i) it does not require very high speed separations in the second dimension as in online LC × LC; (ii) the 2D separation could be carried out using a single 1D-LC instrument; and (iii) high peak capacities can be achieved, although at the cost of a high time of analysis. Nevertheless, offline LC × LC is considerably more prone to sample contamination and losses than online mode, which is of particular concern when dealing with complex samples, such as those of environmental and natural products samples. Online LC × LC, on the other hand, is much more technically challenging than its offline counterpart because it generally requires the use of an automated switching system (e.g., a 6-, 8-, or 10-port valve), which interfaces the first and second dimension (^{2}D) columns, and collects fractions of ^{1}D effluent and injects them into the ^{2}D column. Indeed, most of the efforts and progress achieved in LC × LC research field, have been devoted toward the development of adequate interfaces (modulator) between the two separation dimensions. The modulation interface is really the center piece for a successful LC × LC separation, in parallel to the challenge of combining two solvent systems in order to prevent detrimental effects of ^{1}D effluent into the ^{2}D separation. We refer interested readers to Pirok et al. [19] work for further details on modulation and solvent compatibility issues. It is not surprising that automation offered by online LC × LC systems leads to more accurate, reproductive, repeatable 2D separations and is also less labor intensive than the offline mode. This is of particular interest for environmental research since online LC × LC coupled to high-resolution multichannel detectors offer new opportunities to effectively and efficiently profile and map the entire sample, whose complexity is very difficult to address using 1D-LC or LC-LC methods.

Multidimensional separations by means of capillary electrophoresis

Two-dimensional (2D) electrophoresis has been extensively used for the separation of biological samples. Two-dimensional gel electrophoresis (2D-GE) applied to the separation of proteins is an example of this analytical technique and has preceded the birth of proteomics [29]. Unfortunately, 2D-GE cannot be used for the separation of any other types of analytes beyond proteins and peptides [28]. In fact, when dealing with environmental samples, the available literature dealing with separations based on 2D capillary electrophoresis or the coupling of LC with capillary electrophoresis is not nearly as extensive as those of 2D-LC.

There are two major examples of these types of comprehensive electrophoretic multidimensional systems: (a) those employing a combination of two or more different modes of 2D capillary electrophoresis (2D-CE) and (b) methods coupling liquid chromatography (LC) with capillary electrophoresis (CE), namely LC × CE or CE × LC (also available in the heart-cutting mode), depending on which dimension the separation system is located [28]. Either 2D-CE or LC×CE (or CE×LC) present several advantages (such as increased orthogonality and low analysis time) but also some significant disadvantages, such as the presence of the electric field that complicates the use of regular switching valves for the transfer of the effluent from the ^{1}D to ^{2}D, and the large difference between the two dimensions effluent volumes in LC × CE and LC-CE that have hampered their widespread acceptance as everyday tool for separation and analysis of complex environmental samples. The 2D-CE techniques can be further divided into two major assemblages: (1) the separation in both dimensions is made using the same capillary and (2) two or more capillaries are connected to each other by means of an interface.

Multidimensional separation using a single capillary is achieved by changing buffer characteristics when the sample's analytes are already inside the capillary. This modification will lead to different separation conditions due to difference of selectivity between the initial and final buffers [27]. This technique is simple, straightforward, and it can be used in a regular CE device without the need for interfaces or hyphenation with other equipment. On the other hand, as in 2D-LC, the application of 2D-CE, where two or more capillary columns are employed, is much more challenging. In a similar fashion to 2D-LC, it is also necessary an interface that allows the passage of the sample from the ^{1}D to the ^{2}D. These systems are usually divided according to the method that is used for the transport of the sample from the ^{1}D to the ^{2}D [28, 29]. There are very few examples of the use of 2D-CE methods in the analysis of environmental samples, but some examples dealing with the analysis of compounds of environmental interest (e.g., pharmaceuticals) in urine samples have been published, either using a single [39, 40] or multiple capillary columns [41, 42].

The idea of merging together the versatility of LC systems with the separation power of CE is not new. In fact, the first example of a multidimensional separation where CE was used in one of the separation dimension remotes back to 1948 [43]. In this work, Haugaard and Kroner applied an electrical current to a chromatographic paper, allowing the separation of basic and acidic amino acids in opposite directions, while neutral amino acids remained unaffected. Furthermore, the coupling of LC with CE was later described by Giddings as one of the most powerful 2D separations [35]. In 1990, two articles, both developed by Bushey and Jorgenson, were published describing the first successful use of LC × CE combination to separate proteins [44, 45].

Since the LC and CE are two inherently different techniques, the combination of both mechanisms should be able to provide a higher degree of orthogonality than the 2D-LC ever could, even if the two separation columns are the most uncorrelated possible. Furthermore, the CE separation is usually faster than an LC separation. Nonetheless, its major advantage is also the main reason for the lack of use of multidimensional techniques that couple LC with CE since this higher degree of orthogonality leads to compatibility issues between the two dimensions.

Similar to 2D-LC, multidimensional separations where at least one of the dimensions is CE can be operated either online or offline and, likewise, in heart-cutting or comprehensive modes [26]. Furthermore, comprehensive multidimensional separation using LC × CE (or CE × LC) is much more complex than its heart-cutting counterpart. In fact, due to intrinsic characteristics of the CE separation, it can be argued that the development of an LC × CE (or CE × LC) method is much more complex than that of LC × LC. The electric field existing in the electrophoretic dimension is the major cause of this difficulty since it makes the use of switching valves (like the ones used in LC × LC) incompatible. As such, these valves must be electrically isolated or made of nonmetallic components. To solve this issue, other types of interfaces have been created, including membrane and droplet or gated interfaces. Readers are directed to the outstanding reviews of Ranjbar and coworkers [26] and Grochocki and coworkers [28] for further information on the different interfaces available for LC × CE, CE × LC, and CE × CE. These and other challenges on the LC × CE and CE × LC method development, such as the appreciably higher injection volumes and eluent quantities needed for the LC part of the system, deepen even further the compatibility issues. Consequently, these separation methods are not nearly as widespread as GC × GC or even LC × LC, turning them almost into a niche field in terms of analytical methods. Furthermore, the great majority of studies reported in the literature using these techniques are focused on the separation of proteins, and only very few studies have been published in the analysis of environmental and natural products matrices, including tap water (spiked) [46], Traditional Chinese Medicine [47], and wastewater samples [48].

Targeted vs untargeted analysis in 2D-LC: Finding the best separation conditions

2D-LC is adaptable to both targeted and untargeted analysis, but there are significantly different characteristics between both types of analysis. The concepts of peak capacity and orthogonality, which will be discussed later in more detail, are much more important in untargeted analysis than in targeted analysis. The targeted analysis aims at identifying some known analytes or confirming their presence in a sample. This type of analysis requires some preknowledge of the physicochemical properties of the analytes to be identified, thus meaning that it is confined to a relatively small number of well-studied compounds. Since this type of analysis is very selective, and the analytes of interest are known, the chromatographic conditions can easily be optimized using standard solutions. The targeted analysis thus becomes very useful in environmental studies and attempts to find the best separation conditions, being usually focused in increasing the sensitivity and selectivity of the analytical method to quantitatively determine the analytes of interest. Untargeted analysis, on the other hand, is a nonselective search aiming at the identification of unknown constituents in a sample. In principle, this procedure is carried out without any a priori knowledge and information on the compounds to be identified. As highlighted by Matos et al. [49], it is impossible to achieve a complete untargeted analysis in chromatography, because all the choices made in terms of experimental conditions (e.g., the selection of stationary phases, the gradient and composition of mobile phases, and the detectors used) will be restricted to the scope of the analytical work as well as the range of properties associated to the analytes that can be separated and detected. Thus, finding the best separation conditions in untargeted analysis usually aims to increase the number of analytes that can be successfully detected and identified. Taking into account these constraints, the untargeted analysis can be further classified into two groups, depending on the analytical challenge or environmental problem to be solved: "fully untargeted" (from now onward referred just as untargeted), where there can be a vast number of unknown analytes present in the sample; and "semitargeted," where some specific classes of compounds or some analytes are expected to be found [50]. Obviously, this classification between "fully untargeted," "semitargeted," and "targeted" must be kept in mind when developing new analytical 2D-LC procedures to address a given problem.

Tables 7.1–7.3 summarize some important examples of 2D-LC (either LC-LC or LC × LC) and 2D-CE (either CE-CE or CE × CE) applications for the analysis of environmental samples from an untargeted, semitargeted, and targeted perspective, respectively. As shown in Table 7.1, the most common untargeted 2D-LC application is the profiling of natural products in plant extracts, particularly those used in Traditional Chinese Medicine (e.g., ginseng and other plants extracts). This is due to the complexity of these samples, encompassing hundreds or thousands of constituents with very different properties, and possibly with synergistic effects, where the quality control of these samples is a demanding issue [93]. When addressing the samples of interest from a semitargeted point of view (in Table 7.2), 2D-LC is

Table 7.1 Examples of 2D-LC (either LC-LC or LC × LC) applications for untargeted analysis of environmental samples.

Sample	Analytes	Mode	Separation technique	Detection	Observations	Reference
Zhengtian pills (TCM)	876 compounds	Comprehensive	NPLC × RPLC (2 tandem columns in ^{2}D)	UV (240 nm)	**Extraction method:** soxhlet with ethanol; **Mobile phase:** ^{1}D: *n*-hexane/1,4 dioxane (99.5:0.5), 100 μL/min; ^{2}D: isopropanol/ H_2O (2:98) and methanol, 1 mL/min; **Theoretical peak capacity:** 1740 **Additional information:** two different NPLC columns were tested; various RPLC columns with different lengths were tested; different column temperatures were tested.	[51]
Magnolia-vine (*Schisandra chinensis*)	More than 40 compounds detected, 14 identified	Comprehensive	ILLC × RPLC	UV (254 nm) and MS	**Extraction method:** ultrasound-assisted extraction with n-hexane; **Mobile phase:** ^{1}D:10-mM ammonium acetate solution at 1 mL/min; ^{2}D: H_2O and acetonitrile at 3 mL/min; **Additional information:** ILLC column was compared with other types of columns in a LC × LC system.	[52]
Roots of *Pueraria lobata* and *Pueraria thomsonii*	271 and 254 peaks respectively	Comprehensive and heart-cutting	RPLC × RPLC	DAD and MS	**Extraction method:** ultrasound-assisted extraction with methanol; **Mobile phase:** ^{1}D: methanol and formic acid 0.1% at 100 μL/min; ^{2}D: acetonitrile and formic acid 0.1% at 2.5 mL/min; **Theoretical peak capacity:** 3245 **Effective peak capacity:** 677	[53]

					Practical peak capacity: 1593 and 1776 **Orthogonality:** 68.5% and 76.8% **Additional information:** LC-LC mode was used to remove major compounds.	
Household dust and laundry dryer lint	Plasticizers, flame retardants, pesticides, drug metabolites	Comprehensive	RPLC × RPLC	MS	**Extraction method:** ultrasound-assisted extraction with methanol; **Mobile phase:** ^{1}D: H_2O and acetonitrile at 100 μL/min; ^{2}D: formic acid 0.1% and acetonitrile with 0.1% formic acid at 2 mL/min; **Orthogonality:** 67% (household dust) and 73% (laundry dryer lint) **Additional information:** identity of seven compounds was confirmed and their concentrations were semiquantitative determined (0.4–17.0 μg/g).	[54]
Sewage treatment plant effluents	20 environmental contaminants (pesticides, food additives and pharmaceuticals)	Comprehensive	RPLC × RPLC	UV (290 nm) and MS	**Extraction method:** soxhlet with acetone and methanol; **Mobile phase:** ^{1}D: H_2O and acetonitrile; 100 μL/min; ^{2}D: formic acid 0.1% and acetonitrile with 0.1% of formic acid at 2 mL/min.	[55]
Chinese medicine formula	280 compounds (LC × LC) and 13 minor compounds (mLC-LC)	Comprehensive and multiple heart-cutting	RPLC × RPLC	UV (270 nm) and MS	**Extraction method:** ultrasound-assisted extraction with methanol; **Mobile phase:** ^{1}D: acetonitrile and formic acid 0.1% at 100 μL/min; ^{2}D: solution of ammonium hydroxide 10% and ammonium acetate 10 mM at 2 mL/min; **Theoretical peak capacity:** 2763 **Effective peak capacity:** 710.3 **Practical peak capacity**: 1628 **Orthogonality:** 84.1%	[56]

Continued

Table 7.1 Continued

Sample	Analytes	Mode	Separation technique	Detection	Observations	Reference
					Additional information: different extraction solvents were tested; different stationary phases were tested.	
Dengzhan Shengmai (TCM)	283 compounds (LC × LC) plus 12 pairs of isomer compounds (mLC-LC)	Comprehensive and multiple heart-cutting	RPLC × RPLC and RPLC-Chiral	UV (280 nm) and MS	**Extraction method:** ultrasound-assisted extraction with ethanol; **Mobile phase:** ^{1}D: formic acid 0.1% and methanol with 0.1% of formic acid at 100 μL/min; ^{2}D: formic acid 0.1% and acetonitrile with 0.1% formic acid at 2 mL/min; **Effective peak capacity:** 1123 **Correlation coefficient of the two columns:** 0.414 **Additional Information:** different percentages of ethanol were tested; different columns were tested on both dimensions of LC × LC	[57]
River water	Dissolved organic matter	Heart-cutting	HILIC-HILIC	1D, 2D, and 3D NMR	**Extraction method:** ultrasound-assisted extraction with ethanol; **Mobile phase:** ^{1}D: Acetonitrile and 100 μM deuterated ammonium acetate at 300 μL/min; ^{2}D: acetonitrile and 10 μM deuterated ammonium acetate at 300 μL/min; **Additional information:** LC-LC system was used for sample fraction for 1D, 2D, and 3D NMR.	[58]
Wastewater	53 different analytes (antidepressants,	Multiple heart-cutting	RPLC-RPLC	MS	**Extraction method:** solid-phase extraction; **Mobile phase:** ^{1}D: acetonitrile and formic	[59]

	antibiotics, fungicides, herbicides and others)				acid 0.1% at 20 μL/min; ^{2}D: formic acid 0.1% and methanol with 0.1% formic acid at 500 μL/min; **Practical peak capacity:** 7828 **Orthogonality:** 43%	
Panax ginseng leaves	17 major constituents	Heart-cutting	HILIC-RPLC	DAD and MS	**Extraction method:** ultrasound-assisted extraction with methanol/H_2O followed by liquid-liquid extraction with dichloromethane and butanol; **Mobile phase:** ^{1}D: acetonitrile and isopropanol/H_2O (37.5:62.5) with 0.1% formic acid at 250 μL/min; ^{2}D: H_2O and methanol at 300 μL/min.	[60]
Curcuma (*Curcuma kwangsiensis*)	105 compounds	Comprehensive	RPLC × RPLC	DAD and MS	**Extraction Method:** solid-liquid extraction with methanol; **Mobile phase:** ^{1}D: acetonitrile and formic acid 0.1% at 20 μL/min; ^{2}D: acetonitrile and formic acid 0.1% at 0.7 and 2 mL/min; **Theoretical peak capacity:** 1825 **Effective peak capacity:** 430 **Practical peak capacity:** 1416 **Orthogonality:** 93.2% **Additional information:** four different types of RPLC columns were tested	[61]

^{1} *D*, first-dimension; ^{2} *D*, second-dimension; *Chiral*, chiral liquid chromatography; *ILLC*, immobilized liposome chromatography column; *MS*, mass spectrometry; *NMR*, nuclear magnetic resonance; *NPLC*, normal-phase liquid chromatography column; *RPLC*, reversed-phase liquid chromatography column; *TCM*, Traditional Chinese Medicine.

Table 7.2 Examples of 2D-LC and LC × CE applications for semitargeted analysis of environmental and natural products samples.

Sample	Analytes	Mode	Column set	Detection	Observations	Reference
White and red ginsengs	Ginsenoids	Comprehensive	HILIC × RPLC	MS	**Extraction method:** ultrasound-assisted extraction with Methanol; **Mobile phase:** ^{1}D: 10-mM ammonium formate with 0.2% formic acid and acetonitrile with 0.2% formic acid at 150 μL/min; ^{2}D: formic acid 0.2% and acetonitrile, 0.6 mL/min; **Additional information:** other RPLC column was used as trap column. **Effective peak capacity:** 4392 **Orthogonality:** 55%	[62]
Animal urine	Sulfonamides, β-agonists and hormones	Comprehensive	RPLC × RPLC	DAD and MS	**Mobile phase:** ^{1}D: H_2O/acetonitrile (90:10) with 0.1% formic acid and H_2O/acetonitrile (10:90) with 0.1% formic acid, 40 or 60 μL/min; ^{2}D: H_2O/acetonitrile (90:10) with 0.1% formic acid and H_2O/acetonitrile (10:90) with 0.1% formic acid, 0.8 or 0.35 mL/min; **Limits of detection:** 1–10 μg/L **Additional information:** different RPLC columns were tested on both dimensions; different column temperatures were tested.	[63]
Apiaceous vegetables	Furanocoumarins	Comprehensive	RPLC × RPLC	DAD	**Extraction method:** solid-liquid extraction with H_2O and acetonitrile with 0.1% acetic acid followed by QuEChERS; **Mobile phase:** ^{1}D: 5-mF sodium phosphate and methanol, 250 μL/min; ^{2}D: 20-mM phosphoric acid and acetonitrile, 2.5 mL/min.	[64]

Urban organic aerosols	Water-soluble organic matter and alkaline-soluble organic matter	Comprehensive	HILIC × SEC	UV (254 nm) and FLD (λExc: 240 nm/λEm: 410 nm)	**Extraction method:** extraction from filters with H_2O; **Mobile phase:** ^{1}D: 20 mM of ammonium acetate (pH adjusted to 6.0 with acetic acid) with 10% of acetonitrile, 20 or 17 μL/min; ^{2}D: 20 mM-ammonium hydrogen carbonate with 11% of acetonitrile, 2.5 mL/min.	[65]
Sargassum muticum	Phlorotannins	Comprehensive	HILIC × RPLC	DAD and MS	**Extraction method:** pressurized liquid extraction with ethanol and H_2O; **Mobile phase:** ^{1}D: acetonitrile with 2% acetic acid and methanol/H_2O/acetic acid (95:3:2), 15 μL/min; ^{2}D: formic acid 0.1% and acetonitrile, 3 mL/min; **Theoretical peak capacity:** 1050 and 906.	[66]
Natural organic matter (NOM)	Suwannee River and Pony Lake fulvic acids	Comprehensive	HILIC × SEC	UV (254 nm) FLD (λExc: 240 nm/λ Em: 450 nm) and ELSD	**Mobile phase:** ^{1}D: 20 mM of ammonium acetate (pH adjusted to 6.0 with acetic acid) with 10% or 20% of acetonitrile, 20 μL/min; ^{2}D: 20-mM ammonium hydrogen carbonate with 11% of acetonitrile, 2 mL/min; **Additional information:** different types of columns were tested on both dimensions.	[67]
Atmospheric aerosols	Water-soluble organic matter	Comprehensive	HILIC × SEC	DAD	**Extraction method:** extraction from filters with H_2O; **Mobile phase:** ^{1}D: 20 mM of ammonium acetate (pH adjusted to 6.0 with acetic acid) with 10% of acetonitrile, 20 μL/min; ^{2}D: 20-mM ammonium hydrogen carbonate with 11% of acetonitrile, 2.5 mL/min.	[49]
Wastewater	23 target compounds	Comprehensive	RPLC × RPLC	MS	**Mobile phase:** ^{1}D: formic acid 0.1% and methanol, 0.2 μL/min; ^{2}D: formic acid 0.1% and acetonitrile, 0.040 mL/min.	[68]

Continued

Table 7.2 Continued

Sample	Analytes	Mode	Column set	Detection	Observations	Reference
Wastewater	65 target compounds	Comprehensive	RPLC × RPLC	MS	**Mobile phase:** ^{1}D: formic acid 0.1% and methanol, 0.2 μL/min; ^{2}D: formic acid 0.1% and acetonitrile, 0.040 mL/min; **Orthogonality:** 61%.	[69]
Vegetables and tea	300 pesticides	Heart-cutting	HILIC-RPLC	MS	**Extraction method:** solid-liquid extraction with acetonitrile; **Mobile phase:** ^{1}D: 5-mM ammonium formate solution with 0.1% acetic acid and acetonitrile/H_2O (90:10) with 5-mM ammonium formate solution with 0.1% acetic acid at 200 μL/min;^{2}D: 5-mM ammonium formate solution with 0.1% acetic acid and methanol with 5-mM ammonium formate solution with 0.1% acetic acid 3 mL/min; **Additional information:** trapping column was also used.	[70]
Tap water (spiked)	Haloacetic acids, dalapon, and common inorganic anions	Comprehensive	AX × CE	Capacitively coupled contactless conductivity detection (C^4D)	**Mobile phase:** ^{1}D: 7-mM sodium hydroxide at 0.3 mL/min; ^{2}D: 70 mM (tris (hydroxymethyl) and aminomethane (Tris))/70 mM *N*-cyclohexyl-2-aminoethanesulfonic acid buffer. **Theoretical peak capacity:** 867 **Effective peak capacity:** 498 **Orthogonality:** 68%	[46]

Atmospheric aerosols	Acidic compounds	Comprehensive	SCX × RPLC	UV (210 and 254) and MS	**Extraction method:** ultrasound-assisted extraction with methanol; **Mobile phase:** ^{1}D: acetonitrile and 17.5 mM of acetic acid at 40 μL/min; ^{2}D: acetonitrile and 17.5 mM of acetic acid at 3.5 and 2 mL/min; **Additional information:** two different columns were tested on ^{2}D. Quantitative analysis was performed on various compounds.	[16]

AX, anion exchange chromatography; *CE*, capillary electrophoresis; *ELSD*, evaporative light scattering detector; *FLD*, fluorescence detector; *HILIC*, hydrophilic interaction chromatography column; *MS*, mass spectrometry; *RPLC*, reversed-phase liquid chromatography column; *SCX*, strong cation-exchange chromatography column; *SEC*, size-exclusion chromatography column.

Table 7.3 Examples of 2D-LC (either LC-LC or LC × LC), 3D-LC and 2D-CE (either CE-CE or CE × CE) applications for targeted analysis of environmental and natural products samples.

Sample	Analytes	Mode	Column set	Detection	Observations	Reference
Animal urine	Atrazine (herbicide) + 11 metabolites and hydrolysis products	Heart-cutting	SCX-RPLC	MS	**Extraction method:** offline solid-phase extraction; **Mobile phase:** ^{1}D: formic acid 0.5%, 20-mM ammonium acetate, methanol and methanol with 0.1% of ammonium hydroxide at 1 mL/min; ^{2}D: formic acid 0.5%, 20-mM ammonium acetate, methanol and methanol with 0.1% of ammonium hydroxide at 1 mL/min; **Limits of detection:** 0.1–50 µg/L; **Additional information:** offline solid-phase extraction was optimized by means of experimental design.	[71]
Human urine	Phthalate metabolites	Heart-cutting	RPLC-RPLC	MS	**Mobile phase:** ^{1}D: acetonitrile/acetic acid 1% (10:90), acetonitrile/acetic acid 1% (90:10) and methanol at 0.6 or 0.8 mL/min, ^{2}D: methanol/acetic acid 1% (10:90) at 0.8 mL/min; **Additional information:** quantitative determination of levels of phthalate metabolites in urine samples was performed.	[72]
Peucedani Radix (TCM)	Angular-type pyranocoumarins	Heart-cutting	RPLC-Chiral column	DAD and MS	**Extraction method:** microwave-assisted extraction with ethanol; **Mobile phase:** ^{1}D: formic acid 0.1% and acetonitrile with 0.1% Formic acid at 0.3 mL/min; ^{2}D: acetonitrile/H_2O at 0.7 mL/min; **Additional information:** different extraction temperatures, volumes and ethanol percentages were tested.	[73]

Wastewater and estuarine water	Enantiomers of Omeprazole	Heart-cutting	RPLC-Chiral column	UV (302 nm) and MS	**Mobile phase:** ^1D: H_2O, acetonitrile/H_2O (35:65), and acetonitrile/H_2O (80:20) at 1 mL/min,^2D: acetonitrile/H_2O (35:65) at 1 mL/min; **Limits of detection:** 5.00 μg/L and 0.0250 μg/L for UV and MS detection, respectively; **Additional information:** RPLC column was also a RAM column, used for the exclusion of humic substances; two different RPLC columns were tested.	[74]
Wastewater and river water	Estrogenic chemicals	Heart-cutting	RPLC-RPLC	UV (280 nm) and MS	**Extraction method:** solid-phase extraction; **Mobile phase:** ^1D: H_2O/acetonitrile (95:5); ^2D: 10-mM formic acid and acetonitrile, 10-mM formic acid and methanol, methanol and H_2O or acetonitrile and 10 mM 4-methylmorphline; **Limits of detection:** 0.23–0.91 ng/L; **Additional information:** different pentafluorobenzyl bromide and Dansyl chloride were used as derivatization agents; RAM precolumn was used; quantitative analysis was performed.	[75]
Wastewater	Nine opiates and phenytoin	Selective Comprehensive	RPLC × RPLC, RPLC × RPLC ×RPLC, and RPLC-RPLC	UV (210, 220, and 254 nm)	**Extraction method:** solid-phase extraction; **Mobile phase:** Ex1-^1D:10-mM phosphate buffer and acetonitrile/20-mM ammonium phosphate at 0.7 mL/min; ^2D: acetonitrile and 0.1% trifluoroacetic acid at 3 mL/min; Ex2-^1D: acetonitrile and 10-mM ammonium acetate at 0.5 mL/min ^2D: acetonitrile/10-mM phosphate buffer at 0.5 mL/min ^3D: acetonitrile and 10-mM phosphate buffer at 2 mL/min Ex3-^1D: acetonitrile and 10-mM	[76]

Continued

Table 7.3 Continued

Sample	Analytes	Mode	Column set	Detection	Observations	Reference
					ammonium acetate at 0.5 mL/min; ^{2}D: acetonitrile and acetonitrile with 0.1% formic acid.	
Human urine	Salbutamol, salmeterol, and atenolol	Heart-cutting	HILIC-Chiral	UV (227 nm)	**Extraction method:** solid-phase extraction; **Mobile phase:** ^{1}D: methanol and 2-mM ammonium acetate buffer at 0.5 mL/min; ^{2}D: methanol/acetonitrile/5-mM ammonium acetate (90:5:5) at 0.4 mL/min; **Additional information:** quantitative analysis was performed.	[77]
River water	Fluoroquinolones	Heart-cutting	RPLC-RPLC	FLD (λExc: 294 nm/λEm: 514 nm)	**Extraction method:** online solid-phase extraction; **Mobile phase:** ^{1}D acetonitrile and 50-mM phosphate buffer with 5% acetonitrile at 1.5 mL/min; ^{2}D: acetonitrile and 50-mM phosphate buffer with 5% acetonitrile at 1.5 mL/min; **Limits of detection:** 0.001–0.002 μg/L; **Limits of quantification:** 0.080–0.100 μg/L; **Additional information:** three different RPLC columns were tested on ^{1}D; quantitative analysis was performed.	[78]
River and wastewater	Endocrine disruptors and related compounds	Heart-cutting	RPLC-RPLC	MS	**Extraction method:** solid-phase extraction; **Mobile phase:** ^{1}D H_2O and MeOH or 20-mM ammonium formate and 0.1% acetic acid and MeOH with 20-mM ammonium formate and 0.1% acetic acid at 0.2, 0.5, 1, and 1.75 mL/min ^{2}D: H_2O and MeOH or 20-mM ammonium formate and 0.1% acetic acid and MeOH with 20-mM ammonium formate and 0.1% acetic acid at 0.3 mL/min **Limits of detection:** 0.0038–0.55 ng/L;	[79]

					Additional information: three different columns were tested; quantitative analysis was performed.	
Wastewater, human urine, and river water	Phenytoin, chlorophene, hydrocortisone, and nicosulfuron	Heart-cutting	WCX-RPLC-RPLC	UV (210 nm) and MS	**Extraction method:** solid-phase extraction; **Mobile phase:** ^{1}D: acetonitrile and ammonium acetate buffer; ^{2}D: acetonitrile and phosphate buffer; ^{3}D: acetonitrile and formic acid or phosphate buffer; **Additional information:** quantitative analysis was performed.	[80]
Wastewater	Phenytoin	Selective comprehensive	RPLC × RPLC × RPLC	DAD	**Extraction method:** solid-phase extraction; **Mobile phase:** ^{1}D: acetonitrile and 10-mF phosphoric acid; ^{2}D: acetonitrile and 10-mF phosphoric acid at 0.5 mL/min; ^{3}D: acetonitrile and 10-mF phosphoric acid at 2 mL/min **Additional information:** quantitative analysis was performed.	[81]
Wastewater	Beta-blockers (atenolol, nadolol, metoprolol, bisoprolol, and betaxolol)	Heart-cutting	RPLC-RPLC	FLD (λExc: 230 nm/λEm: 302 nm)	**Extraction method:** solid-phase extraction; **Mobile phase:** ^{1}D: methanol and 25-mF phosphate buffer at 1 mL/min; ^{2}D: methanol and 25-mF phosphate buffer at 1.5 mL/min **Limits of detection:** 0.0020–0.0034 µg/L; **Limits of quantification:** 0.00520–0.0089 µg/L; **Additional information:** several types of columns were tested on both dimensions; quantitative analysis was performed.	[82]
Wastewater	Cationic cardiovascular drugs	Comprehensive	CE × CE	UV (214 nm)	**Running buffer solutions:** ^{1}D: 25-mM sodium acetate with 20% acetonitrile, adjusted to pH 2.8 with concentrated acetic acid; ^{2}D: 50-mM borax, 30-mM sodium dodecyl sulfate and 10 mM β-cyclodextrin with 20% acetonitrile;	[48]

Continued

Table 7.3 Continued

Sample	Analytes	Mode	Column set	Detection	Observations	Reference
					Limits of detection: 0.03–0.1 μg/L; **Additional information:** CE × CE separation combined CZE in ^{1}D and MEKC in ^{2}D.	
Human urine	Propranolol, metoprolol, nimodipine, and nicardipine	Heart-cutting	CE-CE	UV (214 nm)	**Running buffer solutions:** ^{1}D: 20-mM sodium acetate; ^{2}D: 50 mM borax, 30-mM sodium dodecyl sulfate and 10-mM β-cyclodextrin; **Limits of detection:** 0.1–0.3 μg/L; **Additional information:** CE × CE separation combined CZE in ^{1}D and MEKC in ^{2}D.	[41]
Human urine	Quinine	Heart-cutting	CE-CE	DAD	**Running buffer solutions:** ^{1}D: 10-mM sodium acetate, 20-mM acetic acid, 1-mM sodium dihydrogen phosphate, 0.1% methylhydroxyethylcellulose, 10 mM β-alanine, 10-mM acetic acid; ^{2}D: 25 mM β-alanine; 25-mM acetic acid; 0.1% methylhydroxyethylcellulose; **Additional information:** CE × CE separation combined CITP in ^{1}D and CZE in ^{2}D. **Limits of detection:** 8.62 ng/mL; **Limits of quantification:** 28.74 ng/mL.	[42]
TCM	Saponins	Multiple heart-cutting	RPLC-RPLC	DAD	**Extraction method:** ultrasound-assisted extraction with ethanol; **Mobile phase:** ^{1}D: H_2O and methanol at 0.25 mL/min; ^{2}D: acetonitrile and H_2O at 2 mL/min.	[83]
Human urine	Phthalate metabolites	Heart-cutting	RPLC-RPLC	MS	**Mobile phase:** ^{1}D: acetonitrile/acetic acid 1% (10:90), acetonitrile/acetic acid 1% (90:10) and methanol at 0.6 or 0.8 mL/min;	[84]

					^{2}D: methanol/acetic acid 1% (10:90) at 0.8 mL/min; **Additional information:** quantitative analysis was performed.	
Groundwater	Six different herbicides	Heart-cutting	RPLC-RPLC	FLD (λExc: 300 nm/λEm: 400 nm)	**Extraction method:** solid-phase extraction; **Mobile phase:** ^{1}D methanol and H_2O at 1.4 mL/min; ^{2}D methanol and H_2O at 1.4 mL/min; **Limits of detection:** 0.0028–0.00994 μg/L; **Additional information:** several types of columns were tested on both dimensions; quantitative analysis was performed.	[85]
Exhaust particulate matter	1-Nitropyrene	Heart-cutting	RPLC-RPLC	FLD (λExc: 360 nm/λEm: 430 nm)	**Extraction method:** soxhlet extraction with dichloromethane; **Mobile phase:** ^{1}D: methanol and H_2O at 1 mL/min; ^{2}D: methanol and H_2O at 1 or 3 mL/min; **Limits of detection:** 0.3 and 0.01 ng/mL; **Additional information:** reduction column was also used; quantitative analysis was performed.	[11]
Fruits	*o*-Phenylphenol and bitertanol	Heart-cutting	RPLC-RPLC	FLD (λExc: 224 and 257, nm/λEm: 329 and 343 nm)	**Extraction method:** solid-liquid extraction with acetone, dichloromethane, petroleum ether and ethyl acetate; **Mobile phase:** ^{1}D: acetonitrile and 10-mM ammonium acetate at 1 mL/min; ^{2}D: acetonitrile and 10-mM ammonium acetate at 1 mL/min; **Limits of detection:** 0.01–0.05 mg/kg; **Additional information:** quantitative analysis was performed.	[86]
Human urine	4-Nitrophenol and 3-methyl-4-nitrophenol	Heart-cutting	RPLC-RPLC	MS	**Mobile phase:** ^{1}D: acetonitrile and formic acid 0.01% 0.2 mL/min; ^{2}D: acetonitrile and H_2O at 0.2 mL/min;	[87]

Continued

Table 7.3 Continued

Sample	Analytes	Mode	Column set	Detection	Observations	Reference
					Limits of detection: 0.1–20 μg/L; **Additional information:** quantitative analysis was performed	
River water	Perfluorooctane sulfonate	Heart-cutting	RPLC-RPLC	MS	**Mobile phase:** ^{1}D: methanol and 10-mM ammonium acetate at 5 and 1 mL/min; ^{2}D: methanol and 10-mM ammonium acetate at 1 mL/min; **Limits of detection:** 5.35 pg/mL; **Limits of quantification:** 17.86 pg/mL.	[88]
Human urine	Phenol, catechol, and hydroquinone	Heart-cutting	RPLC-RPLC	FLD (λExc: 272, 276, and 289 nm/ λEm: 310, 320, and 332 nm)	**Mobile phase:** ^{1}D: 5-mM sodium citrate buffer, 5-mM sodium citrate buffer/acetonitrile (90:10) or 5-mM sodium citrate buffer/acetonitrile (80:20); ^{2}D: 5-mM sodium citrate buffer/acetonitrile (90:10), 5-mM sodium citrate buffer/acetonitrile (85:15), or 5-mM sodium citrate buffer/ acetonitrile (75:25) **Limits of detection:** 0.05–0.1 mg/L; **Limits of quantification:** 0.2–0.35 mg/L; **Additional information:** quantitative analysis was performed.	[89]
Human urine	Phthalate metabolites	Heart-cutting	RPLC-RPLC	MS and H-NMR	**Mobile phase:** ^{1}D: acetonitrile/acetic acid 1% (10:90), acetonitrile/acetic acid 1% (90:10) and methanol at 0.6 or 0.8 mL/min; ^{2}D: methanol/acetic acid 1% (10:90) at 0.8 mL/min; **Limits of detection:** 0.25–1 μg/L; **Limits of quantification:** 0.5–2 μg/L.	[90]
Air-dust	Dioxins	Heart-cutting	NPLC-RPLC	DAD	**Extraction method:** microwave-assistant solvent extraction with acetone;	[91]

					Mobile phase: ^{1}D: isohexane at 1 mL/min; ^{2}D: dichloromethane at 1 mL/min; **Additional information:** LC-LC used as clean-up method for GC analysis.	
Human urine	Phthalate metabolites	Heart-cutting	RPLC-RPLC	MS	**Mobile phase:** ^{1}D: acetonitrile/acetic acid 1% (10:90), acetonitrile/acetic acid 1% (90:10) and methanol at 0.6 or 0.8 mL/min; ^{2}D: methanol/acetic acid 1% (10:90) at 0.8 mL/min; **Additional information:** quantitative analysis was performed	[92]
Particulate matter	1-Nitropyrene	Heart-cutting	RPLC-RPLC	MS	**Extraction method:** ultrasound-assisted extraction with ethanol and benzene; **Mobile phase:** ^{1}D ethanol and 20-mM acetate buffer at 0.1 or 0.175 mL/min; ^{2}D methanol/H_2O (80:20) with 0.1% formic acid; **Additional information:** quantitative analysis was performed.	[12]

CE, capillary electrophoresis; *Chiral*, chiral liquid chromatography; *CITP*, capillary isotachophoresis; *CZE*, capillary zone electrophoresis; *ELSD*, evaporative light scattering detector; *FLD*, fluorescence detector; *HILIC*, hydrophilic interaction chromatography column; *MEKC*, micellar electrokinetic chromatography; *MS*, mass spectrometry; *NPLC*, normal-phase liquid chromatography column; *RPLC*, reversed-phase liquid chromatography column; *SCX*, strong cation-exchange chromatography column; *SEC*, size-exclusion chromatography column; *WAX*, weak anion-exchange chromatography column.

usually applied into the separation of complex samples such as wastewater, natural organic matter, and plant extracts used in TCM. Nevertheless, to the best of the author's knowledge, there is only one study using LC × CE for the semitargeted analysis of Haloacetic acids, dalapon, and common inorganic anions in tap water matrix [46]. The targeted approach (in Table 7.3) has been mostly used for the separation of phthalate metabolites, pesticides, and drugs in environmental and natural products samples. For this targeted approach, the method of choice has been 2D-LC (either LC-LC or LC × LC) although there are a few reports using 2D-CE (either CE-CE or CE × CE) that have been successfully applied for the separation of pharmaceutical drugs, such as propranolol, quinine, and nicardipine [41, 42, 48].

LC × LC, in particular, can be extraordinarily helpful for the semitargeted and untargeted separation of environmental matrices, allowing acquiring as much information as possible about a given sample. However, in spite of its great potential, the application of LC × LC to environmental samples has not developed that much at this point in time. Online LC × LC coupled to MS detector has been used for semitargeted analysis of wastewater samples, allowing the identification of 23–65 compounds, including analgesics such as Paracetamol and Tramadol, herbicides Diuron and Monuron, Benzotriazole a known Corrosion inhibitor, and antidepressants such as Venlafaxine and Sertraline (Refs. [68, 69] in Table 7.2). LC × LC coupled to three detectors in series (UV, fluorescence detector (FLD), and evaporative light-scattering detector (ELSD)) was also applied to resolve the chemical heterogeneity of Suwannee River fulvic acid standard material and Pony Lake fulvic acid reference material (Ref. [67] in Table 7.2). Due to the complex nature of these samples, incompletely resolved fractions were still portrayed. Nevertheless, in cases where samples separation was accomplished according to different separation mechanisms (i.e., hydrophobicity × molecular weight), it was concluded that smaller molecular weight group fractions seem to be related to a more hydrophobic nature. Following this seminal work with complex natural organic matter, online LC × LC coupled to either a DAD and FLD in series [65] or a single DAD [49] was applied to exploit the compositional changes over a molecular size continuum and associated light-absorption properties of chemically distinct pools of urban organic air particles (i.e., water-soluble organic matter (WSOM) and alkaline-soluble organic matter (ASOM)). The results obtained in these two later studies highlight the potential of MDLC techniques, namely of online LC × LC coupled to high resolution detectors, for unraveling the complexity of the substructures present in complex environmental organic matrices. This fact constitutes a huge advantage in comparison to the traditional 1D-LC. Moreover, even if one tries to replicate these results using only 1D-LC techniques, this would need at least several chromatographic analysis and tedious procedures for the collection of fractions. Table 7.3 also provides several examples of the targeted analysis of environmental samples using 2D-LC separation methods. As already mentioned, due to the specific characteristics of this analytical method, the heart-cutting mode is much more commonly applied than that of comprehensive mode in the targeted separation of environmental samples. The LC-LC technique coupled to different detectors (e.g., DAD, MS, UV, and FLD) has been applied mostly in the direct analysis of wastewater, river water, and urine samples, usually for the determination of specific drugs or drug metabolites [74, 76, 77, 79, 82] and, to a lesser extent, phthalates [72, 84, 90, 92, 94], pesticides [71, 85, 95, 96], and their respective metabolites.

In any chromatographic method, the optimization of the separation conditions is crucial and LC × LC is no exception. Considerable efforts must be devoted to finding the best LC × LC separation conditions because there are many different factors that can significantly influence the final peak capacity. Selection of mobile phase composition in both dimensions and their respective compatibility and flow rates, the type of switching valve and the volume of the sampling loop, as well as the selection of a fit-for-purpose detection system, are important factors to have in mind when developing an LC × LC-based method. Nonetheless, the most important factor when designing an LC × LC method is arguably the selection of the separation mechanisms (i.e., columns) to be employed in both dimensions, taking into account the analytical problem to be answered and whether the selected separation mechanisms are distinct from each other, but compatible at the same time. If successful in reaching this condition, enhanced orthogonality and peak capacity will be achieved. In this regard, the analytical problem to be solved plays, in fact, an important role in column selection. If dealing with a targeted analysis, the orthogonality is not a crucial outcome of the LC × LC procedure. The separation columns in both dimensions can share the same separation mechanisms, as long as they are successful in answering the scientific question set a priori. In the targeted analysis, there are only a few sets of compounds of interest, and a small difference between stationary phases can be enough to reach the desired separation. On the other hand, in semitargeted and untargeted analysis the scenario is completely different and the train of thought must be necessarily different.

In the case of semitargeted analysis, where the main purpose typically encompasses studying different sets of classes of compounds, it is important to consider the chemical characteristics of these analytes. For instance, if the compounds of interest all have low polarity, then it would be a wise choice to use a reversed-phase LC (RPLC) column in one of the dimensions, instead of a normal-phase LC (NPLC) column or even a hydrophilic interaction chromatography (HILIC) column (which can behave as a RPLC or NPLC, depending on mobile phase composition). Furthermore, the use of a RPLC × RPLC system can be suitable for the separation of some specific classes of compounds, such as phenolic compounds in Rooibos plants and Cocoa (Refs. [97, 98] in Table 7.2), as long as the two stationary phases have different properties (e.g., different particle size, composition or different bonded phase). Other clear orthogonal combinations might be more difficult to accomplish, such as HILIC × SEC, SCX × RPLC, and NPLC × RPLC, due to possible mobile phase incompatibility. However, HILIC × SEC was already successfully used for resolving and profiling the chemical heterogeneity of natural organic matter from aquatic [67] and atmospheric matrices [49], whereas SCX × RPLC was employed for separation of low-molar-mass organic acids in different matrices [99] (Table 7.2). NPLC × RPLC is perhaps the least likely practical combination in terms of mobile phase compatibility, although exhibiting a high orthogonality from a theoretical point of view. As shown in Table 7.2, NPLC × RPLC has been applied for the separation of carotenoids [100, 101] (reported theoretical peak capacities of 651 and 986 [101]), and phenolic and stevioside compounds [102] (reported peak capacities of 1850 (practical) and

3468 (theoretical)). This NPLC × RPLC combination was only possible because the ^{1}D encompassed columns either exhibiting characteristic features of HILIC systems (acting as NPLC) [102], or cyano microbore columns that offer normal-phase separations [100, 101].

On the other hand, if the desired type of analysis is untargeted, then using an RPLC × RPLC would probably not be enough to achieve the best orthogonality and peak capacity in the chromatographic separation. However, the most commonly employed combinations in terms of separation mechanisms in natural products and environmental research are RPLC × RPLC, as shown in Table 7.1. In an untargeted analysis, it is questionable at first whether the combination of two or more RPLC columns that would probably have similar separation characteristics will yield a high orthogonality and peak capacity. Similar concerns are also valid when dealing with a combination of any other stationary phases that separate compounds by polarity/hydrophobicity, such as HILIC, RPLC, and NPLC. Nonetheless, if the aim is to achieve the maximum orthogonality with these sorts of columns, then the separation conditions of the selected columns must be as much uncorrelated as possible. However, as shown in Table 7.1, some reports dealing with the untargeted analysis of natural products [53, 56, 57, 61], household dust and laundry dryer lint [54] using RPLC × RPLC, had reported orthogonality values as high as 93%. Nevertheless, an interesting NPLC × RPLC combination is also noteworthy in the untargeted analysis of traditional Chinese medicine [51] (Table 7.1). A silica column was chosen for NPLC in the ^{1}D, and water-soluble nonpolar 1,4-dioxane was selected as mobile phase modifier in NPLC. As a result, 876 peaks were detected, and the total peak capacity reached 1740 [51]. Since NPLC is suitable for group separation, and RPLC exhibits high resolution for less polar compounds, the potential of NPLC × RPLC combination for the analysis of complex environmental samples is enormous and should be further explored.

Updating and trends in peak capacity and orthogonality in 2D-LC

The focal point behind the application of two (or more) separation columns on a chromatographic system is to boost the maximum number of well-resolved peaks, thus raising the number of analytes that can be successfully separated and further identified in just a single chromatographic run. To accomplish this goal in 2D-LC, and particularly in LC × LC, a careful and exhaustive selection of the best combination of separations columns must be accomplished. As discussed previously, when the separation mechanisms in each chromatographic dimension are too alike, no significant benefit will be gained by using more than one dimension, typically because the separation achieved with just one dimension will be essentially the same, or even lower in the subsequent dimensions. In such situations, it may be more adequate to rely on the separation of a conventional 1D-LC system, since there will be no decrease on sensitivity triggered by the extra dilution of analytes in the following chromatographic dimensions. Hence, the ideal combination of columns should be those with entirely uncorrelated separation mechanisms.

Complete orthogonality may be theoretically achieved if the separation mechanisms in every dimension are completely uncorrelated from each other [103]. Orthogonality varies between 0% and 100%, where 100% means that full orthogonality has been reached [104]. In practice, however, it is almost impossible to achieve full orthogonality in LC × LC, because this depends not only on the separation mechanisms, but also on the optimized separation conditions (e.g., mobile phase composition, flow rates, and percentage of organic solvent) and samples characteristics [105]. For example, the combination of two entirely dissimilar separation mechanisms, such as SEC × RPLC, can still exhibit some correlation [105]. On the other hand, when samples are too different, for example, if subjected to a sample preparation step discarding the polar compounds, then orthogonality obtained using the common RPLC × RPLC system will be much higher than it would be if the polar compounds were still present on the sample. If we consider the concept of orthogonality in a more graphic way, a chromatogram where peaks are more dispersed in the 2D space is thought to be more orthogonal compared to a situation where peaks are positioned nearer the diagonal of the 2D space (Fig. 7.3A and F, respectively). It should also be noted that orthogonality is a characteristic specific of each sample. Two different separation mechanisms combined in a multidimensional system that exhibit high levels of orthogonality for a given sample may not be satisfactory for attaining an orthogonal separation of other samples. Therefore, chromatographic separation conditions should be extensively studied and optimized in order to ensure a successful separation outcome. Fig. 7.3 illustrate the expected consequences of a poor choice of set of separation columns in LC × LC. Although showing low correlation coefficient between the two separation mechanisms, the ^{2}D column in chromatograms B and C in Fig. 7.3 represent two scenarios of an excessive and low interaction, respectively, between the analytes and the stationary phase, which is not wanted for a

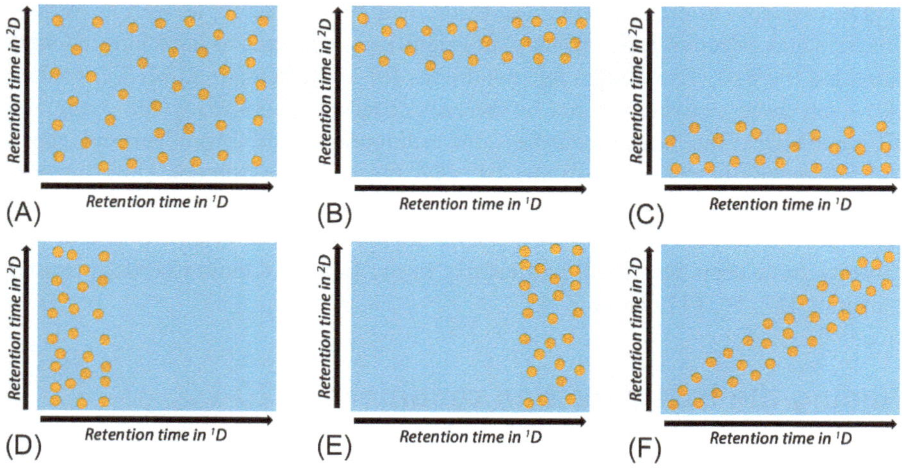

Fig. 7.3 Schematic representation of different 2D-LC separations, displaying different levels of orthogonality (A–E) and extremely low orthogonality (F).

successful LC × LC separation. The same is true in chromatograms D and E in Fig. 7.3, where a rather unfortunate choice of the ^{1}D separation column has a similar result on the overall LC × LC separation.

The peak capacity concept was first defined by Giddings, in 1969, as "the upper limit of resolvable components for a given technique under prescribed conditions" [106]. In MDLC, it is generally accepted that the theoretical peak capacity is merely the product of individual peak capacities of each separation column [35]. Obviously, the theoretical peak capacity value represents the "best case scenario." For example, it infers that no undersampling phenomena arise and, consequently, the resolution reached in one dimension is not lost in the subsequent dimensions. However, the sample transfer, from the ^{1}D to the ^{2}D, will always promote some loss of resolution. Additionally, even though peak capacity should be considered a significant way to assess the success of a separation process, we should also keep in mind that evenly spaced peaks in a 2D chromatogram are rather improbable to be found in any real complex samples. Remarkably, it has been demonstrated that the number of well resolved peaks, in a specific sample, is not likely to be higher than 18% of total peak capacity of a MDLC system [107]. Consequently, the concepts of effective and practical peak capacity were created as criteria that more truthfully estimate the maximum number of peaks that can be successfully separated in a single 2D chromatographic run. The process of calculating the effective peak capacity is fairly easy. Briefly, a correction factor from the undersampling phenomena is applied to the theoretical peak capacity value, leading to a more accurate prediction of the effective peak capacity [108]. On the other hand, the concept of practical peak capacity is rather more complex. As aforementioned, in LC × LC, it is almost impossible to attain full orthogonality; therefore, by default, some areas of a 2D chromatogram will never be occupied by any peak. Obviously, this leads to a decrease in the available 2D chromatographic area where peaks can be separated, which will yield a value for the practical peak capacity that is lower than the theoretical peak capacity [103].

The concepts of orthogonality and theoretical peak capacity, as well as those of practical and effective peak capacity, are significant concepts to have in mind when assessing the best chromatographic conditions for implementing an MDLC method. However, these metrics must not be the only criteria for comparing and optimizing different MDLC methods; nonetheless, the orthogonality and theoretical peak capacity should both be estimated in any LC × LC method development. Unfortunately, these metrics are not often reported in the literature [109] although in the last years some good examples of orthogonality and theoretical, effective, and practical peak capacity have been described, particularly in the field of Chinese herbal medicine screening [53, 56, 61] (Table 7.1).

Finding the best detection conditions in 2D-LC

When dealing with samples with a huge number of analytes by means of conventional 1D-LC, the identification of those analytes by comparison with a mixture of standard compounds stops being a viable choice. Nevertheless, coupling the chromatographic methods with a high-resolution detector, such as an MS detector, has proved to be an excellent way to overcome this situation and achieve the desired analyte identification. Generally, MDLC studies summarized in Tables 7.1–7.3 employ as detection the UV absorbance (either DAD or single UV wavelength), MS, or both detectors in series, with the noticeable exception of two reports on natural organic matter that also use fluorescence and evaporative light scattering detectors (Table 7.2) [65, 67] and two other reports on the drug determination on river and wastewater (Table 7.3) [78, 82]. Although a thorough identification of the analytes is overwhelmingly more common using an MS detector, this does not mean that the use of any other type of detector, namely UV and DAD, becomes pointless. It is possible to acquire rather important information (at the expense of a lesser degree of specificity) using a DAD detector since chromophores with similar mass-to-charge ratio may have distinct absorption maxima, which can be used to differentiate between different molecules [110].

Hence, in MDLC, the use of an MS detector is extremely common, and an improved separation before the MS detection has various advantages when compared to 1D-LC-MS. Some of these advantages include the reduction of matrix effects resulting from coeluting analytes due to the increased separation power of the MDLC system, particularly LC $\times$ LC, higher ionization efficiency, and reduction of ion suppression [34]. On the other hand, the coupling of an MS detector to an LC $\times$ LC system is not as straightforward as in 1D-LC and some specific modifications are essential. In LC $\times$ LC-MS, all mobile phase constituents, particularly in the ^{2}D, must be volatile to be compatible with the MS source [34]. Thus, the ^{2}D separation column must be compatible with mobile phases with a higher percentage of organic solvents, such as acetonitrile or methanol, and volatile additives, such as formic acid and ammonium acetate. In this regard, NPLC columns, or any other type of separation column that does not meet these requirements, can hardly be used as the last separation dimension before the MS detector.

Another noteworthy prerequisite when coupling LC $\times$ LC to an MS detector is the sampling rate. It has been established by Murphy and coworkers [111] that the sampling rate of the ^{1}D mobile phase has a significant influence on the resolution achieved in the ^{2}D, leading to the suggestion that each peak in the ^{1}D should be sampled at least three times. Hence, in LC $\times$ LC, the flow rate in the ^{2}D is typically orders of magnitude higher than the one used in the ^{1}D. The extremely high flow rates in the ^{2}D (up to 5 mL/min) are completely incompatible with any sort of MS detection. In the case of electrospray ionization (ESI), the maximum flow rate should be, at most, 1 mL/min, while in atmospheric-pressure chemical ionization (APCI) mode the maximum flow rate can reach 2 mL/min. In practice, though, these flow rate values should be much lower than these values allowed [69, 112]. The most common way to solve this issue in LC $\times$ LC-MS is to use a flow splitter before the MS detector. This solution allows the detection of the analytes in the MS but also seriously decreases the sensitivity. A more sophisticated solution to this problem is the miniaturization of the entire LC $\times$ LC system, which is an approach commonly used in proteomics but has also been effectively employed by Haun and coworkers in wastewater analysis [69] (Table 7.2). The focal point of this latter study was the construction of a miniaturized

2D-LC system, coupled to a quadrupole/time-of-flight MS detection, for wastewater profiling without the need to split the flow [69]. To reduce ^{2}D analysis time, elevated temperature and pressure conditions, as well as a stationary phase of superficially porous sub-3-μm were employed. Besides the reduction of the total time of analysis, the miniaturization of the LC × LC system also promotes a much lower solvent consumption. Using a standard mixture of 99 target analytes, the miniaturized LC × LC-MS system allowed the detection and identification of 65 these standard compounds in the wastewater samples. Despite obvious advantages over other LC × LC-MS systems, the work of Haun et al. [69] displays a real possibility of an unwarranted reduction in system sensitivity when compared to a 1D-LC-MS system, due to the dilution effect of the ^{2}D. It must also be considered that MS identification does not strictly need a complete chromatographic separation of the analytes. To assess these major issues, one additional work was published later by the same group of authors, using the same miniaturized LC × LC-MS system and aiming at comparing its sensitivity to that of a conventional 1D-LC-MS system for the analysis of wastewater samples [68]. It was stated that the absolute intensity of the analytical signal in the LC × LC-MS system was 10 times lower than that of the 1D-LC-MS system. Nevertheless, as the signal-to-noise ratio was only around 1.5 times lower, the decrease of analytical signal only leads to a small decrease in sensitivity when using the miniaturized LC × LC-MS. Furthermore, the number of compounds successfully identified was always higher in the miniaturized LC × LC-MS system. Still, the main issue with miniaturization of the LC × LC-MS system is perhaps its cost, particularly due to the use of expensive nanoLC pumps. Nonetheless, the simple reduction of the mobile phase flow rate in the ^{2}D to the range of 1–2 mL/min, is likely to show some beneficial effects. Even if this means that a flow splitter is still necessary, at least the mobile phase in the ^{2}D does not have to be split as much as in other LC × LC-MS systems, which will lead to an enhanced analytical sensitivity.

Conclusions

Over the last two decades, the advent of more efficient systems and analytical methodologies based on online MDLC separations has become a clear trend for the analysis of complex samples, such as natural products, proteins, and polymers. Surprisingly, and despite the advent of MDLC-based methods, the use of LC × LC methods is yet to be fully exploited in environmental research. In the case of the 2D-CE (CE × CE, LC × CE, CE × LC, and corresponding heart-cutting modes), the lack of application of these methods is even more noticeable. On the other hand, the importance of these multidimensional separation techniques, namely LC × LC, is not just related with the separation of sample constituents, but also with the rapidly evolving field of analytical instrumentation, which has developed more sophisticated detectors capable of providing a higher discrimination power. It is also clear from a vast assortment of studies available in the literature that the use of hyphenated 1D-LC techniques (e.g., LC-MS and LC-NMR, Fig. 7.1) has been able to provide new insights on the compositional and structural features of exceedingly complex samples, with a level of detail that would be considered technically impossible not long ago. Yet, the continuous development and use of MDLC, namely online LC × LC, has shown the potential to provide a more complete and deeper knowledge into the structural complexity of environmental samples, despite quite a few technological challenges that still need to be overcome to achieve its full capability.

Most environmental studies using MDLC have focused either on screening or identification of small sets of compounds. The use of LC × LC in environmental research has also been focused on the global sample characterization for achieving a heuristic knowledge of the structural composition and interrelationships between different components within the samples. Hence, it is with no surprise that pivotal developments in the stationary phase technology aiming at reaching increasing levels of orthogonality are still required, specifically for attaining a more complete understanding of the interactions that occur between samples' constituents and the stationary phases. Also, it is of the utmost importance to decrease or even eliminate the confounding effects due to these interactions occurring in the chromatographic separation. This will be particularly demanding, but it will be compulsory for the identification of the interactions that really occur within environmental samples. If a clear idea of the mechanisms that occur in the separation process is lacking, it will never be possible to fully assess the complexity of an environmental sample, simply because the decoding of the complexity of such samples cannot be accomplished following a separation process equally demanding and complex. However, if able to solve these methodological challenges, MDLC, and particularly LC × LC-based methods should be seen as promising tools for advancing environmental research and achieve a deeper level of knowledge and know-how within this field.

Acknowledgments

Thanks are due to FCT/MCTES for the financial support to CESAM (UID/AMB/50017/2019) and project AMBIEnCE (PTDC/CTA-AMB/28582/2017), through national funds (OE). FCT/MCTES is also acknowledged for a Research Fellowship within the framework of project AMBIEnCE. FCT/MEC and the European Social Fund are also acknowledged for a PhD grant (PD/BD/142931/2018) and an Investigator FCT Contract (IF/00798/2015).

References

[1] R. Consden, A.H. Gordon, A.J. Martin, Qualitative analysis of proteins: a partition chromatographic method using paper, Biochem. J. 38 (3) (1944) 224–232.
[2] F. Erni, R.W. Frei, Two-dimensional column liquid chromatographic technique for resolution of complex mixtures, J. Chromatogr. A 149 (1978) 561–569.
[3] E.A. Hogendoorn, C.E. Goewie, P. van Zoonen, Application of HPLC column-switching in pesticide residue analysis, Fresenius J. Anal. Chem. 339 (6) (1991) 348–356.
[4] C. Aguilar, F. Borrull, R.M. Marcé, Determination of pesticides by on-line trace enrichment-reversed-phase liquid chromatography-diode-array detection and confirmation by particle-beam mass spectrometry, Chromatographia 43 (11) (1996) 592–598.

[5] J.V. Sancho, et al., Rapid determination of glufosinate, glyphosate and aminomethylphosphonic acid in environmental water samples using precolumn fluorogenic labeling and coupled-column liquid chromatography, J. Chromatogr. A 737 (1) (1996) 75–83.
[6] F. Hernández, et al., New method for the rapid determination of triazine herbicides and some of their main metabolites in water by using coupled-column liquid chromatography and large volume injection, J. Chromatogr. A 778 (1) (1997) 171–181.
[7] C. Hidalgo, J.V. Sancho, F. Hernández, Trace determination of triazine herbicides by means of coupled-column liquid chromatography and large volume injection, Anal. Chim. Acta 338 (3) (1997) 223–229.
[8] F. Hernández, et al., Coupled-column liquid chromatography applied to the trace-level determination of triazine herbicides and some of their metabolites in water samples, Anal. Chem. 70 (15) (1998) 3322–3328.
[9] K. Hayakawa, et al., A high-performance liquid chromatographic system equipped with on-line reducer, clean-up and concentrator columns for determination of trace levels of nitropolycyclic aromatic hydrocarbons in airborne particulates, Anal. Chim. Acta 445 (2) (2001) 205–212.
[10] N. Tang, et al., Improvement of an automatic HPLC system for nitropolycyclic aromatic hydrocarbons: removal of an interfering peak and increase in the number of analytes, Anal. Sci. 19 (2) (2003) 249–253.
[11] T. Murahashi, F. Tsuruga, S. Sasaki, An automatic method for the determination of carcinogenic 1-nitropyrene in extracts from automobile exhaust particulate matter, Analyst 128 (11) (2003) 1346–1351.
[12] J.P. Miller-Schulze, et al., Analysis of 1-nitropyrene in air particulate matter standard reference materials by using two-dimensional high performance liquid chromatography with online reduction and tandem mass spectrometry detection, J. Chromatogr. A 1167 (2) (2007) 154–160.
[13] E.A. Hogendoorn, et al., Strategies in using analytical restricted access media columns for the removal of humic acid interferences in the trace analysis of acidic herbicides in water samples by coupled column liquid chromatography with UV detection, Anal. Chem. 71 (6) (1999) 1111–1118.
[14] E.A. Hogendoorn, et al., Microwave assisted solvent extraction and coupled-column reversed-phase liquid chromatography with UV detection: use of an analytical restricted-access-medium column for the efficient multi-residue analysis of acidic pesticides in soils, J. Chromatogr. A 938 (1–2) (2001) 23–33.
[15] J. Martínez Fernández, et al., Application of restricted-access media column in coupled-column RPLC with UV detection and electrospray mass spectrometry for determination of azole pesticides in urine, Chromatographia 53 (9–10) (2001) 503–509.
[16] J. Pól, et al., Comprehensive two-dimensional liquid chromatography–time-of-flight mass spectrometry in the analysis of acidic compounds in atmospheric aerosols, J. Chromatogr. A 1130 (1) (2006) 64–71.
[17] P.J. Marriott, P. Schoenmakers, Z.Y. Wu, Nomenclature and conventions in comprehensive multidimensional chromatography—an update, LC GC Europe 25 (5) (2012) 266–275.
[18] J.L. Cao, et al., Application of two-dimensional chromatography in the analysis of Chinese herbal medicines, J. Chromatogr. A 1371 (2014) 1–14.
[19] B.W.J. Pirok, D.R. Stoll, P.J. Schoenmakers, Recent developments in two-dimensional liquid chromatography: fundamental improvements for practical applications, Anal. Chem. 91 (1) (2019) 240–263.

[20] X. Wang, S. Buckenmaier, D. Stoll, The growing role of two-dimensional LC in the biopharmaceutical industry, J. Appl. Bioanal. 3 (2017) 120–126.
[21] D.R. Stoll, P.W. Carr, Two-dimensional liquid chromatography: a state of the art tutorial, Anal. Chem. 89 (1) (2017) 519–531.
[22] X. Zhang, et al., Multi-dimensional liquid chromatography in proteomics—a review, Anal. Chim. Acta 664 (2) (2010) 101–113.
[23] P.Q. Tranchida, et al., Potential of comprehensive chromatography in food analysis, TrAC Trends Anal. Chem. 52 (2013) 186–205.
[24] H. Wang, S. Hanash, Multi-dimensional liquid phase based separations in proteomics, J. Chromatogr. B Anal. Technol. Biomed. Life Sci. 787 (1) (2003) 11–18.
[25] P.F. Brandão, A.C. Duarte, R.M.B.O. Duarte, Comprehensive multidimensional liquid chromatography for advancing environmental and natural products research, TrAC Trends Anal. Chem. 116 (2019) 186–197.
[26] L. Ranjbar, J.P. Foley, M.C. Breadmore, Multidimensional liquid-phase separations combining both chromatography and electrophoresis—a review, Anal. Chim. Acta 950 (2017) 7–31.
[27] F.J. Kohl, L. Sánchez-Hernández, C. Neusüß, Capillary electrophoresis in two-dimensional separation systems: techniques and applications, Electrophoresis 36 (1) (2015) 144–158.
[28] W. Grochocki, M.J. Markuszewski, J.P. Quirino, Multidimensional capillary electrophoresis, Electrophoresis 36 (1) (2015) 135–143.
[29] P.A. Kler, D. Sydes, C. Huhn, Column-coupling strategies for multidimensional electrophoretic separation techniques, Anal. Bioanal. Chem. 407 (1) (2015) 119–138.
[30] P.R. Kootstra, H.A. Herbold, Automated solid-phase extraction and coupled-column reversed-phase liquid chromatography for the trace-level determination of low-molecular-mass carbonyl compounds in air, J. Chromatogr. A 697 (1) (1995) 203–211.
[31] D. Matějíček, Multi heart-cutting two-dimensional liquid chromatography–atmospheric pressure photoionization-tandem mass spectrometry method for the determination of endocrine disrupting compounds in water, J. Chromatogr. A 1231 (2012) 52–58.
[32] D. Matejicek, On-line two-dimensional liquid chromatography-tandem mass spectrometric determination of estrogens in sediments, J. Chromatogr. A 1218 (16) (2011) 2292–3000.
[33] J.C. Giddings, in: H.J. Cortes (Ed.), Multidimensional Chromatography, Marcel Dekker, New York, 1990, pp. 1–27.
[34] F. Cacciola, et al., Comprehensive two-dimensional liquid chromatography coupled to mass spectrometry: fundamentals, method development and applications, in: A. Cappiello, P. Palma (Eds.), Comprehensive Analytical Chemistry, Elsevier, 2018, pp. 81–123 (Chapter 4).
[35] J. Calvin Giddings, Two-dimensional separations: concept and promise, Anal. Chem. 56 (12) (1984). 1258A–1268A, 1270A.
[36] J.T.V. Matos, R.M.B.O. Duarte, A.C. Duarte, Trends in data processing of comprehensive two-dimensional chromatography: state of the art, J. Chromatogr. B Anal. Technol. Biomed. Life Sci. 910 (2012) 31–45.
[37] K.M. Pierce, et al., Review of chemometric analysis techniques for comprehensive two dimensional separations data, J. Chromatogr. A 1255 (2012) 3–11.
[38] L. Mondello, et al., Quantification in comprehensive two-dimensional liquid chromatography, Anal. Chem. 80 (14) (2008) 5418–5424.
[39] C. Kukusamude, S. Srijaranai, J.P. Quirino, Stacking and separation of neutral and cationic analytes in interface-free two-dimensional heart-cutting capillary electrophoresis, Anal. Chem. 86 (6) (2014) 3159–3166.

[40] X. Zhang, Z. Zhang, Heart-cut two-dimensional separation method via hyphenation of micellar electrokinetic capillary chromatography and capillary zone electrophoresis using analyte focusing by micelle collapse, J. Chromatogr. B Anal. Technol. Biomed. Life Sci. 879 (19) (2011) 1641–1646.
[41] Z. Zhang, X. Zhang, F. Li, The multi-concentration and two-dimensional capillary electrophoresis method for the analysis of drugs in urine samples, Sci. China Chem. 53 (5) (2010) 1183–1189.
[42] P. Mikuš, et al., 2D capillary electrophoresis hyphenated with spectral detection for the determination of quinine in human urine, J. Chromatogr. Sci. 50 (10) (2012) 849–854.
[43] G. Haugaard, T.D. Kroner, Partition chromatography of amino acids with applied voltage, J. Am. Chem. Soc. 70 (6) (1948) 2135–2137.
[44] M.M. Bushey, J.W. Jorgenson, Automated instrumentation for comprehensive two-dimensional high-performance liquid chromatography of proteins, Anal. Chem. 62 (2) (1990) 161–167.
[45] M.M. Bushey, J.W. Jorgenson, Automated instrumentation for comprehensive two-dimensional high-performance liquid chromatography/capillary zone electrophoresis, Anal. Chem. 62 (10) (1990) 978–984.
[46] L. Ranjbar, et al., Online comprehensive two-dimensional ion chromatography × capillary electrophoresis, Anal. Chem. 87 (17) (2015) 8673–8678.
[47] X. Zhang, et al., Comprehensive two-dimensional capillary LC and CE for resolution of neutral components in traditional Chinese medicines, J. Sep. Sci. 24 (5) (2001) 385–391.
[48] Z.X. Zhang, M.Z. Zhang, S.S. Zhang, Online preconcentration and two-dimensional separation of cationic compounds via hyphenation of capillary zone electrophoresis with cyclodextrin-modified micellar electrokinetic capillary chromatography, Electrophoresis 30 (11) (2009) 1958–1966.
[49] J.T.V. Matos, et al., Profiling water-soluble organic matter from urban aerosols using comprehensive two-dimensional liquid chromatography, Aerosol Sci. Technol. 49 (6) (2015) 381–389.
[50] B.L. Milman, I.K. Zhurkovich, The chemical space for non-target analysis, TrAC Trends Anal. Chem. 97 (2017) 179–187.
[51] Y. Wei, et al., A comprehensive two-dimensional normal-phase × reversed-phase liquid chromatography based on the modification of mobile phases, J. Chromatogr. A 1216 (44) (2009) 7466–7471.
[52] S. Wang, et al., Comprehensive two-dimensional high performance liquid chromatography system with immobilized liposome chromatography column and monolithic column for separation of the traditional Chinese medicine *Schisandra chinensis*, Anal. Chim. Acta 713 (2012) 121–129.
[53] X. Qiao, et al., Separation and detection of minor constituents in herbal medicines using a combination of heart-cutting and comprehensive two-dimensional liquid chromatography, J. Chromatogr. A 1362 (2014) 157–167.
[54] X. Ouyang, et al., Non-target analysis of household dust and laundry dryer lint using comprehensive two-dimensional liquid chromatography coupled with time-of-flight mass spectrometry, Chemosphere 166 (2017) 431–437.
[55] X. Ouyang, et al., Comprehensive two-dimensional liquid chromatography coupled to high resolution time of flight mass spectrometry for chemical characterization of sewage treatment plant effluents, J. Chromatogr. A 1380 (2015) 139–145.
[56] X. Qiao, et al., A chemical profiling solution for Chinese medicine formulas using comprehensive and loop-based multiple heart-cutting two-dimensional liquid

chromatography coupled with quadrupole time-of-flight mass spectrometry, J. Chromatogr. A 1438 (2016) 198–204.

[57] N. Sheng, et al., Chiral separation and chemical profile of Dengzhan Shengmai by integrating comprehensive with multiple heart-cutting two-dimensional liquid chromatography coupled with quadrupole time-of-flight mass spectrometry, J. Chromatogr. A 1517 (2017) 97–107.

[58] G.C. Woods, M.J. Simpson, A.J. Simpson, Oxidized sterols as a significant component of dissolved organic matter: evidence from 2D HPLC in combination with 2D and 3D NMR spectroscopy, Water Res. 46 (10) (2012) 3398–3408.

[59] S. Stephan, et al., Contaminant screening of wastewater with HPLC-IM-qTOF-MS and LC+LC-IM-qTOF-MS using a CCS database, Anal. Bioanal. Chem. 408 (24) (2016) 6545–6555.

[60] B. Ji, et al., Application of fractionized sampling and stacking for construction of an interface for online heart-cutting two-dimensional liquid chromatography, J. Chromatogr. A 1466 (2016) 199–204.

[61] W. Zhou, et al., On-line comprehensive two-dimensional liquid chromatography tandem mass spectrometry for the analysis of *Curcuma kwangsiensis*, Talanta 186 (2018) 73–79.

[62] H. Zhang, et al., A multidimensional analytical approach based on time-decoupled online comprehensive two-dimensional liquid chromatography coupled with ion mobility quadrupole time-of-flight mass spectrometry for the analysis of ginsenosides from white and red ginsengs, J. Pharm. Biomed. Anal. 163 (2019) 24–33.

[63] M.H. Blokland, et al., Multiclass screening in urine by comprehensive two-dimensional liquid chromatography time of flight mass spectrometry for residues of sulphonamides, beta-agonists and steroids, Food Addit. Contam. Part A Chem. Anal. Control Expo. Risk Assess. 35 (9) (2018) 1703–1715.

[64] D.W. Cook, et al., Comparison of multivariate curve resolution strategies in quantitative LCxLC: application to the quantification of furanocoumarins in apiaceous vegetables, Anal. Chim. Acta 961 (2017) 49–58.

[65] A.S. Paula, et al., Two chemically distinct light-absorbing pools of urban organic aerosols: a comprehensive multidimensional analysis of trends, Chemosphere 145 (2016) 215–223.

[66] L. Montero, et al., Anti-proliferative activity and chemical characterization by comprehensive two-dimensional liquid chromatography coupled to mass spectrometry of phlorotannins from the brown macroalga *Sargassum muticum* collected on North-Atlantic coasts, J. Chromatogr. A 1428 (2016) 115–125.

[67] R.M. Duarte, A.C. Barros, A.C. Duarte, Resolving the chemical heterogeneity of natural organic matter: new insights from comprehensive two-dimensional liquid chromatography, J. Chromatogr. A 1249 (2012) 138–146.

[68] J. Leonhardt, et al., A comparison of one-dimensional and microscale two-dimensional liquid chromatographic approaches coupled to high resolution mass spectrometry for the analysis of complex samples, Anal. Methods 7 (18) (2015) 7697–7706.

[69] J. Haun, et al., Online and splitless NanoLC × CapillaryLC with quadrupole/time-of-flight mass spectrometric detection for comprehensive screening analysis of complex samples, Anal. Chem. 85 (21) (2013) 10083–10090.

[70] S. Kittlaus, et al., Development and validation of an efficient automated method for the analysis of 300 pesticides in foods using two-dimensional liquid chromatography–tandem mass spectrometry, J. Chromatogr. A 1283 (2013) 98–109.

[71] Z. Kuklenyik, et al., Two-dimensional high performance liquid chromatography separation and tandem mass spectrometry detection of atrazine and its metabolic and hydrolysis products in urine, J. Chromatogr. B 901 (2012) 1–8.

[72] M. Kasper-Sonnenberg, et al., Levels of phthalate metabolites in urine among mother–child-pairs—results from the Duisburg birth cohort study, Germany, Int. J. Hyg. Environ. Health 215 (3) (2012) 373–382.
[73] Y.L. Song, et al., Qualitative analysis and enantiospecific determination of angular-type pyranocoumarins in Peucedani Radix using achiral and chiral liquid chromatography coupled with tandem mass spectrometry, J. Chromatogr. A 1338 (2014) 24–37.
[74] J.C. Barreiro, et al., A column-switching method for quantification of the enantiomers of omeprazole in native matrices of waste and estuarine water samples, Talanta 82 (1) (2010) 384–391.
[75] G.W. Lien, C.Y. Chen, G.S. Wang, Comparison of electrospray ionization, atmospheric pressure chemical ionization and atmospheric pressure photoionization for determining estrogenic chemicals in water by liquid chromatography tandem mass spectrometry with chemical derivatizations, J. Chromatogr. A 1216 (6) (2009) 956–966.
[76] S.R. Groskreutz, et al., Selective comprehensive multidimensional separation for resolution enhancement in high performance liquid chromatography. Part II: applications, J. Chromatogr. A 1228 (2011) 41–50.
[77] Y. Yang, et al., Chiral determination of salbutamol, salmeterol and atenolol by two-dimensional LC-LC: application to urine samples, Chromatographia 75 (23–24) (2012) 1365–1375.
[78] M.D.G. García, et al., Determination of (fluoro)quinolones in environmental water using online preconcentration with column switching linked to large sample volumes and fluorescence detection, J. Sep. Sci. 35 (7) (2012) 823–831.
[79] M. Gorga, M. Petrovic, D. Barceló, Multi-residue analytical method for the determination of endocrine disruptors and related compounds in river and waste water using dual column liquid chromatography switching system coupled to mass spectrometry, J. Chromatogr. A 1295 (2013) 57–66.
[80] S.W. Simpkins, et al., Targeted three-dimensional liquid chromatography: a versatile tool for quantitative trace analysis in complex matrices, J. Chromatogr. A 1217 (49) (2010) 7648–7660.
[81] H.P. Bailey, S.C. Rutan, D.R. Stoll, Chemometric analysis of targeted 3DLC-DAD data for accurate and precise quantification of phenytoin in wastewater samples, J. Sep. Sci. 35 (14) (2012) 1837–1843.
[82] P.P. Vázquez, et al., Determination of five beta-blockers in wastewaters by coupled-column liquid chromatography and fluorescence detection, Anal. Chim. Acta 666 (1–2) (2010) 38–44.
[83] C.L. Yao, et al., Simultaneous quantitation of five *Panax notoginseng* saponins by multi heart-cutting two-dimensional liquid chromatography: method development and application to the quality control of eight Notoginseng containing Chinese patent medicines, J. Chromatogr. A 1402 (2015) 71–81.
[84] H. Fromme, et al., Occurrence and daily variation of phthalate metabolites in the urine of an adult population, Int. J. Hyg. Environ. Health 210 (1) (2007) 21–33.
[85] A.R. Mughari, et al., Coupled-column liquid chromatography combined with postcolumn photochemical derivatization and fluorescence detection for the determination of herbicides in groundwater, J. Sep. Sci. 30 (5) (2007) 673–678.
[86] T. Zamora, et al., Determination of fungicide residues in fruits by coupled-column liquid chromatography, J. Sep. Sci. 27 (9) (2004) 645–652.
[87] F. Hernández, J.V. Sancho, O.J. Pozo, An estimation of the exposure to organophosphorus pesticides through the simultaneous determination of their main metabolites in urine by liquid chromatography-tandem mass spectrometry, J. Chromatogr. B Anal. Technol. Biomed. Life Sci. 808 (2) (2004) 229–239.

[88] M. Takino, S. Daishima, T. Nakahara, Determination of perfluorooctane sulfonate in river water by liquid chromatography/atmospheric pressure photoionization mass spectrometry by automated on-line extraction using turbulent flow chromatography, Rapid Commun. Mass Spectrom. 17 (5) (2003) 383–390.
[89] G. Marrubini, et al., Direct analysis of phenol, catechol and hydroquinone in human urine by coupled-column HPLC with fluorimetric detection, Chromatographia 62 (1–2) (2005) 25–31.
[90] H.M. Koch, L.M. Gonzalez-Reche, J. Angerer, On-line clean-up by multidimensional liquid chromatography–electrospray ionization tandem mass spectrometry for high throughput quantification of primary and secondary phthalate metabolites in human urine, J. Chromatogr. B 784 (1) (2003) 169–182.
[91] A.C. Van Beuzekom, et al., Fast sample preparation involving MASE and coupled column normal phase liquid chromatography for the rapid trace analysis of dioxins in air-dust samples from fire catastrophe emissions, Talanta 63 (5) (2004) 1183–1191.
[92] H.M. Koch, et al., Internal exposure of the general population to DEHP and other phthalates—determination of secondary and primary phthalate monoester metabolites in urine, Environ. Res. 93 (2) (2003) 177–185.
[93] D.Z. Yang, et al., Multidimensional information-based HPLC technologies to evaluate traditional Chinese medicine, J. Chromatogr. Sci. 51 (7) (2013) 716–725.
[94] R. Preuss, H.M. Koch, J. Angerer, Biological monitoring of the five major metabolites of di-(2-ethylhexyl)phthalate (DEHP) in human urine using column-switching liquid chromatography–tandem mass spectrometry, J. Chromatogr. B 816 (1) (2005) 269–280.
[95] M.M. Galera, et al., Large-volume direct injection for determining naphthalene derivative pesticides in water using a restricted-access medium column in RPLC-LC with fluorescence detection, Chromatographia 60 (9–10) (2004) 517–522.
[96] C. Hidalgo, et al., Improved coupled-column liquid chromatographic method for the determination of glyphosate and aminomethylphosphonic acid residues in environmental waters, J. Chromatogr. A 1035 (1) (2004) 153–157.
[97] T. Beelders, et al., Comprehensive two-dimensional liquid chromatographic analysis of rooibos (*Aspalathus linearis*) phenolics, J. Sep. Sci. 35 (14) (2012) 1808–1820.
[98] M. Muller, A.G.J. Tredoux, A. de Villiers, Predictive kinetic optimisation of hydrophilic interaction chromatography × reversed phase liquid chromatography separations: experimental verification and application to phenolic analysis, J. Chromatogr. A 1571 (2018) 107–120.
[99] J.T.V. Matos, R.M.B.O. Duarte, A.C. Duarte, A simple approach to reduce dimensionality from comprehensive two-dimensional liquid chromatography coupled with a multichannel detector, Anal. Chim. Acta 804 (2013) 296–303.
[100] F. Cacciola, et al., Application of comprehensive two-dimensional liquid chromatography for carotenoid analysis in red mamey (*Pouteria sapote*) fruit, Food Anal. Methods 9 (8) (2016) 2335–2341.
[101] P. Dugo, et al., Comprehensive normal-phase × reversed-phase liquid chromatography coupled to photodiode array and mass spectrometry detection for the analysis of free carotenoids and carotenoid esters from mandarin, J. Chromatogr. A 1189 (1) (2008) 196–206.
[102] F. Cacciola, et al., Employing ultra high pressure liquid chromatography as the second dimension in a comprehensive two-dimensional system for analysis of *Stevia rebaudiana* extracts, J. Chromatogr. A 1218 (15) (2011) 2012–2018.
[103] Z. Liu, D.G. Patterson Jr., M.L. Lee, Geometric approach to factor analysis for the estimation of orthogonality and practical peak capacity in comprehensive two-dimensional separations, Anal. Chem. 67 (21) (1995) 3840–3845.

[104] M. Gilar, et al., Orthogonality of separation in two-dimensional liquid chromatography, Anal. Chem. 77 (19) (2005) 6426–6434.
[105] I. François, K. Sandra, P. Sandra, Comprehensive liquid chromatography: fundamental aspects and practical considerations—a review, Anal. Chim. Acta 641 (1–2) (2009) 14–31.
[106] J.C. Giddings, Generation of variance, "theoretical plates," resolution, and peak capacity in electrophoresis and sedimentation, Sep. Sci. 4 (3) (1969) 181–189.
[107] R.A. Shellie, P.R. Haddad, Comprehensive two-dimensional liquid chromatography, Anal. Bioanal. Chem. 386 (3) (2006) 405–415.
[108] M.R. Filgueira, et al., Improving peak capacity in fast online comprehensive two-dimensional liquid chromatography with post-first-dimension flow splitting, Anal. Chem. 83 (24) (2011) 9531–9539.
[109] H. Malerod, E. Lundanes, T. Greibrokk, Recent advances in on-line multidimensional liquid chromatography, Anal. Methods 2 (2) (2010) 110–122.
[110] I.N. Papadoyannis, H.G. Gika, Peak identification with a diode array detector, in: J. Cazes (Ed.), Encyclopedia of Chromatography, Marcel Dekker, Inc., 2004.
[111] R.E. Murphy, M.R. Schure, J.P. Foley, Effect of sampling rate on resolution in comprehensive two-dimensional liquid chromatography, Anal. Chem. 70 (8) (1998) 1585–1594.
[112] L. Mondello, P.Q. Tranchida, K.D. Bartle, Introduction. in: D.M. Desiderio, N.M. M. Nibbering, L. Mondello (Eds.), Comprehensive Chromatography in Combination with Mass Spectrometry, 2011. https://doi.org/10.1002/9781118003466 ch1.

Multidimensional gas chromatography for environmental exposure measurement

8

Cátia Martins, Carina Pedrosa Costa, Sílvia M. Rocha
Department of Chemistry & QOPNA/LAQV-REQUIMTE, University of Aveiro, Aveiro, Portugal

Introduction

Along life, humans are exposed to a wide range of different environmental exposures/stressors, which can be of physical, chemical, and/or social nature. These exposures may change over the lifetime, and their interaction with the individual characteristics (e.g., sex, health status, genetics, among others) determine the individuals' susceptibility to diseases. Exposome encompasses the human exposures (internal and external) during lifetime and how they are related to individuals' health and well-being [1]. General (e.g., urban, rural or industrialized environments, climate, social classes, education) and specific (e.g., life style, environmental toxicants, diet, smoking habits, physical activity, occupation) factors contribute for the external exposome; while internal exposome measures individual or groups of chemicals through the body fluids analysis that may be possible biomarkers of diseases/exposures (e.g., metabolism, sex, physical activity, age, endocrine factors, gut microflora, lipid oxidation, oxidative stress) (Fig. 8.1) [2, 3]. Nowadays, exposome tends to be explored and monitored using top-down approaches that measure exposures and internal processes through sampling individuals' body fluids (e.g., urine, blood, breath); rather than bottom-up approaches, in which external exposures are directly analyzed through environmental samples (e.g., water, air, foods), without consideration of the individuals' internal processes.

Snapshots of the individual exposome may be accessed over specific lifetimes, and its combination allows an approximation of the full exposome. In the last years, several consortia were established around the world in order to have in-depth knowledge about exposome, being the most representative: HERCULES Exposome Research Center (funded by the National Institute of Environmental Health Sciences, United States) [4], EXPOsOMICS (funded by the European Commission) [5], and HELIX—Human Early Life Exposome (funded by the European Commission) [6].

Multidimensional Analytical Techniques in Environmental Research. https://doi.org/10.1016/B978-0-12-818896-5.00008-9

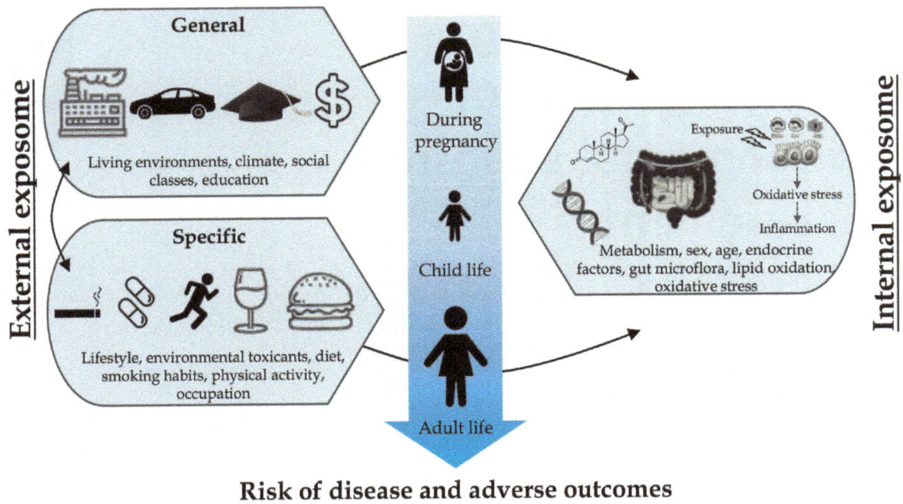

Fig. 8.1 External and internal factors that influence exposome, over human' lifetime, and allows to evaluate disease risks or adverse outcomes.
Modified from S. Dagnino, Unravelling the exposome: conclusions and thoughts for the future, in: S. Dagnino, A. Macherone (Eds.), Unraveling the Exposome, first ed., Springer International Publishing AG, Switzerland, part of Springer Nature, 2019, pp. 425–437, and it was designed using resources from Freepik.com.

The total and comprehensive determination of exposome represents a huge challenge due to its dynamic nature. Nowadays, a wide range of applications related to the study of environmental exposures rely upon the application of gas chromatographic-based methods to address an ever-growing demand to provide useful chemical information. The emergence of new molecules, or new uses of known molecules, as well as increased knowledge about the harmful health effects of various pollutants, have promoted the development of more powerful instrumental and methodological approaches to keep pace with the wealth of complex samples that require analysis. Moreover, despite the number of biologically active chemicals is extremely high (ca. 10^{60}), only one small portion is present in a considerably amount, and other small portion has impact as real chemical exposures [7]. To address these challenges, advances in gas chromatographic instrumentation have evolved from one-dimensional-gas chromatography (1D-GC) and heart cutting approaches (GC-GC), to a range of instrumentation devices mentioned as multidimensional gas chromatography (MDGC). Comprehensive two-dimensional (2D) gas chromatography (GC×GC) represents the main type of MDGC that has gained implementation in several fields, namely in the study of environmental exposures [8–12]. Thus, this chapter aims to present a critical review of the separation capabilities of the analytical platforms based on the comprehensive 2D chromatography, which properly respond to the identification and/or quantitation of a wide range of analytes, sometimes found only in trace amounts, in such complex matrices, as the body fluids. Also, it is discussed the importance of this technique for untargeted studies, meaning that the analytes of interest are either known or unknown beforehand. This chapter aims to cover the fundamental theory, while presenting recent advances and future trends for the application of the GC × GC-based methodologies in environmental research.

Sample preparation toward gas chromatographic analysis

In order to understand environmental exposures, a strategy for the analytes determination should be carefully planned and a workflow should be constructed that includes the sequence of all steps that handle with the sample, i.e., sampling, extraction, instrumental analysis, data processing, and interpretation, is recommended (Fig. 8.2). Environmental research represents itself huge challenges and complications to which questions associated with the study of body fluids may be added, namely the complexity and high variability of body fluids composition, containing a wide range of chemical structures in a wide dynamic concentration range. Thus, multiple sources of variability from sample may be considered, namely, collection and storage conditions, beyond the analysis methods (extraction and chromatographic ones), workflows, reagents, among others.

Although this chapter is focused on the instrumental analysis using GC × GC-based techniques, there are several considerations about sampling and extraction that must be emphasized once they can contribute for misinterpretations of the data. The sampling and extraction steps should be carefully established in order to generate analytes related data that may be useful for comprehension of environmental exposures and/or to determine specific analytes.

The analysis of body fluids' composition is a valuable and powerful approach to understand the environmental exposures once reflects their effects in human's metabolism. Despite its high potential, the characterization of body fluids to evaluate environmental exposures using advanced gas chromatographic techniques is not common, however, serum, urine, and breath have been already used [8–11, 13, 14]. Body fluid selection should reflect the studied exposure, and noninvasive sampling (e.g., urine,

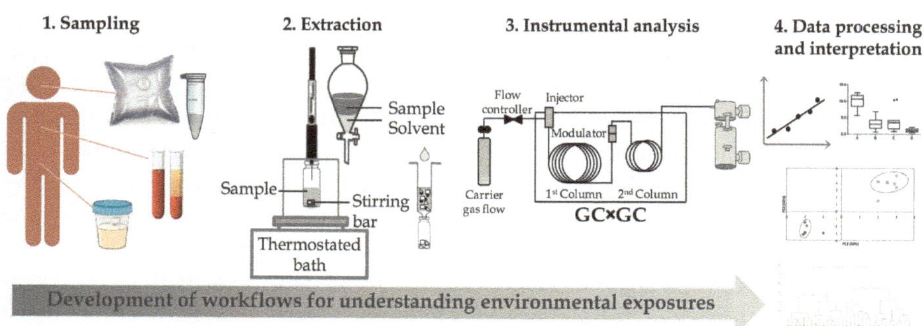

Fig. 8.2 Development of workflows for understanding environmental exposures: From sampling, to extraction, instrumental analysis and data processing, and interpretation.

exhaled breath) can be one factor to be considered on the body fluid selection. Furthermore, after sample collection, it is crucial to have a metabolism-quenching step, where it is recommended to store body fluids' samples at −80°C. The storage should be as fast as possible in order to not comprise the obtained results, due to possible alterations in the sample composition until the samples are frozen [15].

Moreover, it is of huge importance that the number of samples to be collected must be representative of the population under study (i.e., population subjected to a given exposure type), and each sample itself must be also representative even in size/amount. Furthermore, independent samples may be preferred than analytical replicates. In fact, individuals' large number allows to obtain reliable exposure-phenotype interactions. Another factor that may be considered is to have quality control (QC) samples, i.e., pool of aliquots from the body fluids of different individuals that allow the monitoring of the data acquisition performance [13]. Noteworthy, the number of samples is dependent of the strategy used, i.e., target approaches study one or a small group of analytes and require an higher number of samples, while untargeted studies have an enlarged set of analytes, and consequently may involve less samples [16]. However, considering the latter type of analysis, it is important to ensure that the number of samples under study and respective replicates must allow the application of chemometric methods and the generation of robust results.

The selection of extraction techniques prior to gas chromatographic analysis is based mainly on the physical-chemical properties of the analytes, also the matrix composition must be taken into consideration. For instance, solid-phase microextraction (SPME) has been used for direct extraction of small and volatile metabolites from breath and urine, namely alkanes and aldehydes associated with lipid peroxidation [14, 17, 18]. SPME, as a solvent free and user-friendly technique, has been also used for a wide range of applications. Lamani et al. (2015) [11] developed a methodology based on in situ derivatization SPME combined with GC×GC-qMS (quadrupole-mass spectrometry) for a comprehensive fingerprinting of aromatic amines in urine. Due to the high polarity of amino compounds, urine complexity and prevalence of conjugated anilines, pretreatment steps such as acidic hydrolysis, liquid-liquid extraction (LLE), and derivatization of amines to their corresponding aromatic iodine compounds are necessary, prior to the derivatives enrichment by headspace SPME.

On the other hand, for profiling urine metabolome of residents living near oil refineries and coal-fired power plants, which includes the prospection of a wide range of chemical families (benzenoids, hydrocarbons, lipids and lipid-like molecules, organic acids and derivatives, organoheterocyclic compounds, organonitrogen compounds, organooxygen compounds, phenylpropanoids and polyketides, and organosulfur compounds), the sample preparation for gas chromatographic analysis involved the depletion of urea via treatment with urease, protein precipitation with methanol, and trimethylsilyl derivatization [10, 13].

Due to their potential adverse outcomes in the population, some pollutants such as polychlorinated biphenyls (PCBs) or organochlorine pesticides (OCPs), brominated flame retardants, and polybrominated diphenyl ethers (oxides) (PBDEs or PBDPOs) have been determined in serum, and the solid-phase extraction (SPE) was applied to their extraction, before chromatographic analysis [8, 9]. The elution of the compounds of interest was accomplished using nonpolar solvents.

Finally, it is important to highlight that samples' composition can be significantly changed at any of the sample handling stages if there is not performed an adequate procedure. Due to the high relevance of this issue, recent studies have aimed to improve knowledge about degradation/changes on samples' profile [12, 19–28]. For instance, analytes may be loss and/or modified during the extraction steps, namely during solvent extracts concentration before analysis, storage before analysis, cross-contamination with storage, and/or analysis materials, among others. Hence, the extraction protocol should ensure that analytes are extracted in a representative way, minimizing analytes degradation or losses. The compliance of these indications must be considered to avoid discrepancies in the results obtained from different laboratories, as well as to produce results that meet the legal requirements.

The role of the multidimensional gas chromatographic in the assessment of environmental exposures

Principles of GC×GC and instrumental advances

Although 1D-GC is widely used in the qualitative and quantitative analysis of a wide range of samples and fields, providing high-quality analytical data, sometimes their complexity exceeds the separation capacity of a single chromatographic column, producing overloaded chromatograms. Several untargeted and targeted challenges have been identified to monitor the bioaccumulation of many pollutants, some of them require the detection of trace pollutants and the respective degradation products, or the determination of many analytes in a single analysis.

Due to the complexity of biological samples, such as the case of the body fluids, peaks' co-elution might occur, which difficult compounds' identification and quantification. Therefore, in order to increase chromatographic resolution and peak capacity, there have been developed independent techniques, namely GC×GC that arises as a powerful and high-throughput solution [12, 29, 30], which plays a key role in the determination of a wide range of analytes. The well separation of chromatographic peaks in complex matrices, high peak resolution [29, 31], as well as the possible enantiomeric recognition are important features of GC×GC [12].

In the last decades, significant improvements have been observed in the field of the chromatography, such as development of a wide set of GC columns' stationary phases (namely, oxygen-resistant stationary phases, extended, both the lower and upper temperature limits, or novel ionic liquids-based stationary phases), enhancement of the instrumental technologies (e.g., new types of modulators, new detectors, such as high-resolution mass spectrometers), and more powerful data analysis through new software and/or algorithms [12, 32]. Also, instrumentation has evolved from 1D-GC to multidimensional gas chromatography (MDGC), namely comprehensive 2D gas chromatography (GC×GC) that has been successfully applied to environmental research in a wide range of applications, such as in soil and sediments, water, biological matrices, and has been also extended to pollutants and impurities in various matrixes and applications, including fire debris, electronics waste, cosmetics, plastics, cigarette smoke, and reference materials [12, 33–39]. These examples illustrate that the use of GC×GC allowed to fulfill legal requirements and/or provided novel data useful in the understanding current environmental challenges. For instance, Tranchida et al. [35] developed a targeted chromatographic methodology that allowed the determination of 54 fragrance contact allergens, in one single analysis, in concentrations significantly lower than the legal limits (10 ppm) imposed by the European Directive 2003/15/EC. These allergens were reported by the Scientific Committee on Consumer Safety (SCCS) as being skin sensitizers, and their presence in cosmetics should be provided in products labels. Also, untargeted GC×GC-ToFMS analysis allowed to identify novel dichlorodiphenyltrichloroethane (DDT)-related compounds in bottlenose dolphins that are not detect in current biomonitoring studies [40]. This study provides huge information for a more complete understanding of the transportation and fate of these molecules with a biological system then previously known.

In general, gas chromatographic processes involve the analytes' separation through their partitioning (dispersion and specific interactions) between two immiscible phases, namely the mobile phase (e.g., helium, hydrogen, or nitrogen) and the stationary phase [41]. GC×GC comprises two orthogonal separation mechanisms that combines the use of two GC columns (Fig. 8.3), which are coated with specific stationary phases ruled by different properties, namely volatility, chirality, or polarity. The columns' combination most commonly used is a nonpolar as ^{1}D column (separation ruled by volatility, generally 30 m, and by the analytes carbon length) and a polar as ^{2}D column (separation ruled by polarity, usually 1–2 m) [30]. Depending on the physicochemical characteristics of the analytes under study, different sets of GC columns may be used, but the set including 5% of phenylmethylpolysiloxane or equivalents as a nonpolar column and a polyethylene glycol or equivalents as a polar column is currently used. Nevertheless, it is possible to use a reverse-phase column set, which consists in a polar phase in the first dimension, and a nonpolar/mid-polar phase in the second dimension, as reported in the analysis of organic acids from urines of infants with inborn metabolism errors [42]. The selection of the appropriate column set should be one of the first step to taken into consideration once it will contribute for an effective analytes' separation.

A specific interface, called modulator, makes the connection between the two GC columns. The modulator is responsible for the transfer of small portions of the eluate

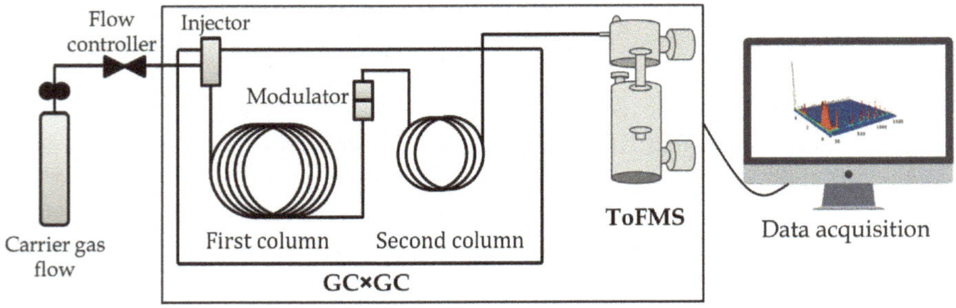

Fig. 8.3 Schematic representation of comprehensive two-dimensional gas chromatographic system coupled with time-of-flight mass spectrometer (GC×GC-ToFMS).

(2–8 s) from the primary ^{1}D column, and then to reinject into the secondary ^{2}D column, preserving the integrity of ^{1}D separation, once each peak is modulated several times. Different categories of modulators are commercially available, namely valve-based, flow, and thermal types. Either on valve-based and flow modulation, the collection of the eluate is made by a loop or channel; while in thermal modulation, the eluate is trapped using temperature. The cryomodulator is the most widely used type, where a fast cooling of the eluate occurs using a cryogenic jet (for instance with liquid nitrogen) followed by a fast heating applied by a hot-gas jet, this allows the immobilization and then remobilization of the compounds [8–10, 12, 29, 31]. Narrow peaks are produced once the collect fractions are no bigger than ¼ of the width peak, thus they are "flash" separated before the end of the modulation time. However, wrap-around phenomenon might occur, when the separation of an analyte exceeds the modulation time, and it is not finished before the next modulation. This phenomenon may promote co-elutions, interfering in the precise quantification of the compounds [29, 31, 43], being necessary to formerly optimize the modulation time to avoid wrap around.

The GC×GC equipment include a detector, such as a flame ionization detector (FID, which is cheaper and appropriate for quantitative analysis) or a mass spectrometer (MS). MS detectors have higher sensitivity and sensibility, and they are more effective on the analytes' identification due to the mass spectrum fragmentation pattern and potential comparison with MS databases. However, the peaks generated by GC×GC are very narrow, with typically width at half height of 0.1 s or less, and several data points are needed to record these peaks, which imply higher data acquisition rates (ca. 100 full-mass-range spectra per second) that are only possible with time-of-flight mass spectrometer (ToFMS) [29, 44]. Furthermore, reliable spectra deconvolution of overlapped peaks is possible due to full-mass spectra acquisition and continuity (same ion abundance ratios is observed in all the point of the chromatographic peak, for the different masses in the spectrum) of ToFMS, even at trace levels [41, 45].

GC×GC data analysis

Once the GC×GC data are collected, experts ultimately intend to identify and quantify analytes (in either a targeted or untargeted fashion) to address a variety of goals in the context of the experimental design. Targeted and untargeted analysis refers to when the analytes of interest are either known or unknown beforehand, respectively [12].

The complexity of the data generated by a GC×GC-ToFMS analysis can be handled through specific software that allows the data processing, which commonly comprises different algorithms to perform a set of operations, namely the signal deconvolution and automated peak finding, letting to perform the acquisition, processing, and data report. The software's also create and allow the visualization of the GC×GC chromatograms, in a contour plot or a three-dimensional (3D) plot [44] (Fig. 8.4).

The accurate analyte identification is only confirmed through the co-injection of authentic standards; however, they can be expensive or not commercially available.

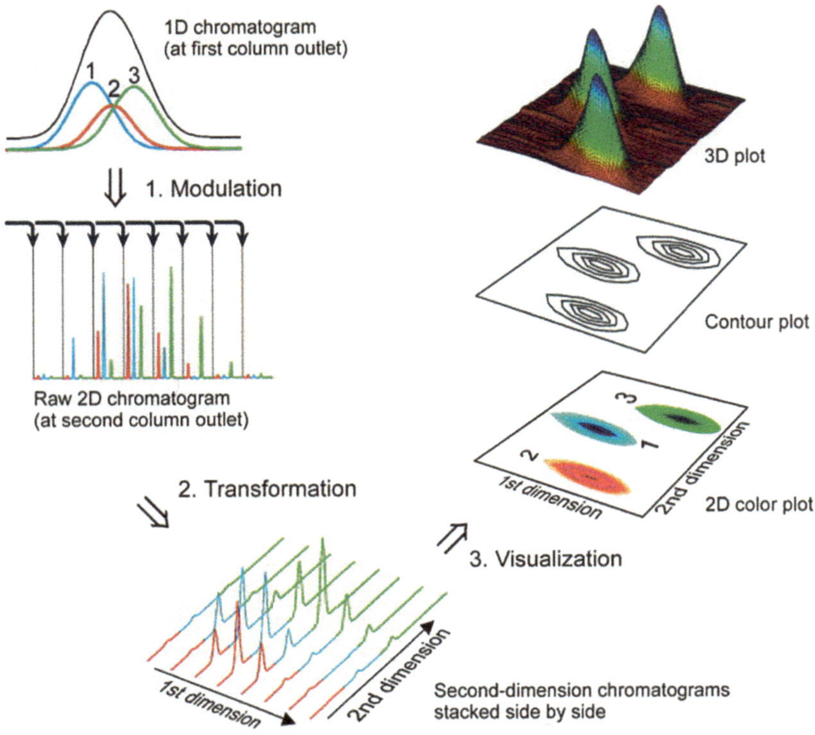

Fig. 8.4 GC × GC generation and visualization: Large series of high-speed second-dimension chromatograms are transformed to form a two-dimensional chromatogram that is usually performed using specific software. The visualization can be made in a contour plot or in a 3D plot.
Reproduced with permission from J. Dallüge, J. Beens, U.A.T. Brinkman, Comprehensive two-dimensional gas chromatography: a powerful and versatile analytical tool, J. Chromatogr. A 1000 (1–2) (2003) 69–108, Copyright (2003), with permission from Elsevier.

For that reason, the use of the commercial or open access mass spectra databases can be decisive tools for compounds' putative identification [46, 47]. Even so, the utilization of certified reference materials can be of huge importance if it is considered the implementation or utilization of methodologies that are in agreement with the legal policies. Certified reference materials are important to guarantee the quality assurance and quality control tools, for instance to validate analytical methodologies or to perform instruments calibration [48].

The confidence of the compounds identification may be improved using other strategies, namely using linear retention indices (LRI) values. Basically, a *n*-alkanes series (typically ranging from C_6 to C_{20}) is injected in the GC (using the GC×GC ovens temperature programs and modulator conditions established for the samples analysis), and retention time of each *n*-alkane is recorded. An LRI value of 100 times the number of carbons is attributed to each *n*-alkane. Then, compounds' retention time is normalized using the retention times of the adjacent eluting *n*-alkanes, through the van Den Dool and Kratz equation [49, 50]. The calculated LRI can be compared with LRIs available in the literature or in the open sources databases, in which they were achieved with similar GC column to the column used experimentally in the first dimension. The LRIs are not dependent of several GC instrumental features (e.g., carrier gas type and velocity, column' diameter and length, film thickness, among others), which allows the comparison between LRIs values acquired in different equipment and laboratories [41, 51]. The criteria described above are also commonly used for 1D-GC, but GC×GC provides an extra powerful tool for analytes identification, namely the formation of structured chromatograms. The orthogonal separation mechanism of GC×GC (promoted by the combination of a nonpolar with a polar column) lets the 2D spatial distribution of the compounds ruled by their chemical properties, i.e., chemically related compounds are organized in the same 2D chromatographic space, determining a 2D "chemical map." This feature is particularly helpful in the achievement of reliable identifications, especially in the analysis of complex matrices like body fluids, and when chemical standards are absence. It can also simplify the data analysis, and consequently decreases the analysis time [8, 14, 18, 31].

Fig. 8.5A shows an example of a structured chromatogram obtained from urine of a healthy individual that highlights one of the most peculiar characteristics of the GC×GC comparing with 1D-GC. For instance, it can be observed that linear aliphatic hydrocarbons (alkanes) show the lower retention time on the second dimension (${}^2t_R$) once they are nonpolar analytes, contrary to acids that have higher polarity, and therefore exhibited the highest ${}^2t_R$. Moreover, it can also be observed the chromatographic zones (bands and clusters) where it is possible to find analytes from a certain chemical family, in which the functional group globally contributes for their display on the chromatogram as follows (with the increase of the ${}^2t_R$): aliphatic alkanes < alkenes ~ aromatic alkanes < ketones ~ aldehydes < alcohols < acids. As terpenic compounds include different chemical families (alcohols, aldehydes, hydrocarbons, and ketones), they are shown within a rectangle arranged along the first (between ca. 600 and 1600 s) and second dimensions (between ca. 0.540 and 1.640 s). This complex chemical composition of urine provides relevant data about the humans' global physiological status [18]. Indeed, if individuals are in contact with the cigarette smoke (specific external exposure), it may provoke numerous metabolic alterations (internal exposome) that will lead to pathophysiological lungs disorders. These effects can be quantified in urine through the aldehydes and alkanes, which are known as being products of oxidative stress [18]. Also, the detected terpenoids may be associated to the humans' diet, another specific external exposure.

Fig. 8.5B exhibits blow-up of a part of the previous contour plot (Fig. 8.5A) and highlights one of the advantages of GC×GC that allows the separation of analytes with similar volatility through the secondary ^{2}D column. For instance, decanal (e) and verbenone (f) have the similar volatility (${}^1t_R - 948$ s), but they were separated by the ^{2}D column once they present different polarities (${}^2t_R - 0.56$ and 0.88 s, respectively). Another important feature of the GC×GC-ToFMS is related with its sensitivity and spectral quality of the acquired data, which are crucial in the analytes' identification and quantification, particularly for those in trace amounts as can be

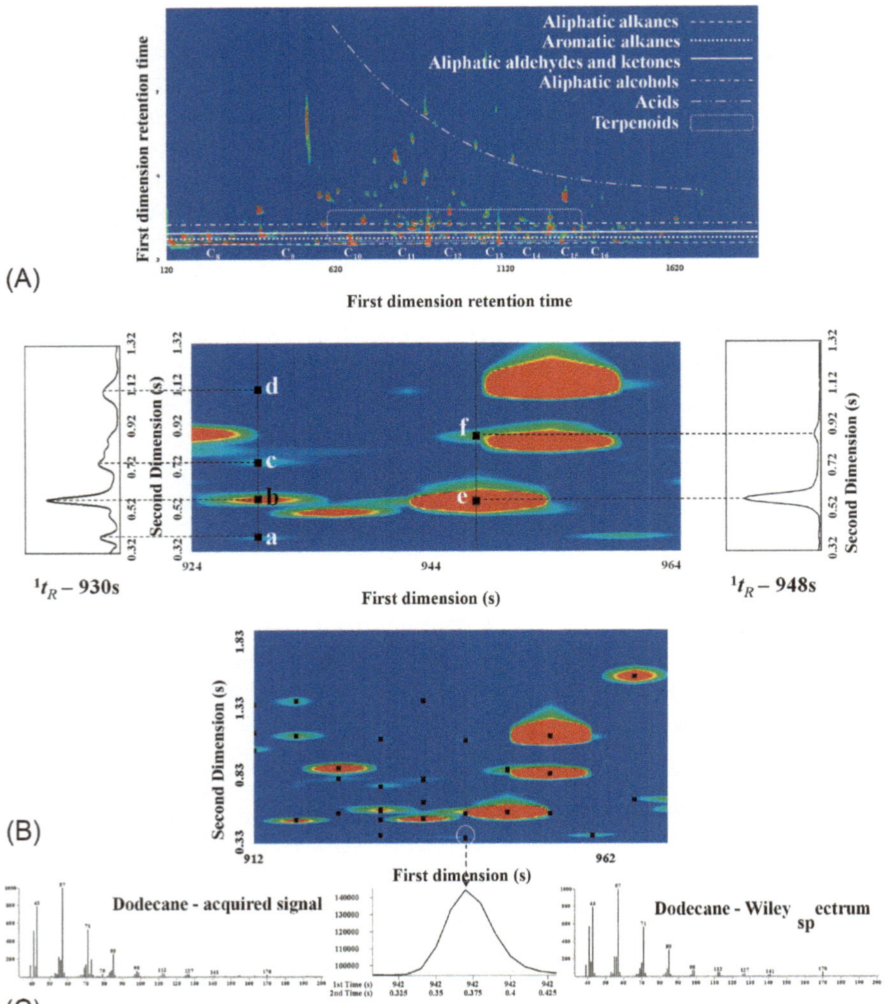

Fig. 8.5 (A) GC×GC-ToFMS total ion chromatogram contour plot of urine from a healthy adult. The *n*-alkanes series (C_8–C_{16}) was superimposed on the contour plot. Bands and cluster formed by structurally related compounds are indicated. (B) Blow-up of a part of the previous contour plot showing the separation of 1-dodecene (a), 2-decanone (b), linalool (c), α-phellandren-8-ol (d), decanal (e), and verbenone (f). (C) The identification of dodecane, one trace metabolite found in urine, whose GC×GC peak had 100ms-wide that was acquired at a mass spectral of 125 spectra/s.

Reproduced with permission from S.M. Rocha, M. Caldeira, J. Carrola, M. Santos, N. Cruz, I.F. Duarte, Exploring the human urine metabolomic potentialities by comprehensive two-dimensional gas chromatography coupled to time of flight mass spectrometry, J. Chromatogr. A 1252 (2012) 155–163, Copyright (2012), with permission from Elsevier.

found in body fluids samples. In fact, narrow peaks are obtained due to the peaks' cryo-focusing in the GC×GC modulator; and ToFMS acquire a continuous full-range mass spectrum, allowing getting high acquisition rates, which promote the spectral quality. An practical example can be observed in Fig. 8.5C, in which dodecane (trace urine metabolite) GC×GC peak, with 100ms-wide, was detected and identified at a mass spectral acquisition of 125 spectra/s, and had a mass spectrum similarity of 937/1000, compared with Wiley database.

GC×GC-ToFMS data may be acquired and processed through different approaches, namely by full-scan (scanning in a *m/z* range), single or multiple ion monitoring (SIM or MIM—data acquisition using specific *m/z* as diagnostic ions), and/or ion extraction mode (IEC—use of specific *m/z* as diagnostic ions to process MS data) [46, 47]. Full-scan acquisition mode is commonly used to acquire GC×GC data in exposome studies [8–11, 14, 18]. Nevertheless, IEC may be used to minimize the contribution of co-eluted peaks (allow a better visualization of a certain chemical family on the chromatographic space), and therefore to increase the target compounds' peak area. For instance, Rocha et al. [18] showed the use of several diagnostic ions (*m/z* 43, 56, 57, 71, and 85) to highlight lipid oxidation metabolites, such as aliphatic alkanes and aldehydes, in a chromatogram from human' urine, which helped to simplify the data obtained, and also reduced the analysis time. Qualifier ions can also help in analytes' identification, as Lamani et al. [11] show through the use of a specific iodine fragment ion (*m/z* 127) to identify aromatic amines in humans' urine, once this specific fragment ion is related with the in situ derivatization SPME procedure applied to samples.

Untargeted studies usually presuppose the use of the chromatographic area to estimate the analytes' amount, once the goal is to measure and compare as many analytes as possible [10, 14, 18]. For instance, by profiling urine metabolome in children and elderly residents living near a petrochemical complex, based on chromatographic area of each derivatized metabolite extracted using the unique mass [10], it is possible to link their internal exposure to oxidative stress biomarkers, which may be traced to multiple air toxics emitted from specific sources of oil refineries and coal-fired power plants. On the other hand, targeted or semitargeted studies imply an accurate identification and quantification of a defined set of analytes, therefore, acquired raw data may be normalized through the use of internal (or external) standards, and analytes' amount can be determined by external calibration curves or standard addition method. The selection of the internal standard is dependent on the analytical methodology and the target compounds. In fact, pollutants' determination commonly use $^{13}C_{12}$ labeled internal standards [8, 9], which are the recommended type of standards once is assured an equal response factor of the analytes. Moreover, it is important to mention that pollutants quantification is usually performed by isotope dilution (i.e., a known amount of an isotopically enriched compound is added to the sample under study). Also, there are several directives that should be followed in order to fulfill the legal criteria for pollutants quantification [52–55], in which there are several important guidelines, regarding: the analytical methodology that should be used; the number of replicates; the criteria for identification of interest' analytes (including the standards and qualifier ions criteria); the limits of detection and quantification of the methodology; the procedure to perform calibration curves; the accuracy, recovery, and precision of the methodology; among others.

Applications toward current challenges

Environmental exposures are very dynamic and variable throughout humans' lifetime, being still quite challenging to measure the exposome in a comprehensive way. The complete coverage of the humans' exposome implies a parallel analysis through different analytical platforms and/or complementary assays. On the other hand, current challenges also comprise the simultaneously determination of wide range of chemicals, namely persistent pollutants with potential adverse impact on human well-being and health. In order to highlight the real benefits of the use of GC×GC, representative research studies of these current challenges were selected.

An example of the exposome assessment of a population that lives near oil refineries and coal-fired power plants using different analytical methodologies was illustrated by Chen et al. [10]. In fact, this study combines the impact of external and internal exposures in the humans' exposome, through the analysis of different analytes in urine from children and elderly residents (high and low exposure, according to the living distance to the petrochemical complex). The general external exposures included the living distance to the petrochemical complex (oil refinery and coal-fired power plant), the road area surrounding homes, the ambient' concentrations of vanadium and several polycyclic aromatic hydrocarbons (PAH). External exposures, namely one PAH and several heavy metals, were assessed in urine through high-performance liquid chromatography and inductively coupled plasma mass spectrometry (ICP-MS), respectively. Furthermore, four oxidative stress biomarkers were analyzed in urine by liquid chromatography-tandem mass spectrometry (LC-MS/MS). An untargeted urinary metabolomics approach was also performed using GC×GC-ToFMS, in which 405 and 391 potential metabolite peaks were identified in each individual' urine from children and elderly, respectively.

These different domains of information were processed through different statistical approaches, namely: partial least squares discrimination analysis (PLS-DA) to identify exposures-related potential metabolites, and the models were validated using permutation and cross-validation test; Pearson's correlation test to assess the association of the oxidative stress biomarkers; Student's *t*-test or Wilcoxon Mann Whitney test were used to analyze continuous variables, while Chi-squared test or Fisher's exact test were used to discrete variables, regarding the basic characteristics and external exposures; also an analysis of covariance (ANCOVA) test was performed to compare the urine exposure biomarker concentration between low and high exposure groups, adjusting several parameters (e.g., age, gender, smoking, alcohol consumption, among others).

These statistical procedures were combined in a meet-in-the-middle approach (Fig. 8.6), in which was possible to correlate the different exposures (external and internal) and their impact on the biological pathways, particularly in the urine metabolic profiles and oxidative stress biomarkers. GC×GC-ToFMS played a key role to unveil the urine metabolic profiles of this population, which allowed understanding the metabolic pathways that are disrupted under these environmental exposures, and consequently founding putative intermediate biomarkers that are involved in pathways potentially affected by these exposures, namely the tryptophan metabolism and

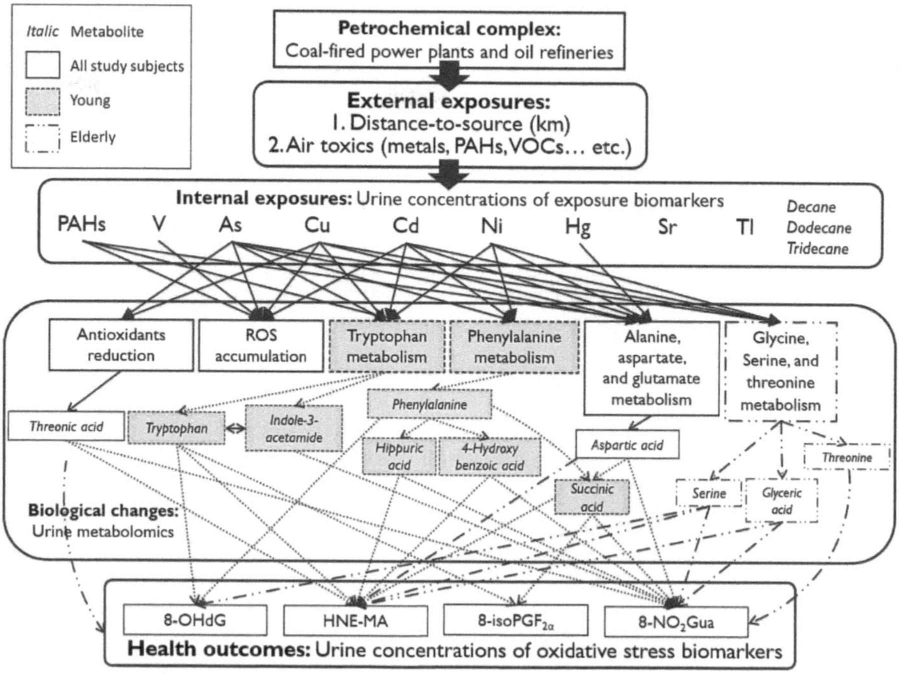

Fig. 8.6 Meet-in-the-middle approach that correlates the different exposures (external and internal) and their impact on the biological pathways within a population exposed to petrochemical air pollution. The urine metabolic profile changes and oxidative stress biomarkers were also correlated, according to the age-specific biological pathways. Reproduced with permission from C.H.S. Chen, T.H. Yuan, R.H. Shie, K.Y. Wu, C.C. Chan, Linking sources to early effects by profiling urine metabolome of residents living near oil refineries and coal-fired power plants, Environ. Int. 102 (2017) 87–96, Copyright (2017), with permission from Elsevier.

phenylalanine metabolism in children subjects; glycine, serine, and threonine metabolism in elderly subjects; and alanine, aspartate, and glutamate metabolism in both age groups. Chen et al. [10] also demonstrated that the air toxics from this industrial emission source had huge impact on population' health, once they showed to have higher levels of the urine exposure biomarkers and changes in the urine metabolic profiles that are related with increased oxidative stress (namely oxidative and nitrative DNA damage, and lipid peroxidation).

The persistent organic pollutants (POPs) represent a huge environmental concern, once they do not breakdown easily, and consequently accumulate in the environment, posing a risk of harm to human health [55]. Indeed, most of the POPs identified until now were already banned or have restricted use, being polychlorinated biphenyls (PCBs) and organochlorine pesticides (OCPs) the most studied chemical pollutants. Also, there have been appearing emergent persistent toxic compounds, such as polybrominated diphenyl ethers (oxides) (PBDEs or PBDPOs) that is a class of brominated flame retardants (BFRs). The determination of these different classes of pollutants is still quite challenging, once the determination of each one requires different instrumental parameters and GC columns. Focant et al. [8] proposed a strategy that intended to overcome this drawback and allowed a simultaneous determination of several classes of pollutants based on GC×GC-IDToFMS (isotope dilution). This methodology was applied for the simultaneous quantification of PCB, OCPS, and BFRs in human serum and milk, using reference ^{13}C-labeled internal standards.

The performance of the methodology based on GC×GC-IDToFMS was compared and tested against to the reference method using a GC-isotope dilution high-resolution mass spectrometry (IDHRMS) [8]. This targeted approach based on GC×GC-IDToFMS had a huge advantage once it allowed determining 59 pollutants (Fig. 8.7) in a single analysis, while 3 separate injections were required to perform their quantification in the GC-IDHRMS. This fact is due to the enlarged GC×GC chromatographic space that increases the peak capacity due to the analytes' separation in two dimensions, and the separation of potential interfering compounds. Also, the detection limits (LOD) were similar between the two methodologies since the 1D-GC was performed using a high-mass resolution detector when comparing to a low-mass resolution as ToFMS, namely the instrumental LOD ranged between

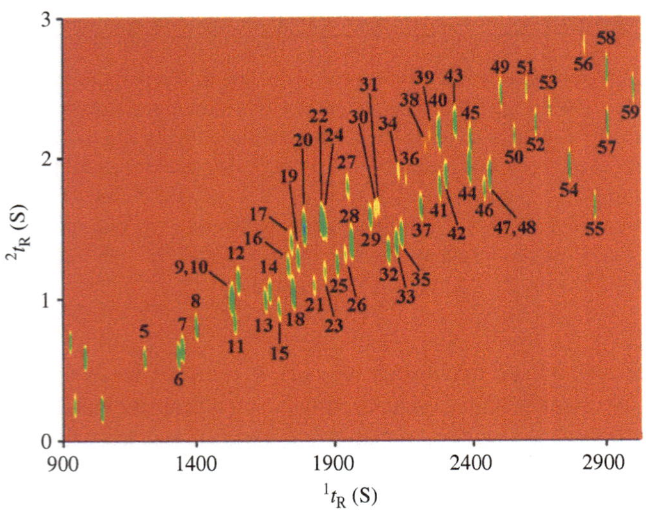

Fig. 8.7 GC×GC total ion contour plot of 59 reference standard pollutants. The chromatographic parameters for the separation of the 59 standards were selected to ensure a minimum of co-elutions in an acceptable time scale of 50 min (1.2 compound/min). The peak number assignment is available in the original paper [8].
Reproduced with permission from J.F. Focant, A. Sjödin, W.I. Turner, D.G. Patterson, Measurement of selected polybrominated diphenyl ethers, polybrominated and polychlorinated biphenyls, and organochlorine pesticides in human serum and milk using comprehensive two-dimensional gas chromatography isotope dilution time-of-flight mass spectrometry, Anal. Chem. 76 (21) (2004) 6313–6320, Copyright (2004) American Chemical Society.

0.5 and 10 pg/μL and the method LOD ranged between 1 and 15 pg/μL. GC×GC-IDToFMS-based methodology showed to be a well-suited tool for the determination of these persistent and emerging pollutants and can be further enlarged to other toxicants (e.g., toxaphenes, dibenzo-*p*-dioxins, among others).

Specific external exposures, such as the cigarette smoking can promote a significant risk factor for cancer development, either for smokers and nonsmokers. In fact, there has been reported an increased risk of cancer, coronary heart disease, and respiratory illnesses due to exposure of nonsmokers to second-hand tobacco smoke [56]. These exposures may promote several metabolic alterations that can be assessed in the body fluids, through different instrumental and methodological approaches.

Table 8.1 comprises the comparison between the performance of 1D-GC and GC×GC for the determination of two classes of compounds in different body fluids: alkanes, metabolites related to oxidative stress, and aromatic amines, harmful components of cigarette smoke. Exhaled breath was the body fluid used to the alkanes determination [57, 58], while the aromatic amines were determined in urine [11, 56]. GC×GC-ToFMS proved to be more efficient for this targeted approach, for instance, the number of identified analytes per analysis was lower for 1D-GC comparing to GC×GC-ToFMS determinations: 13 and 41 in alkanes, and 6 and 16 in aromatic amines, respectively. Moreover, the LOD range was significantly lower (41–85 pg) for the alkanes determined by GC×GC-ToFMS [58], instead of the ca. 1800 pg registered for GC-qMS [57]. Regarding the aromatic amines determination in urine, the LOD range was also lower for GC×GC-ToFMS (3.8–24.4 pg/mL) [11], comparing to 1.8–112.2 pg/mL registered for GC-MS/MS [56]. These examples

Table 8.1 Comparison between the performance (number of identified analytes per analysis and LOD range) of 1D-GC and GC×GC in the analysis of alkanes and aromatic amines in different body fluids (exhaled breath and urine).

Body fluid	Analytes' class	Sample extraction	Chromatographic analysis	N° of identified analytes per analysis	LOD range[a]	Reference
Exhaled breath	Alkanes	Bio-VOC tube followed by SPME[b]	GC-qMS[c]	13	ca. 1800 pg	[57]
		Bio-VOC tube followed by thermal desorption tube (trap)	GC×GC-ToFMS[d]	41	41–85 pg	[58]
Urine	Aromatic amines	SPE[e]	GC-MS/MS	6	1.8–112.2 pg/mL	[56]
		SPME	GC×GC-ToFMS	16	3.8–24.4 pg/mL	[11]

[a] This range includes the LODs determined for several components of each chemical family.
[b] Solid-phase microextraction.
[c] Gas chromatography-quadrupole-mass spectrometry.
[d] Comprehensive two-dimensional gas chromatography-time-of-flight mass spectrometry.
[e] Solid-phase extraction.

show the pertinence and utility of GC×GC-ToFMS as a powerful analytical tool, which by one side allows to have in-depth comprehension of the body fluids composition, and on the other side is able to detect and quantify target compounds that may usually be present in trace amounts, but that can act as real chemical exposures or have impact on humans' metabolism.

Concluding remarks and future trends

Exposome measurement represents a huge challenge due to the amount of data and metadata that is generated in the analysis of all exposures along lifetime. Therefore, exposome' measurement and characterization requires multidisciplinary approaches, using multiple analytical technologies and tools. MDGC may have a key role in the understanding of exposome; in particular, GC×GC is currently being used in top-down approaches that implies the exposome' study through the analysis of body fluids. GC×GC-ToFMS presents several advantages over the conventional 1D-GC, namely shorter run times, lower limits of detection, enhanced resolution and peak capacity, and higher mass sensitivity and selectivity due to the peak focusing in the modulator [29, 43, 59]. Furthermore, signal-to-noise ratio is improved for GC×GC compared to 1D-GC [44]. Therefore, GC×GC-ToFMS can provide in-depth information about the body fluids composition, either in the amount of achieved data but also in the detection of analytes that may be present in trace amounts. This higher sensitivity and selectivity could be an important feature, for instance in the detection and quantification of certain pollutants, whose legal quantification parameters must be more easily achieved. However, the high costs associated to the equipment acquisition, operation, and maintenance are some of the GC×GC disadvantages. Novel cryogen-free versions of the modulators are already available, which contribute to the reduction of the operation costs. Moreover, expert technician is required due to the complexity of the instrumentation, which generates quite complex data. The alignment, integration, and processing of the generated data is still a challenging and time-consuming task [29]. In fact, data processing and storage are still a limitation, being open to new developments of bioinformatics tools that allow a rapid, easily, and also a comprehensive analysis of different domains of information for the interpretation of achieved signals from the chromatographic systems, which sometimes cannot be interpreted. The recent equipment include integrated software platform that acquires data, controls all hardware and analyzes and reports results with a high level of automation, tailored to reduce the operator's work. The creation of databases may potentially help researchers in the unknown analytes' identification. Also, it is expected in the near future, the development of fastest and miniaturized solutions (even portable devices) that will lower the equipment's production cost [60] and will allow real-time monitoring of exposures, and consequently will be routinely used in the labs.

Despite the huge potential, GC×GC is still few explored in top-down approaches. Indeed, the metabolites concentration is highly conditioned by the body fluid selection, once size, weight, or volume of the individual may significantly affect the relative quantification results. Body fluids normalization is required due to the greater diversity of metabolite structures that can be detected. In fact, this is a challenging issue once there is no standard method for directly measuring the total metabolites amount in a single way [61]. Furthermore, the high-dimensional data produced by GC×GC can be a great input in exposome studies, allowing the correlation between metabolites and biological pathways, which can posterior lead to network construction. Nevertheless, these data must be complemented with other level of information's, including other instrumental analysis or databases in order to allow biological interpretation. For that, it is crucial to have common scientific language and quality controls for a successful collaboration and also for interlaboratory assays, e.g., methodological standard procedures for sample collection, storage time, the use of reference materials, or use of methodologies. The combination of the comprehensive monitoring of body fluids' composition at different omic levels, and the current developments and use of sensor-based technologies may contribute, in the future, to perceive the risk assessment in patients [62]. These outcomes will contribute for a step forward into personalized medicine, enabling a better understanding of the impact of environmental exposures on human health.

Acknowledgments

Thanks are due to FCT/MEC for the financial support to the QOPNA (UID/QUI/00062/2019) and LAQV-REQUIMTE (UIDB/50006/2020), through national funds and where applicable cofinanced by the FEDER, within the PT2020 Partnership Agreement.

References

[1] C.P. Wild, Complementing the genome with an "exposome": the outstanding challenge of environmental exposure measurement in molecular epidemiology, Cancer Epidemiol. Biomark. Prev. 14 (8) (2005) 1847–1850.

[2] C.P. Wild, The exposome: from concept to utility, Int. J. Epidemiol. 41 (1) (2012) 24–32.

[3] M.T. Smith, C.M. McHale, R. de la Rosa, Using exposomics to assess cumulative risks from multiple environmental stressors, in: S. Dagnino, A. Macherone (Eds.), Unraveling the Exposome, first ed., Springer International Publishing AG, part of Springer Nature. Switzerland, 2019, pp. 3–22.

[4] M.M. Niedzwiecki, G.W. Miller, The exposome paradigm in human health: lessons from the emory exposome summer course, Environ. Health Perspect. 125 (6) (2017) 1–4.

[5] P. Vineis, M. Chadeau-Hyam, H. Gmuender, J. Gulliver, Z. Herceg, J. Kleinjans, M. Kogevinas, S. Kyrtopoulos, M. Nieuwenhuijsen, D.H. Phillips, N. Probst-Hensch, A. Scalbert, R. Vermeulen, C.P. Wild, The exposome in practice: design of the EXPOsOMICS project, Int. J. Hyg. Environ. Health 220 (2) (2017) 142–151.

[6] M. Vrijheid, R. Slama, O. Robinson, L. Chatzi, M. Coen, P. van den Hazel, C. Thomsen, J. Wright, T.J. Athersuch, N. Avellana, X. Basagaña, C. Brochot, L. Bucchini, M. Bustamante, A. Carracedo, M. Casas, X. Estivill, et al., The human early-life exposome (HELIX): project rationale and design, Environ. Health Perspect. 122 (6) (2014) 535–544.

[7] T. Athersuch, Metabonomics, in: S. Dagnino, A. Macherone (Eds.), Unraveling the Exposome, first ed., Springer International Publishing AG, part of Springer Nature. Switzerland, 2019, pp. 147–181.

[8] J.F. Focant, A. Sjödin, W.I. Turner, D.G. Patterson, Measurement of selected polybrominated diphenyl ethers, polybrominated and polychlorinated biphenyls, and organochlorine pesticides in human serum and milk using comprehensive two-dimensional gas chromatography isotope dilution time-of-flight mass spectrometry, Anal. Chem. 76 (21) (2004) 6313–6320.
[9] D. Megson, J.F. Focant, D.G. Patterson, M. Robson, M.C. Lohan, P.J. Worsfold, S. Comber, R. Kalin, E. Reiner, G. O'Sullivan, Can polychlorinated biphenyl (PCB) signatures and enantiomer fractions be used for source identification and to age date occupational exposure? Environ. Int. 81 (2015) 56–63.
[10] C.H.S. Chen, T.H. Yuan, R.H. Shie, K.Y. Wu, C.C. Chan, Linking sources to early effects by profiling urine metabolome of residents living near oil refineries and coal-fired power plants, Environ. Int. 102 (2017) 87–96.
[11] X. Lamani, S. Horst, T. Zimmermann, T.C. Schmidt, Determination of aromatic amines in human urine using comprehensive multi-dimensional gas chromatography mass spectrometry (GCxGC-qMS), Anal. Bioanal. Chem. 407 (1) (2015) 241–252.
[12] S.E. Prebihalo, K.L. Berrier, C.E. Freye, H.D. Bahaghighat, N.R. Moore, D.K. Pinkerton, R.E. Synovec, Multidimensional gas chromatography: advances in instrumentation, chemometrics, and applications, Anal. Chem. 90 (2018) 505–532.
[13] E.C.Y. Chan, K.K. Pasikanti, J.K. Nicholson, Global urinary metabolic profiling procedures using gas chromatography-mass spectrometry, Nat. Protoc. 6 (10) (2011) 1483–1499.
[14] M. Caldeira, R. Perestrelo, A.S. Barros, M.J. Bilelo, A. Morête, J.S. Câmara, S.M. Rocha, Allergic asthma exhaled breath metabolome: a challenge for comprehensive two-dimensional gas chromatography, J. Chromatogr. A 1254 (2012) 87–97.
[15] M.A. Fernández-Peralbo, M.D.L. de Castro, Preparation of urine samples prior to targeted or untargeted metabolomics mass-spectrometry analysis, Trends Anal. Chem. 41 (2012) 75–85.
[16] M.D.L. de Castro, F. Priego-Capote, The analytical process to search for metabolomics biomarkers, J. Pharm. Biomed. Anal. 147 (2018) 341–349.
[17] C.I. Mack, B. Egert, E. Liberto, C.H. Weinert, A. Bub, I. Hoffmann, C. Bicchi, S. E. Kulling, C. Cordero, Robust markers of coffee consumption identified among the volatile organic compounds in human urine, Mol. Nutr. Food Res. 63 (10) (2019) 1–12.
[18] S.M. Rocha, M. Caldeira, J. Carrola, M. Santos, N. Cruz, I.F. Duarte, Exploring the human urine metabolomic potentialities by comprehensive two-dimensional gas chromatography coupled to time of flight mass spectrometry, J. Chromatogr. A 1252 (2012) 155–163.
[19] K.D. Nizio, M. Ueland, B.H. Stuart, S.L. Forbes, The analysis of textiles associated with decomposing remains as a natural training aid for cadaver-detection dogs, Forensic Chem. 5 (2017) 33–45.
[20] M. Chen, S. Kim, J.E. Park, H.J. Jung, J. Hur, Structural and compositional changes of dissolved organic matter upon solid-phase extraction tracked by multiple analytical tools, Anal. Bioanal. Chem. 408 (23) (2016) 6249–6258.
[21] P. Armstrong, K.D. Nizio, K.A. Perrault, S.L. Forbes, Establishing the volatile profile of pig carcasses as analogues for human decomposition during the early postmortem period, Heliyon 2 (2) (2016) e00070.
[22] S.L. Forbes, A.N. Troobnikoff, M. Ueland, K.D. Nizio, K.A. Perrault, Profiling the decomposition odour at the grave surface before and after probing, Forensic Sci. Int. 259 (2016) 193–199.
[23] K.A. Perrault, T. Rai, B.H. Stuart, S.L. Forbes, Seasonal comparison of carrion volatiles in decomposition soil using comprehensive two-dimensional gas chromatography—time of flight mass spectrometry, Anal. Methods 7 (2015) 690–698.

[24] K.A. Perrault, K.D. Nizio, S.L. Forbes, A comparison of one-dimensional and comprehensive two-dimensional gas chromatography for decomposition odour profiling using inter-year replicate field trials, Chromatographia 78 (15–16) (2015) 1057–1070.
[25] P.H. Stefanuto, K. Perrault, S. Stadler, R. Pesesse, M. Brokl, S. Forbes, J.F. Focant, Reading cadaveric decomposition chemistry with a new pair of glasses, ChemPlusChem 79 (6) (2014) 786–789.
[26] S. Stadler, P.H. Stefanuto, M. Brokl, S.L. Forbes, J.F. Focant, Characterization of volatile organic compounds from human analogue decomposition using thermal desorption coupled to comprehensive two-dimensional gas chromatography-time-of-flight mass spectrometry, Anal. Chem. 85 (2) (2013) 998–1005.
[27] P.H. Stefanuto, K.A. Perrault, S. Stadler, R. Pesesse, H.N. Leblanc, S.L. Forbes, J. F. Focant, GC × GC-TOFMS and supervised multivariate approaches to study human cadaveric decomposition olfactive signatures, Anal. Bioanal. Chem. 407 (16) (2015) 4767–4778.
[28] S. Stadler, P.H. Stefanuto, J.D. Byer, M. Brokl, S. Forbes, J.F. Focant, Analysis of synthetic canine training aids by comprehensive two-dimensional gas chromatography-time of flight mass spectrometry, J. Chromatogr. A 1255 (2012) 202–206.
[29] L. Mondello, P. Tranchida, P. Dugo, G. Dugo, Comprehensive two-dimensional gas chromatography-mass spectrometry: a review, Mass Spectrom. Rev. 27 (2008) 101–124.
[30] P.J. Marriott, S. Chin, B. Maikhunthod, H. Schmarr, S. Bieri, Multidimensional gas chromatography, Trends Anal. Chem. 34 (2012) 1–21.
[31] P.J. Marriott, S.T. Chin, B. Maikhunthod, H.G. Schmarr, S. Bieri, Multidimensional gas chromatography, Trends Anal. Chem. 34 (2012) 1–20.
[32] J.W. Wong, D.G. Hayward, K. Zhang, Gas chromatography-mass spectrometry techniques for multiresidue pesticide analysis in agricultural commodities, in: I. Ferrer, E. M. Thurman (Eds.), Comprehensive Analytical Chemistry, first ed., Elsevier B.V., Oxford, 2013, pp. 3–22.
[33] K.L. Organtini, A.L. Myers, K.J. Jobst, J. Cochran, B. Ross, B. McCarry, E.J. Reiner, F. L. Dorman, Comprehensive characterization of the halogenated dibenzo-p-dioxin and dibenzofuran contents of residential fire debris using comprehensive two-dimensional gas chromatography coupled to time of flight mass spectrometry, J. Chromatogr. A 1369 (2014) 138–146.
[34] M. Ubukata, K.J. Jobst, E.J. Reiner, S.E. Reichenbach, Q. Tao, J. Hang, Z. Wu, A.J. Dane, R.B. Cody, Non-targeted analysis of electronics waste by comprehensive two-dimensional gas chromatography combined with high-resolution mass spectrometry: using accurate mass information and mass defect analysis to explore the data, J. Chromatogr. A 1395 (2015) 152–159.
[35] P.Q. Tranchida, M. Maimone, F.A. Franchina, T.R. Bjerk, C.A. Zini, G. Purcaro, L. Mondello, Four-stage (low-)flow modulation comprehensive gas chromatography-quadrupole mass spectrometry for the determination of recently-highlighted cosmetic allergens, J. Chromatogr. A 1439 (2016) 144–151.
[36] H.E. Toraman, T. Dijkmans, M.R. Djokic, K.M. Van Geem, G.B. Marin, Detailed compositional characterization of plastic waste pyrolysis oil by comprehensive two-dimensional gas-chromatography coupled to multiple detectors, J. Chromatogr. A 1359 (2014) 237–246.
[37] X. Wang, J. Guo, J. Shang, L. Ding, G. Zhao, F. Xie, Y. Jia, Y. Qin, Y. Yu, L. Chen, S. Zhang, Determination of nitroalkanes in mainstream cigarette smoke by heart-cutting multidimensional gas chromatography system coupled with mass spectrometry detection, J. Chromatogr. A 1424 (2015) 118–126.

[38] C. Manzano, E. Hoh, S.L.M. Simonich, Improved separation of complex polycyclic aromatic hydrocarbon mixtures using novel column combinations in GC × GC/ToF-MS, Environ. Sci. Technol. 46 (14) (2012) 7677–7684.
[39] X. Li, X. Dai, X. Yin, M. Li, Y. Zhao, J. Zhou, T. Huang, H. Li, Impurity analysis of pure aldrin using heart-cut multi-dimensional gas chromatography-mass spectrometry, J. Chromatogr. A 1277 (2013) 69–75.
[40] S.A. Mackintosh, N.G. Dodder, N.J. Shaul, L.I. Aluwihare, K.A. Maruya, S.J. Chivers, K. Danil, D.W. Weller, E. Hoh, Newly identified DDT-related compounds accumulating in Southern California bottlenose dolphins, Environ. Sci. Technol. 50 (22) (2016) 12129–12137.
[41] R.L. Grob, E.F. Barry (Eds.), Modern Practice of Gas Chromatography, fourth ed., John Wiley & Sons, Inc., New Jersey, 2004.
[42] K.A. Kouremenos, J. Pitt, P.J. Marriott, Metabolic profiling of infant urine using comprehensive two-dimensional gas chromatography: application to the diagnosis of organic acidurias and biomarker discovery, J. Chromatogr. A 1217 (1) (2010) 104–111.
[43] P.J. Marriott, T. Massil, H. Hügel, Molecular structure retention relationships in comprehensive two-dimensional gas chromatography, J. Sep. Sci. 27 (15–16) (2004) 1273–1284.
[44] J. Dallüge, J. Beens, U.A.T. Brinkman, Comprehensive two-dimensional gas chromatography: a powerful and versatile analytical tool, J. Chromatogr. A 1000 (1–2) (2003) 69–108.
[45] T. Górecki, O. Panić, N. Oldridge, Recent advances in comprehensive two-dimensional gas chromatography (GC × GC), J. Liq. Chromatogr. Relat. Technol. 29 (7–8) (2006) 1077–1104.
[46] K. Dettmer, P.A. Aronov, B.D. Hammock, Mass spectrometry-based metabolomics, Mass Spectrom. Rev. 26 (2007) 51–78.
[47] B.L. Milman, General principles of identification by mass spectrometry, Trends Anal. Chem. 69 (2015) 24–33.
[48] ILAC, ILAC-G9:2005—Guidelines for the Selection and Use of Reference Materials, Australia, 2005.
[49] H. van Den Dool, P.D. Kratz, A generalization of the retention index system including linear temperature programmed gas—liquid partition chromatography, J. Chromatogr. A 11 (1963) 463–471.
[50] B.d.'A. Zellner, C. Bicchi, P. Dugo, P. Rubiolo, G. Dugo, L. Mondello, Linear retention indices in gas chromatographic analysis: a review, Flavour Fragr. J. 23 (2008) 297–314.
[51] V.I. Babushok, Chromatographic retention indices in identification of chemical compounds, Trends Anal. Chem. 69 (2015) 98–104.
[52] EPA, Method 8082A—Polychlorinated Biphenyls (PCBs) by Gas Chromatography, 2007.
[53] The European Commission, Comission Regulation (EU) No 589/2014 of 2 June 2014 laying down methods of sampling and analysis for the control of levels of dioxins, dioxin-like PCBs and non-dioxin-like PCBs in certain foodstuffs and repealing Regulation (EU) No 252/2012, Off. J. Eur. Union L164 (2014) 18–40.
[54] The European Commission, Regulation (EU) 2019/1021 of the European Parliament and of the Council of 20 June 2019 on persistent organic pollutants (recast), Off. J. Eur. Union L169 (850) (2019) 45–77.
[55] Science for Environment Policy, Future Brief: Persistent Organic Pollutants: Towards a POPs-Free Future, Future Brief 19. Brief produced for the European Commission DG Environment, 2017.

[56] S. Mazumder, R.A. Ahamed, E. McGahee, L. Wang, T.H. Seyler, A new automated method for the analysis of aromatic amines in human urine by GC-MS/MS, J. Anal. Toxicol. 43 (1) (2019) 25–35.
[57] D. Poli, P. Carbognani, M. Corradi, M. Goldoni, O. Acampa, B. Balbi, L. Bianchi, M. Rusca, A. Mutti, Exhaled volatile organic compounds in patients with non-small cell lung cancer: cross sectional and nested short-term follow-up study, Respir. Res. 6 (1) (2005) 71.
[58] M.K. Das, S.C. Bishwal, A. Das, D. Dabral, A. Varshney, V.K. Badireddy, R. Nanda, Investigation of gender-specific exhaled breath volatome in humans by GCxGC-TOF-MS, Anal. Chem. 86 (2) (2014) 1229–1237.
[59] J.V. Seeley, S.K. Seeley, Multidimensional gas chromatography: fundamental advances and new applications, Anal. Chem. 85 (2) (2013) 557–578.
[60] T. Dymerski, Two-dimensional gas chromatography coupled with mass spectrometry in food analysis, Crit. Rev. Anal. Chem. 48 (4) (2018) 252–278.
[61] Y. Wu, L. Li, Sample normalization methods in quantitative metabolomics, J. Chromatogr. A 1430 (2016) 80–95.
[62] S. Dagnino, Unravelling the exposome: conclusions and thoughts for the future, in: S. Dagnino, A. Macherone (Eds.), Unraveling the Exposome, first ed., Springer International Publishing AG, part of Springer Nature. Switzerland, 2019, pp. 425–437.

Synchrotron radiation-based spatial methods in environmental biogeochemistry

9

Aakriti Sharma, Dean Hesterberg
Department of Crop and Soil Sciences, North Carolina State University, Raleigh, NC, United States

Introduction

The advent of synchrotron-radiation (SR)-based analytical techniques such as X-ray absorption near-edge structure (XANES) spectroscopy, extended X-ray absorption fine structure (EXAFS) spectroscopy, synchrotron X-ray diffraction (SXRD), scanning transmission X-ray microscopy (STXM), and X-ray computed microtomography (CMT) have led to major scientific advances in environmental biogeochemistry [1–14]. The key reason is that synchrotron radiation (SR) techniques allow biogeochemists to delve more deeply into the complexity of fine-scale processes that drive element cycling. The advancement of SR techniques in terms of high flux density and coherence, low divergence, sensitive detector systems, and novel sample preparation techniques [10, 15] have provided scientists with powerful tools to probe the elemental spatial distributions and chemical speciation of nutrients and contaminants in complex environmental samples. These advanced state-of-the-art techniques have created possibilities to explore complex biogeochemical systems at different spatial scales, ranging from millimeter to nanometer scales, providing a unique perspective at each scale [16–18]. Fig. 9.1 illustrates a range of connected spatial scales related to soil and agricultural systems, highlighting the macro (e.g., millimeter) to molecular length scales that are typically targeted by SR techniques to derive a mechanistic understanding of the biogeochemical processes. The impact of SR and other analyses would be enhanced by improved upscaling models to connect the small-scale information to the practical management of land, water, and air in the context of global biogeochemical cycling. The complexities of biogeochemical systems present an analytical paradox related to scaling. That is, analyses at large scale (global, landscape, etc.) only infer the underlying drivers of the integrated processes at the scale that can be managed, whereas fine-scale (millimeter, micrometer, nanometer) analyses provide information about underlying mechanisms that control processes operating at the large scales—but a focus on the finer-scale system details tends to lose the larger-scale context. Therefore, analytical techniques that probe attributes of these

Multidimensional Analytical Techniques in Environmental Research. https://doi.org/10.1016/B978-0-12-818896-5.00009-0

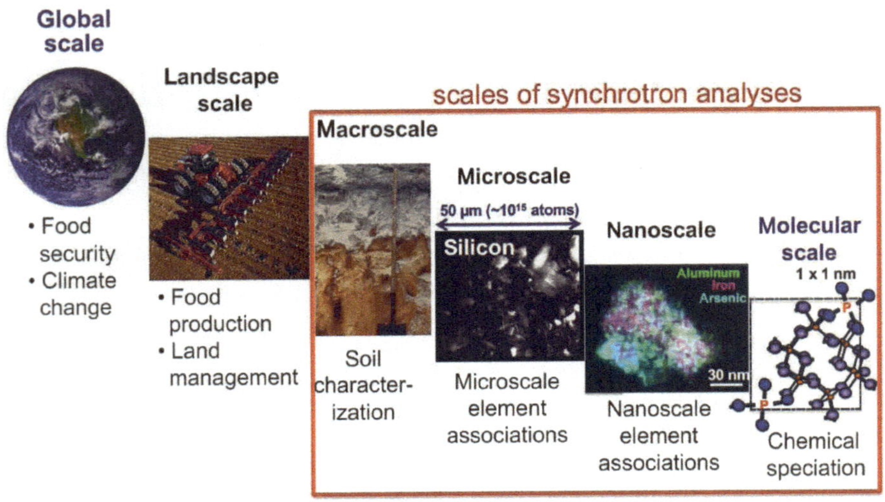

Fig. 9.1 Overview of different length scales of soil systems. Complex biogeochemical processes occur in soil systems that span from the global-to-molecular scale. A *red* box shows spatial scales at which different synchrotron methods are applied to gain a mechanistic understanding of effective land management at the landscape-to-global scale.
Modified from D. Hesterberg, M. Ricker, J. LeBeau, J. Guinness, M. Polizzotto, The NSF 2026 Idea Machine Competition, 2019. https://nsf2026imgallery.skild.com/.

systems at different spatial scales could ideally be combined to integrate knowledge about physical, chemical, and biological processes and their relative impacts on outcomes observed at a larger scale. Many research studies in biogeochemical systems are conducted, often independently, at different spatial scales. However, integrating analytical information from SR and other spatially resolving analytical techniques are complicated by differences in spatial resolution and spatial registry needed to overlay information.

Quantitatively upscaling micro, nano, and molecular-scale information to macro, landscape, and global scales is a major challenge because of the large-scale differences—the volume of planet earth is $\sim 10^{50}$ orders of magnitude greater than that of a single anion of oxygen (O^{2-}), the dominant element in the Earth's crust [19]. Moreover, there are overlying layers of complexity of biogeochemical systems with increasing spatial scale [20]. Although highly resolved spectroscopic techniques are critical for elucidating underlying reaction mechanisms of nutrients and contaminants in biogeochemical systems, the utility of these advanced analyses is somewhat marginalized by this lack of connectivity across spatial scales. Even establishing relationships between outcomes of measurements taken from millimeter-to-nanometer scales should improve our approaches for connecting such data to larger spatial scales. The goal of this chapter is to describe a number of SR techniques in the context of spatial scale, including the advantages and limitations of these techniques in relation to their impacts on the management of biogeochemical systems.

Overview of synchrotron radiation methods

Synchrotrons are large electron accelerators that produce broad-spectrum radiation. The generated radiation can be monochromized to select specific wavelengths, mainly within the hard X-ray, tender X-ray, soft X-ray, ultraviolet-visible (UV-vis), and infrared ranges. The broad-spectrum radiation or monochromized radiation can also be focused to micron-to-nanometer-sized beams. Focused beams are used for a variety of imaging techniques, including two-dimensional (2D) and three-dimensional (3D) chemical imaging, tomographic imaging, scattering, diffraction, Bragg coherent diffraction imaging (CDI), and ptychography [8, 21–28] to obtain chemical, physical, biological, crystallographic, and morphological details about biogeochemical systems. A number of synchrotron methods typically used in environmental biogeochemistry studies and key information provided by these methods are listed in Table 9.1. A summary of each technique with some examples is given later. We generally discuss these techniques in the context of microscale spatial resolution, but many of the techniques were first applied to macroscale samples (e.g., millimeter-scale) and have evolved into submicron-to-nanometer scale resolutions.

Table 9.1 List of synchrotron radiation-based methods applied in environmental biogeochemistry studies and key information provided by each method.

Synchrotron methods	Acronyms	Information provided	Selected references
X-ray absorption spectroscopy	XAS		
X-ray absorption near edge structure	XANES	Valence state, coordination geometry	[3, 8, 9, 29–31]
Extended X-ray absorption fine structure	EXAFS	Local molecular structure	[20, 32–35]
X-ray fluorescence microprobe	XRF	Chemical imaging	[3, 36–38]
Synchrotron X-ray diffraction	SXRD	Crystal structure, crystallinity	[6, 39–42]
Scanning transmission X-ray microscopy-Near-edge X-ray absorption fine structure	STXM-NEXAFS	Chemical imaging Local molecular structure	[4, 7, 11, 43, 44]
X-ray standing wave	XSW	Surface structure	[45–47]
X-ray fluorescence microtomography	XFMT	3D chemical imaging	[24, 48, 49]
Confocal X-ray fluorescence imaging	CXFI	3D chemical imaging	[50–52]

Continued

Table 9.1 Continued

Synchrotron methods	Acronyms	Information provided	Selected references
Ptychography	–	Structural anomalies; 3D imaging	[22, 25, 28]
Coherent diffraction imaging	CDI	3D shape, strain structure, crystal defects	[26, 27, 53]
Small-angle X-ray scattering	SAXS	Particle sizes and particle shapes; nucleation processes	[54–57]
Wide-angle X-ray scattering	WAXS	Crystalline phases; crystal structure	[58–61]
X-ray photoemission spectroscopy	XPS	Oxidation state; local chemical environment	[62–66]
Infrared spectromicroscopy	IRMS	Organic and inorganic chemical speciation; molecular structure	[67–70]

X-ray absorption spectroscopy

Among different synchrotron-based spectroscopic techniques, X-ray absorption spectroscopy (XAS) has been widely used to determine the chemical speciation of elements in environmental samples [3, 8, 9, 29, 30, 32, 33, 71–75]. Speciation is important for deepening our insights into binding/release mechanisms, chemical transformations, bioavailability, and toxicity. The power of XAS lies in its element specificity. It can provide information on oxidation states and local molecular coordination, regardless of the complexity of the matrix [5, 6, 8, 76, 77]. Two XAS techniques, XANES and EXAFS spectroscopies, involve measuring the absorption coefficient of a sample as a function of the incident X-ray energy scanned across the absorption edge of a given element of interest [78]. Features in the spectra around the edge, in the near-edge region above the edge energy, and in the extended (EXAFS) region out to >800 eV above the edge are related to the binding energies of the electrons in atoms of the target element, electronic transitions into bound or continuum states, and scattering of the photoelectrons from coordinating atoms around an excited atom [5, 32, 33, 77, 78]. The energies and shapes of these features are sensitive to oxidation states and average, local molecular bonding environments around these atoms [71–73, 79, 80]. Depending on the size of the X-ray beam used, XAS spectra can be collected from sample areas ranging from square millimeters to square nanometers in lateral dimensions. For example, a typical X-ray beam for bulk-sample analysis is 5–15 mm wide x ~1 mm high, whereas examples of focused beams may range from $5 \times 5\,\mu m^2$ down to $5 \times 5\,nm^2$ [81, 82]. The actual volume of the sample from which the absorption or fluorescence signal arises depends on the thickness of the sample, the penetration depth of the X-ray beam, and the escape depth of fluorescence electrons when measuring in fluorescence mode.

One of the advantages of XAS is that it allows for in-situ analyses of environmental samples, which helps to understand the biogeochemical processes under more representative environmental conditions, assuming that the samples do not change under the X-ray beam. Moreover, many biogeochemical processes such as redox reactions, adsorption, precipitation, etc. occur rapidly during the initial reaction, where a significant portion of the reaction process occurs before the first measurement could be made if conventional batch and flow techniques are to be used. The use of quick-XAS allows the study of rapid processes in real-time at sub-second time scales, providing information about chemical kinetics and reaction mechanisms, and giving insights into how initial reactions influence succeeding processes [5].

X-ray fluorescence microprobe

Many environmental and geochemical applications have used 2D chemical imaging of a host element in the context of sample matrix elements, either in fluorescence (e.g., micro X-ray fluorescence—μ-XRF) or transmission mode (e.g., scanning transmission X-ray microscopy—STXM). The energy used for imaging depends on the elements of interest, with the greatest sensitivity generally being achieved at energy at or above a feature of the absorption edge of a given element. For example, imaging at the arsenic K-edge (11,867 eV) with a focused X-ray beam of 12–13 keV would be sensitive to As, but also elements with edges at lower energies (e.g., Ge, Ga, Zn, Cu, Ni, Co, Fe, Mn, Cr, V, etc.), particularly in fluorescence mode.

μ-XRF is a powerful complement to μ-XAS as it provides information about the spatial distribution and relative concentrations of one or more elements of interest (e.g., trace elements) along with matrix elements, which can be correlated to determine spatial colocalization. This information is useful for interpreting μ-XAS data by identifying the elements that are associated with the element of interest. Moreover, the advent of fast detector systems and rapid (fly-)scanning techniques has led to the development of μ-XANES imaging [18, 75, 83] based on μ-XRF maps of hundreds to thousands of pixels collected at different energies across the absorption edge. This technique demonstrates the complementarity between μ-XRF and μ-XAS as it allows for the extraction of μ-XANES spectra from every pixel of the map and enables speciation analysis in relation to chemical composition and elemental distributions. Furthermore, the use of fast detector systems in μ-XANES imaging potentially offers an advantage of preserving sample integrity by reducing sample exposures to milliseconds per energy image rather than minutes for a typical energy scan on a single point.

Synchrotron X-ray diffraction

Synchrotron X-ray diffraction (SXRD) is another complementary technique that provides more definitive information about crystalline soil minerals. Similar to conventional laboratory XRD (LXRD), SXRD works on the principle of Bragg's law. In XRD analysis, constructive or destructive interferences occurring when scattered waves are in phase or out of phase lead to peaks in a scanned diffraction pattern, or bright (high intensity) spots on an aerial detector. Bragg's law relates positions of these peaks to the atomic structure of minerals and other crystalline materials [84]. As shown in Fig. 9.2, SXRD offers better sensitivity and resolution of diffraction peaks than conventional laboratory XRD because of high flux, tunable well-defined wavelength, and better collimation of SR [85], which also improves the identification of minor minerals (Fig. 9.2).

Lateral spatial resolution of the synchrotron X-rays incident on the sample enables one to reduce the heterogeneity of the sample in the diffraction volume. The availability of high brilliance synchrotron sources and the progress in achromatic focusing optics have created possibilities to map in-situ grain orientations, crystalline phase distribution, and strain/stress tensors at a microscopic length scale [86]. The development of scanning X-ray microdiffraction has allowed the collection of mineral species maps from a complex, heterogeneous geochemical matrix [6]. SXRD is a powerful complement to other SR techniques to obtain greater specificity of element bonding in biogeochemical systems [6, 14, 39, 40, 86, 87] as illustrated in Fig. 9.3.

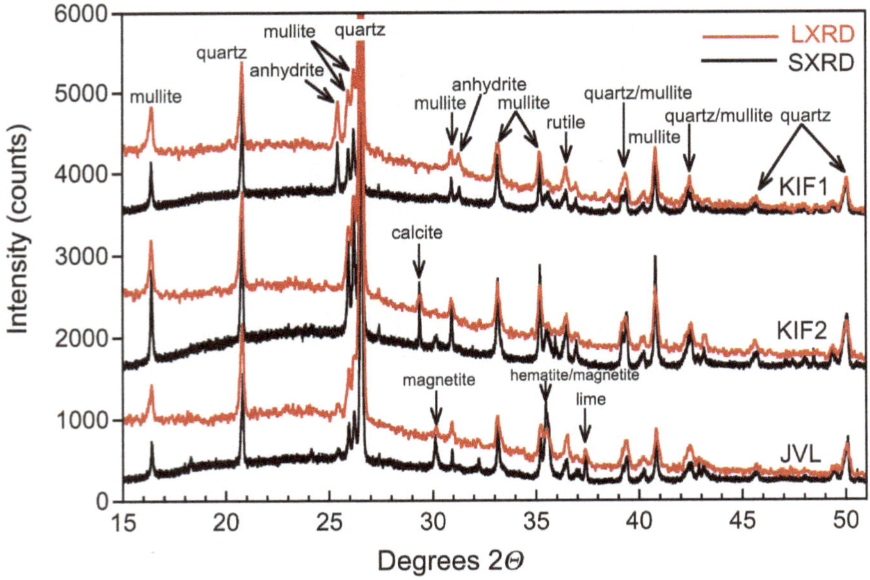

Fig. 9.2 Stacked laboratory (LXRD) and synchrotron (SXRD) X-ray diffraction patterns for coal fly ash samples comparing the sensitivity of these two methods. Generally, diffraction peaks obtained from SXRD are stronger and more well resolved than those from LXRD, especially for minor phases such as magnetite.

Modified and reprinted with permission from N. Rivera, N. Kaur, D. Hesterberg, C.R. Ward, R.E. Austin, O.W. Duckworth, Energy Fuel 29 (2015) 954. Copyright (2015) American Chemical Society.

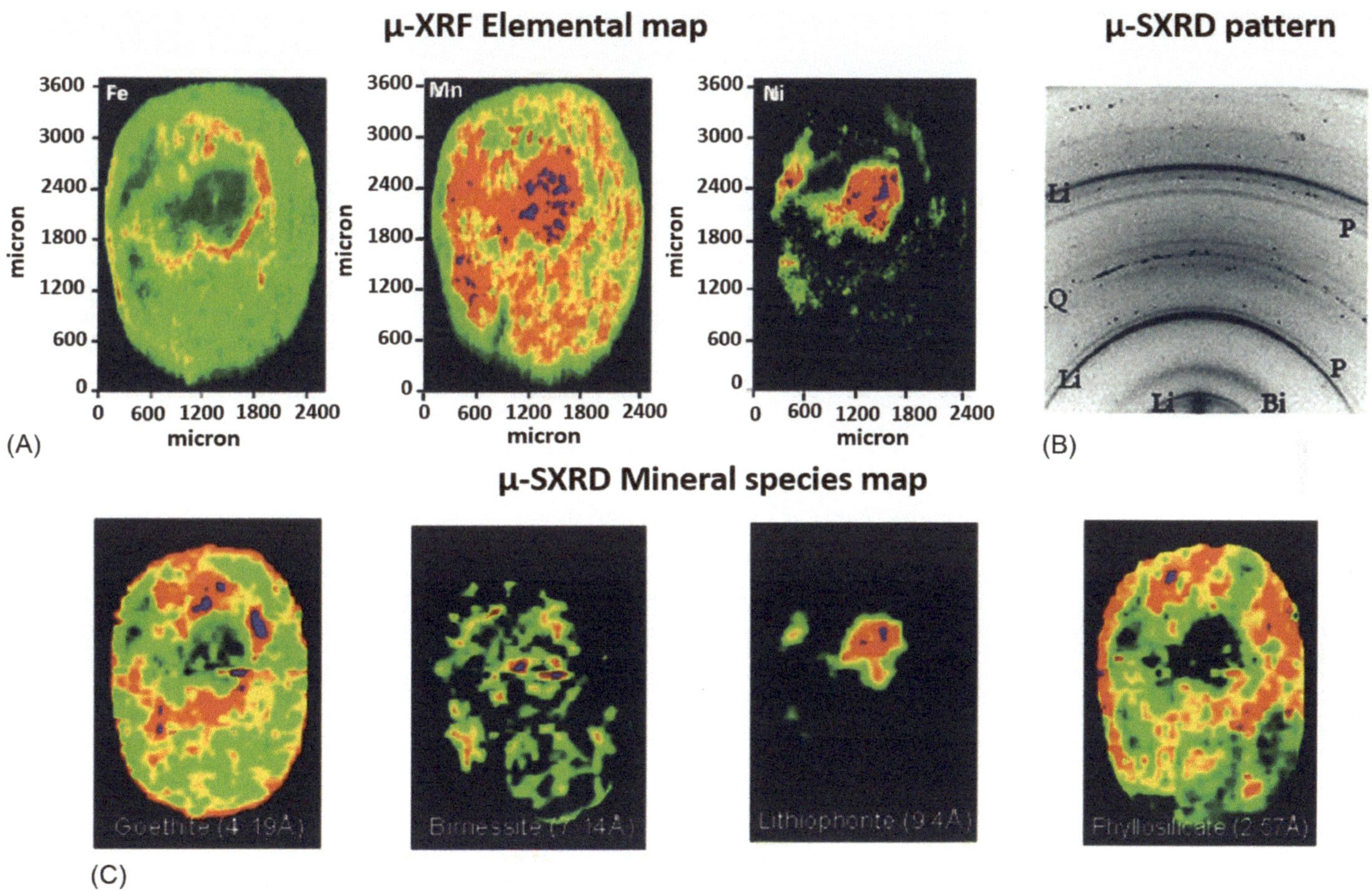

Fig. 9.3 See figure legend in next page

(continued)

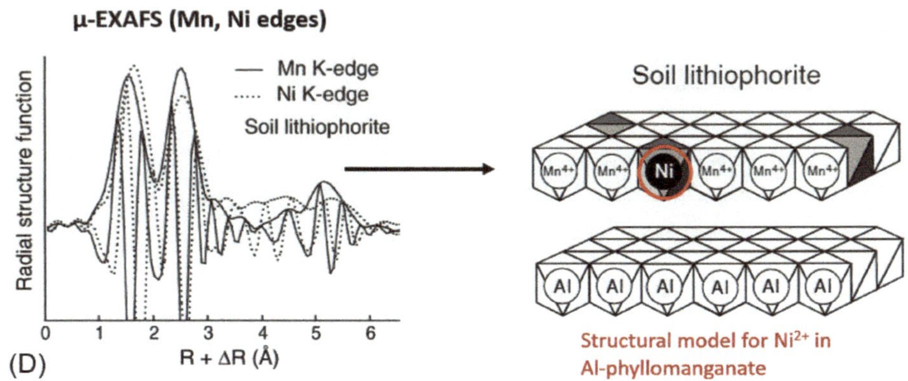

Fig. 9.3, cont'd Integration of spatially resolved synchrotron techniques—μ-XRF (A), μ-SXRD (B–C), and μ-EXAFS (D)—to analyze Ni sequestration in soil ferromanganese nodule. Such integrated approaches increase the specificity of chemical speciation analyses by constraining the possible species of trace elements in complex, heterogeneous geochemical systems.
Reproduced from A. Manceau, N. Tamura, M.A. Marcus, A.A. MacDowell, R.S. Celestre, R.E. Sublett, G. Sposito, H.A. Padmore, Am. Mineral. 87 (2002) 1494, with permission from the Mineralogical Society of America.

Scanning transmission X-ray microscopy

Scanning transmission X-ray microscopy (STXM) is a chemical imaging method that generates a microscopic image of a raster-scanned sample by detecting the intensity of transmitted X-rays from a focused beam. STXM typically uses a monochromatic soft X-ray beam in the energy range of 100–2200 eV; and the energy is scanned across X-ray absorption edges to generate near-edge X-ray absorption fine structure (NEXAFS) spectra. Moreover, sequences of images at tens to hundreds of different photon energies can also be acquired [88] to generate NEXAFS spectra for each pixel. Similar to (μ-)XANES spectra, NEXAFS spectra are generated by absorption of photons through excitation of core electrons to higher unoccupied energy states [89, 90]. The energy range of soft X-rays in STXM allows for accessing the 1s absorption edges of light elements such as C, N, O, Na, Mg, Al, Si, and P; and 2p absorption edges of elements of environmental relevance such as Fe, Mn, Cr, As, Zn, Ni, Ti, and Ca (Fig. 9.4). STXM-NEXAFS is commonly used to determine organic carbon distribution and its association with other elements (i.e., organo-mineral interactions) [1, 7, 43, 91–96]. Other applications include mapping metal species in hydrated organic samples such as biofilms or cell-mineral aggregates formed by bacteria [11, 97], and mapping polymer microstructures [98]. Obst et al. [44] reviewed the applications of STXM in biogeochemical environments.

X-ray standing wave

X-ray standing waves (XSW) are generated by the interference between the coherent X-ray beam impinging on a sample and the reflected X-rays. Unlike several other techniques that employ the interference of X-rays, XSW utilizes the inelastic scattering excited by the X-ray interference field [45, 46]. As the photon energy of the incident wave meets the Bragg condition, the XSW field above the surface changes in a characteristic way. The emitted photoelectrons, Auger electrons, or fluorescence that are generated by inelastic scattering processes can be used as element-specific signals that reflect the position of the adsorbate atom [45, 99]. Knowing the exact pattern and position of an X-ray interference field and being able to manipulate it in space provides information about the real-space distribution of inelastically scattering atoms [46]. The XSW phase is directly linked to the substrate reflecting lattice planes or interfaces, and thereby can be used to directly determine the positions of atoms of different elements relative to these substrate planes [45, 46]. In environmental biogeochemistry, the XSW methods are widely applied to investigate surface, interface, and thin-film structures. The short-period standing waves produced from the Bragg reflection of a perfect crystal have been used to probe the locations of ions within crystals and within the electrical double layer at a mineral/solution interface [45, 100], and to probe the positions of sorbed ions within calcite and at the calcite surface [101–103]. Similarly, long-period standing waves generated at grazing incidence have been used to study the reactivity of mineral surfaces in contact with metalloid bearing aqueous solutions [47]. The advantages of XSW lies in its ability to produce high-resolution,

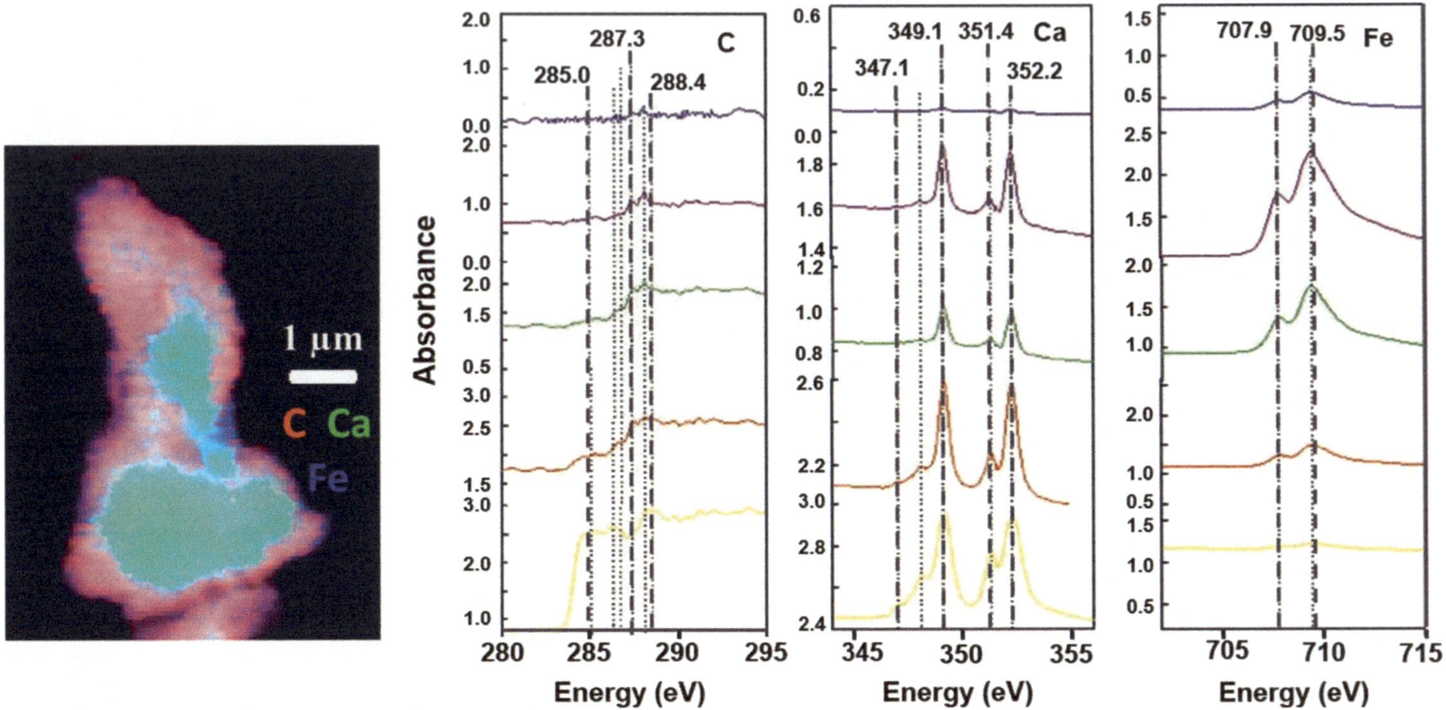

Fig. 9.4 Color-coded optical density map of C *(red)*, Ca *(green)*, and Fe *(blue)* generated using STXM imaging from a sample containing ferrihydrite and natural dissolved organic matter *(left)*; and C (1s), Ca (2p), and Fe (2p) spectra generated from STXM-NEXAFS analysis of the interface between a black carbon particle, other soil carbon and mineral matter *(right)*. This figure shows the applicability of STXM in analyzing both low- and high-Z elements. Figure on the left is reprinted with permission from T. Sowers, D. Adhikari, J. Wang, Y. Yang, D.L. Sparks, Environ. Sci. Technol. (2018). Copyright (2018) American Chemical Society. Figure on the right is reprinted from J. Lehmann, D. Solomon, Organic Carbon Chemistry in Soils Observed by Synchrotron-Based Spectroscopy, Elsevier. 2010, p. 289. Copyright (2010) with permission from Elsevier.

element-specific atomic structure information at the mineral/water interface and determine 3D lattice locations of bulk impurity atoms and surface adsorbates [99].

X-ray transmission and fluorescence microtomography

X-ray (transmission) microtomography (XMT) and X-ray fluorescence microtomography (XFMT) are 3D methods for, respectively, characterizing the physical and chemical structures of a sample. XMT results in 3D reconstructions of X-ray transmission cross sections collected while rotating a sample by 180 degree. XFMT is a variant of X-ray fluorescence imaging in which intact samples are imaged repeatedly while being rotated to create 3D chemical images. In XMT, interfaces between phases or structures of different absorption cross sections are segmented to show physical relationships between sample components. For example, Keyes et al. [104] used synchrotron XMT to visualize the 3D spatial relationships between plant root hairs and structured soil to enhance the modeling of phosphate uptake by roots. XFMT allows the reconstruction of virtual cross sections of elemental distribution from 2D sinograms [24, 105, 106]. In other words, 3D tomography involves acquiring 2D images over a given area of interest from multiple angles and an analytical step that reconstructs a 3D image of the sample [48]. The advantage of XFMT over 2D X-ray fluorescence imaging is that it allows the visualization of the composition or internal structure of a sample without physically sectioning it. However, since the fluorescent X-ray beam must pass through the sample thickness to the detector throughout the angular rotation of the sample, this technique provides element-specific information on limited sample volumes, depending on the fluorescence energy of the imaged element and absorption cross section of the matrix. Other biogeochemical applications of synchrotron XMT and XFMT include analyses of pore structure in relation to soil amendments affecting water retention and flow [107], spatial distributions and chemical speciation of metals in different plant parts [108–110], analyses of hydrated biological samples such as roots [24, 49, 104, 111–113], imaging biofilms in porous media [114–116], and identification of minerals [117, 118]. Besides these applications, synchrotron microtomography has also been used as a complement to other analytical methods. For example, XRF microtomography was combined with micro-XANES to determine Zn distribution and speciation within different parts of a rocket plant (*Eruca vesicaria* L. *Cavalieri*) grown in polluted soil with or without compost amendment (Fig. 9.5) [106].

Confocal X-ray fluorescence imaging

Another variant of X-ray fluorescence (XRF) imaging is confocal X-ray fluorescence imaging (CXFI), which, like XFMT, allows the visualization of the distribution of elements in three dimensions. Whereas XFMT provides a means to collect 3D chemical images especially on thinner samples, CXFI can also be applied to thick samples. Optical sectioning of a thick sample is enabled by the confocal volume or probe volume created by the intersection of incident and detection foci [50, 105, 119]. In general, the CXFI method involves confocal excitation with compound refractive lens

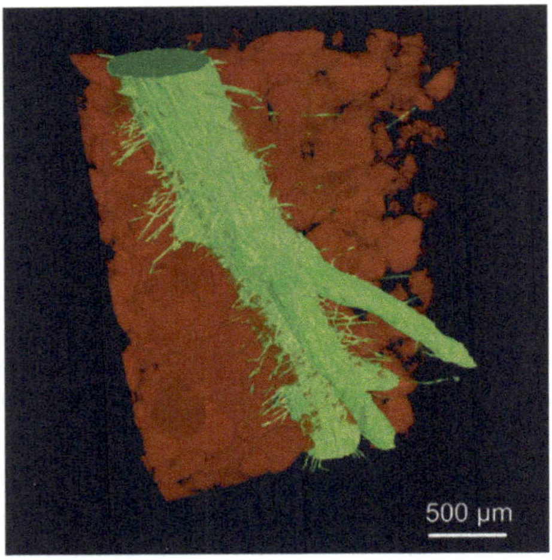

Fig. 9.5 Three-dimensional synchrotron XMT image of a seminal root of Triticum aestivum, including lateral roots and root hairs in a soil matrix.
Reprinted from S.D. Keyes, K.R. Daly, N.J. Gostling, D.L. Jones, P. Talboys, B.R. Pinzer, R. Boardman, I. Sinclair, A. Marchant, T. Roose, High-resolution synchrotron imaging of wheat root hairs growing in soil and image-based modeling of phosphate uptake, New Phytol. 198 (2013) 1023, with permission from John Wiley & Sons. © 2013 The Authors. New Phytologist © 2013 New Phytologist Trust.

systems and confocal detection optics, which enables depth resolution. Traditionally a polycapillary optic, which uses reflective glass capillaries, is used to selectively transmit X-ray fluorescence from sample to the detector. However, the polycapillary depth resolution deteriorates with energy and thus a collimating channel array (CCA), which defines an energy independent confocal volume, is preferred for high depth resolution (Fig. 9.6) [50]. Relevant applications of CXFI in biogeochemical research include identification of minerals [118], and analyses of elemental distribution in plants [120–122] and geologic materials [119, 123]. Moreover, CXFI is also used in combination with XFMT to obtain compositional and morphological information [51, 124].

Coherent diffraction imaging

Coherent diffraction imaging (CDI) using coherent synchrotron X-rays is a high-resolution spatial technique for constructing a real-space image of an irradiated sample via an inverse fast Fourier transform of a reciprocal-space image of diffracted X-rays collected on an aerial detector. Miao et al. [26] reviewed many CDI methods, including plane-wave CDI, Bragg CDI, ptychography, Fresnel CDI, and reflection CDI. Third- and especially fourth-generation synchrotrons (and free-electron lasers) produce a sufficiently high proportion of coherent X-rays to perform CDI imaging. Plane-wave and Bragg CDI utilize a coherent X-ray source that is larger than the reconstructed sample to produce electron density images or crystal-structural dislocations of whole objects. Analogous to these techniques, ptychography uses a focused X-ray beam to achieve even higher spatially resolved structural details of a sample by

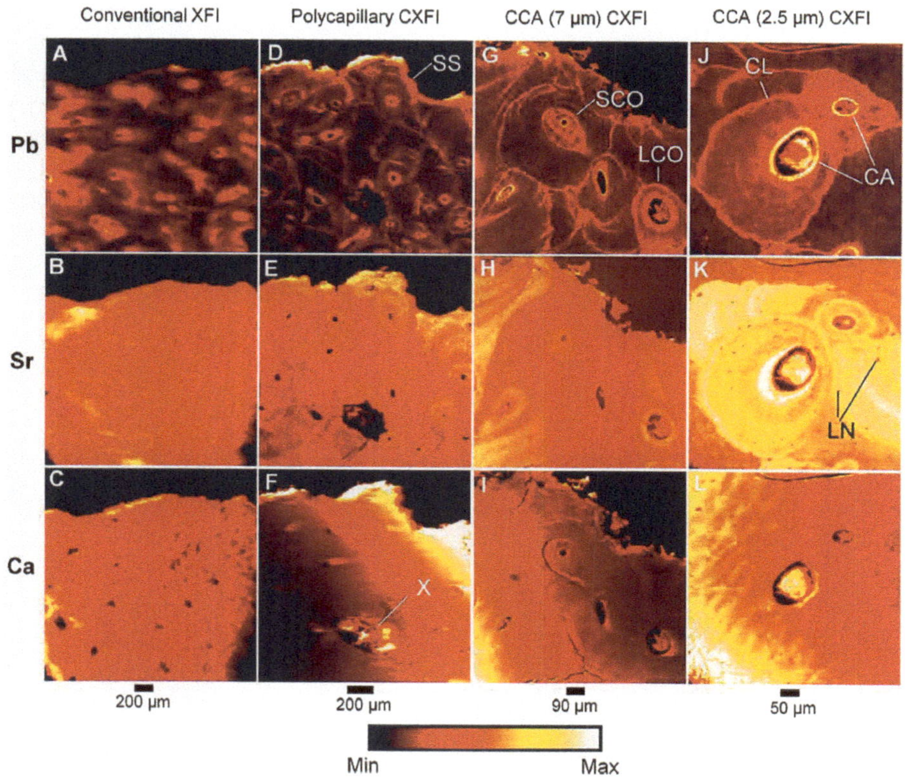

Fig. 9.6 Comparison of different optical configurations used in X-ray fluorescence imaging, showing elemental maps for Pb, Sr, and Ca collected on a thin rib section (A–C) of archaeological human bone for conventional XFI and the same fibula section (D–L) for all CXFI images. X-ray fluorescence imaging by employing CCA optics provided better spatial resolution than conventional and polycapillary-based detector optics and equivalent spatial resolution for low- and high-energy X-ray fluorescence.

reconstructing a real-space image from a series of overlapping images collected by rastering the sample through a focused beam on a 2D grid. Fresnel CDI reconstructs sample features by analysis of the Fresnel diffraction pattern generated from a curved wave, and reflection CDI measures surface topography from the interference pattern of reflected coherent X-ray waves [26]. A key advantage of CDI techniques compared with other spatially resolved synchrotron X-ray techniques is the high spatial resolution achieved for either noncrystalline (e.g., plane-wave CDI, ptychography) or crystalline (Bragg CDI, Fresnel CDI) samples, even without focusing optics [26, 27].

High-resolution CDI imaging techniques have been used to complement other common synchrotron X-ray techniques to characterize micro- to nanoscale structural and chemical complexities in biogeochemical systems. For example, Badger et al. [125] used Bragg CDI imaging in combination with μ-XRF and WAXS to characterize the shapes and crystal-structural displacements of biogenic Ca-oxalate monohydrate (whewellite) crystals in maize tissue (Fig. 9.7). Electron-density isosurface images (Fig. 9.7A), which were achieved with spatial resolutions down to 50 nm, depending

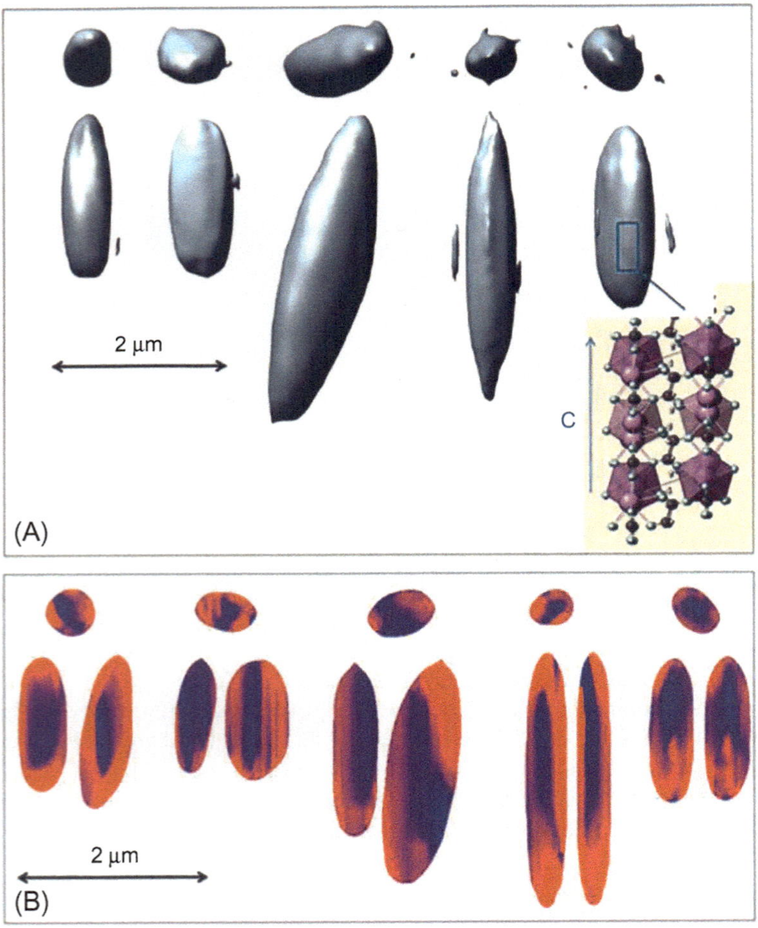

Fig. 9.7 Bragg CDI images of biogenic Ca-oxalate monohydrate (whewellite) crystals in maize. (A) top and side-view isosurface renderings of electron densities of 2–5 μm long whewellite crystals (atomic structure shown in inset) reconstructed from 3D Bragg CDI analysis of a diffraction peak at ~3.65 Å, with minimum spatial resolutions of 50–100 nm; (B) unit-cell displacements imaged with angstrom-scale resolution for orthogonal slices through the centers of the particles shown in (A), where the color scale ranges from *blue* (0 radians) to *red* (π radians, reflecting a half unit cell displacement from ideal position) back to *blue* (2π radians). Reproduced from J. Badger, J. Lal, R. Harder, H. Inouye, S.C. Gleber, S. Vogt, I. Robinson, L. Makowski, Sci. Rep. 3 (2013) 2843.

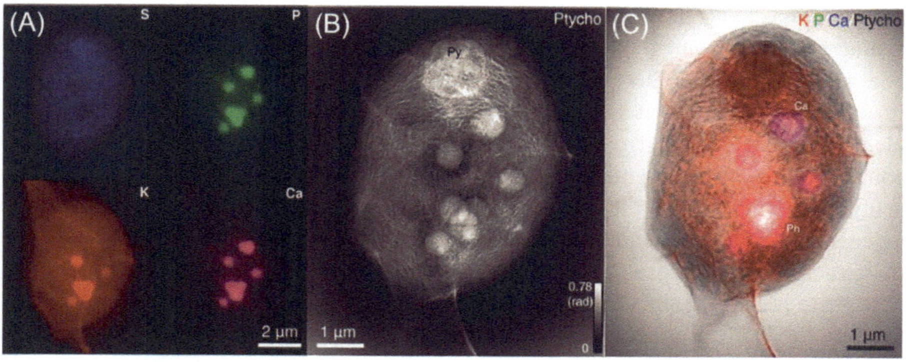

Fig. 9.8 Images of a Chlamydomonas algal cell under cryogenic conditions. (A) Micro-X-ray fluorescence (μ-XRF) chemical image with sub-100-nm spatial resolution showing concentrated regions of sulfur, phosphorus, potassium, and calcium; (B) ptychography image of electron density with sub-20-nm spatial resolution showing cellular substructure, including a pyrenoid (Py); and (C) combined image of the simultaneously collected μ-XRF and ptychography data indicating the presence of a polyphosphate body (Ph) with complexed calcium and a potassium-depleted body (Ca).
Reproduced from J. Deng, D.J. Vine, S. Chen, Q. Jin, Y.S. Nashed, T. Peterka, S. Vogt, C. Jacobsen, Sci. Rep. 7 (2017) 445.

on the diffracted signal intensity, show particle shapes along with details such as terminating sharp points. Analysis of the detailed structure of the selected Bragg reflection with angstrom-scale spatial precision also gives information on unit-cell displacements within the crystal structure. Images of these displacements (Fig. 9.7B) showed sharp boundaries (blue/red) between crystal domains of different phases, often along an axis from the center to the periphery of the whewellite particles (see details in Badger et al. [125]).

As another example, Deng et al. [25] combined synchrotron X-ray ptychography with μ-XRF chemical imaging of frozen algal cells to enhance the identity of cellular substructure (Fig. 9.8). The μ-XRF images themselves show distinct regions enriched in different elements (Fig. 9.8A), but their relationship to the cellular structure is lacking at the ~100-nm spatial resolution. Ptychography produced higher resolution (~20 nm) images of cellular substructure (Fig. 9.8B) but lacked chemical information. However, overlaying the μ-XRF and ptychography images (Fig. 9.8C) indicated the presence of subcellular bodies enriched in Ca-pyrophosphate complexes with varied potassium contents (Fig. 9.8, details in Deng et al. [25]). These examples illustrate that high-resolution CDI imaging is a powerful tool for characterizing both inorganic (minerals, non-crystalline solids) and biological components of biogeochemical systems, especially when used in combination with other spatially resolved synchrotron X-ray techniques.

Small-/wide-angle X-ray scattering

Small-angle X-ray scattering (SAXS) is a scattering method in which a sample is irradiated by X-rays and the elastically scattered X-rays are recorded at very low angles, typically between 0 and 5 degree to produce a spatially averaged intensity distribution [126]. The angular range contains information about the particle sizes and average particle shapes, characteristic distances of partially ordered materials, and pore sizes, in the nanometer size range. The applications of SAXS include the study of nucleation processes as well as the size distribution of colloidal particles. For example, Shaw et al. [127] have used small- and wide-angle X-ray scattering methods in combination with differential scanning calorimetry/thermogravimetric (TGA/DSC) analysis to understand the dehydration, structural collapse, and recrystallization mechanisms of hydrated calcium silicate minerals. Masion et al. used SAXS to study the structure of an aggregate—the hydrolysis product of Fe(III) in the presence of phosphate [128] and silicate [129] ligands. Similarly, Carrado et al. [54] used SAXS to study the nucleation and growth of silane-derived synthetic clay minerals, and Hu et al. [55] used SAXS to study effects of Al on Fe(III) (hydr)oxide nucleation, growth, and Ostwald ripening. Likewise, the distribution of colloidal particle size has been investigated using SAXS by Wagner et al. [130], Nakamura et al. [131], Cordoba et al. [56], and Garcia-Diez et al. [57]. Goertz et al. [132] showed that particle-size distributions of two colloidal, nonaggregated suspensions were more reliable when the SAXS method was used compared to transmission electron microscopy (TEM) and analytical ultracentrifuge (AUC). An advantage of SAXS over these other two techniques was that SAXS allowed determination of the mean diameter of the agglomerates and the primary particles [133].

Another complementary analytical technique to SAXS is wide-angle X-ray scattering (WAXS). WAXS is also one of the scattering methods which analyzes Bragg peaks scattered at wide angles, i.e., $2\theta > 1$ degree from nanometer-sized structures. WAXS is distinct from SAXS in that the distance from the sample to the detector is shorter and thus diffraction maxima are observed at larger angles. WAXS has been used to determine crystallinity and molecular structure [34, 58, 59, 134–136], and in combination with SAXS has been used to study the shape, size, and structure of colloidal particles [60, 61] and nucleation processes [137, 138].

X-ray photoelectron spectroscopy

In X-ray photoelectron spectroscopy (XPS), a monochromatic X-ray beam is used to excite core electrons from a specimen and the kinetic energies of all emitted electrons are measured to determine their binding energies (BE) [62]. The BE is characteristic for a given element and is also sensitive to its oxidation state and the local chemical environment surrounding the atom, which allows the differentiation of chemical species of an element in an unknown sample. XPS is a surface sensitive technique because only photoelectrons from the near surface region are detected due to their small penetration depth. To avoid any surface contamination by adsorbates from the ambient atmosphere, XPS experiments require ultrahigh vacuum conditions. XPS has been extensively used to probe the surface composition and electronic structure of environmental samples. The range of environmental applications of XPS includes studies of sorption processes [62–65, 139], the interaction of water with metal oxide surfaces [66, 140–142], and oxidation of metal sulfide surfaces [143–145].

Synchrotron-based infrared spectromicroscopy

Infrared spectroscopy is one of the vibrational spectroscopic techniques that rely on the absorption of infrared (IR) light with different wavelengths by the sample compounds. Absorption of IR light occurs when distortions of molecular structural units change the dipole moment [67, 146]. These transitions result in vibrations of molecular bonds. The peaks in the IR spectra represent the absorption of IR light by specific molecular moieties at specific frequencies due to stretching, bending, rocking, and wagging vibrations [67]. Most inorganic and organic compounds in the environment are IR active, and this technique can be used to provide information about molecule structure within bulk solids.

Unlike conventional IR spectroscopy, IR radiation emitted by the synchrotron storage ring has high brightness that allows generating diffraction-limited spatially resolved infrared images. Synchrotron-based IR spectroscopy or infrared spectromicroscopy (IRMS) provides the opportunity to examine low-frequency adsorbate-adsorbent vibrational bands and identification of chemical species at interfaces [147]. IR spectroscopy is especially sensitive to vibrations of O—H, C—H, C—O, N—H, and C—N bonds and thus it can be applied to understand the chemistry of biological materials such as living cells [68, 148–151] and organic carbon chemistry [69, 152, 153]. Other applications of IR spectromicroscopy include identification of chemical species [154–157], location, and characterization of contaminants in soils and sediments [158–161], the study of fate and transport of organic contaminants in plants [162–164], and biogeochemical transformation of environmental pollutants [165, 166]. Reviews on the applications of infrared spectromicroscopy to study biogeochemical systems are provided by Holman and Martin [167], Hirschmugl [147], Martin et al. [70], and Holman [68].

In summary, SR techniques have emerged as powerful analytical tools because of their element specificity, high sensitivity, wide tunability, and the ability to perform nondestructive, in-situ analyses of samples, often with minimal sample preparation. The development of spatially resolved SR techniques has allowed geochemists to capitalize on the natural heterogeneity of soils to segregate chemical species into more homogeneous environments. Holistic approaches to integrating multiple SR techniques provide multiple analytical perspectives for greater specificity of analyses of complex biogeochemical systems.

Integration of multidimensional analytical techniques and its challenges

Ideally, spectroscopic information collected at different spatial scales (Fig. 9.1) could be combined to provide a unified analytical assessment of both the context of the chemistry studied at larger scales, e.g., the average speciation of an element, and more specific information at smaller scales including the diversity of minor species of the element. For example, linear-combination fitting analysis of the millimeter-scale (bulk-sample) phosphorus K-edge XANES spectrum for a soil sample included the three phosphate standards that were most frequently included in fits to 13 μ-XANES (~50-nm resolution) spectra collected across a $45 \times 50\,\mu m^2$ area of the sample [16]. However, the μ-XANES fitting analysis included five additional species that were not fit in the bulk sample, indicating an increased sensitivity to minor species, as was also reported by others [2, 6, 43, 62, 168–171].

Because biogeochemical systems are composed of multiple inorganic and organic solids, biomass, water, and gases; no single analytical technique provides a complete picture of the system. To obtain a full representation of the elemental composition, mineralogy, organic matter (including residue and biomass) composition, and element speciation in reactive biogeochemical microsites; ideally one would analyze the full composition of the relevant matrix and nonmatrix (e.g., trace) elements as well as speciation (mineralogy, structural properties, etc.) of each element in the context of the specific research objective. Such analyses would define, for example, the types of phyllosilicate, Al, Fe, and Mn (hydr)oxides present in a given microscale volume, the degree of humification of natural organic matter—including the relative composition of reactive carboxyl and phenolic acid functional groups, and the matrix-dependent speciation of one or more elements of interest which is less abundant. Conceptually, such an approach would involve analysis of the same suite of reactive microsites with multiple, complementary techniques, wherein each technique provides insights on a specific aspect of the system. For this discussion, we refer to reactive microsites/nanosites as the micro/nanoscale volume of a biogeochemical system probed by a focused spectroscopic analysis [3, 20].

As illustrated in Fig. 9.3, Manceau et al. [6] used three complementary synchrotron X-ray techniques—μ-XRF for element spatial distributions, μ-XRD for mineralogical distributions, and μ-EXAFS for molecular coordination—to determine with a high degree of specificity the molecular coordination of Ni with Mn oxide (lithiophorite) in manganese nodules. Microscale spatial resolution was essential to analyze the Mn-oxide phase that accumulated most of the Ni independently of an Fe-(hydr)oxide phase in the sample. In other words, if the Mn and Fe phases were spatially colocalized on the micron scale, it would be challenging to separate the sorbing (host) mineral phase(s) for Ni^{2+} ions using μ-XRF analysis.

Ideally, each complementary analytical technique probes the same volume of sample or micro/nanosites, but this is generally not the case. For synchrotron X-ray analysis, the sampled volume depends on the spot size of the X-ray beam, the energy-dependent depth of penetration of the X-ray beam, and especially the element-specific escape depth of the fluorescence signal (for fluorescence mode measurements).

Spot size of the X-ray beam: The spot size is defined by the lateral spatial resolution of the analytical technique—e.g., μ-XRF imaging, μ-XRD, or μ-CMT. For example, order-of-magnitude greater spatial resolution can be obtained by ptychography, a full-field computational imaging technique, than conventional μ-XRF imaging with the same beam size.

Depth of penetration and escape depth of fluorescence X-rays: Sample transmission or absorption calculations such as those represented in Fig. 9.9 are useful for designing spatially resolved experiments involving μ-/n-XRF imaging, μ/n-XCT

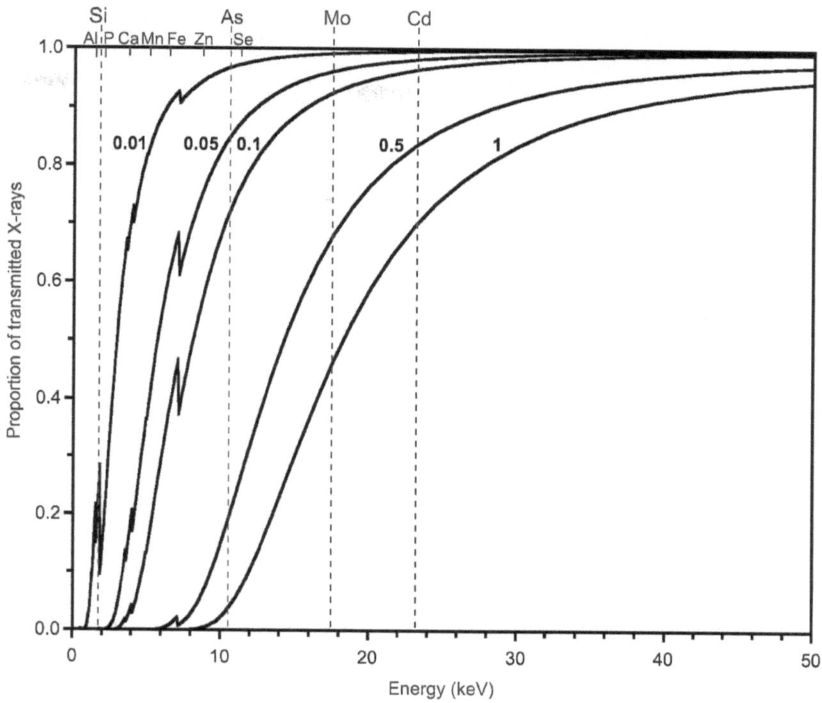

Fig. 9.9 Proportions of X-rays transmitted through a soil matrix of different thicknesses between 0.01 and 1 mm (numbered curves) as a function of energy. Vertical lines and labeled ticks denote Kα fluorescence energies of selected elements of relevance to biogeochemical systems. These transmission curves were generated from data in the NIST database of physics (https:/physics.nist.gov/PhysRefData/FFast/html/form.html). The following chemical formula for an idealized soil of bulk density 1.3 g cm^{-3} was based on the average mole fractions of elements in the continental crust reported in Schulze [19], scaled by 10^4 to yield whole number stoichiometric coefficients: $Si_{2050}Al_{626}Fe_{186}Mg_{179}Ca_{188}Na_{256}K_{138}Ti_{19}Mn_3P_7H_{291}O_{6057}$.

(computed tomography), or μ/n-XFCT (X-ray fluorescence tomography). This figure shows the proportion of X-rays of varying energy that would be transmitted through a representative soil material based on linear attenuation coefficients. Generally, the proportion of transmitted X-rays increases curvilinearly with increasing energy and decreases in proportion to the sample thickness for a given energy. Also, note discontinuities in these curves reflected as sharp decrease in transmission at absorption edges of more abundant elements in the soil sample. For example, the edges for abundant Al, Si, and Fe are evident at energies of 100–700 eV above the Kα emission lines denoted for these elements in Fig. 9.9.

In μ/n-XRF and XFCT imaging of biogeochemical systems, the spatially resolved fluorescence signals from different elements can be statistically correlated with each other to imply spatial colocalization [3, 12, 36, 37, 172, 173]. However, for samples that are thick relative to the escape depth of the fluorescence X-rays for the composing elements, the intensities of fluorescence signals will be skewed toward the elements of greater fluorescence energy, regardless of the penetration depth of the incident X-ray. As an illustration of this point, consider μ-XRF imaging of the model soil in Fig. 9.9 with an incident X-ray energy of 10 keV. The penetration depth of the X-rays will be approximately 1000 μm, at which point 98% of the X-rays are absorbed and 2% is further transmitted. However, the approximate escape depths of X-ray fluorescence signals (defined here as five absorption lengths) from any Zn, Fe, Mn, Ca, P, Si, and Al atoms irradiated by the incident X-ray beam would, respectively, be 850 (Zn), 480 (Fe), 380 (Mn), 120 (Ca), 26 (P), 34 (Si), and 28 (Al) microns. Note that these depths would change with elemental composition and bulk density of the soil or other biogeochemistry matrix. These biases can be reduced by preparing thinner samples for spatially resolved X-ray imaging [50].

Sample thickness: To help normalize the volume analyzed across elements, thin sections of samples can be prepared. However, thin sections of porous biogeochemical samples such as soils typically range in thickness from 20 to 30 μm [87, 174–179]. For STXM, which is transmission microscopy in the soft X-ray region, geochemical samples are usually prepared with a thickness of $\leq$0.2 μm [1, 4, 91, 180]. However, one can see from Fig. 9.9 that even with samples of 30 μm thick, μ-XRF images or μ-XAS spectra for low-Z elements like Al, Si, and P would not come from the full 30-μm thick sample.

The situation of spatially complementary analyses becomes even more challenging when other analytical techniques with different spatial resolutions are included. For example, magnetic resonance imaging (a derivative of NMR) would be useful for characterizing natural organic matter functional groups, but it has a low spatial resolution, typically in the sub-millimeter range [181]. Similarly, vibrational spectroscopy FTIR microscopy provides additional information on both inorganic and organic components, but it commonly has a spatial resolution (related to the diffraction limit) of 2–25 μm in the mid-infrared region [147]. Specialized synchrotron infrared nanoprobes that couple an infrared beam with an atomic force microscope tip provide nanometer-scale resolution and can be applied to study biological materials and polymers [182]. Destructive analyses such as TOF-SIMS also have micro- to nanoscale lateral spatial resolutions, but angstrom-level depth resolution, similar to UV-vis imaging.

In the context of integrating analytical techniques, data must often be combined from different synchrotron X-ray beamlines to cover a wide energy range that includes absorption edges of matrix elements (e.g., C, Al, Si, S, K, Ca, Mn, and Fe) and other elements of interest (e.g., P, trace-element contaminants) in biogeochemical systems. Analysis of high-Z (e.g., Fe and Mn), intermediate Z (e.g., P, S, Ca), and low-Z (e.g., C) elements requires hard, tender, and soft X-rays, respectively. The optical systems of different beamlines are usually optimized in these different energy ranges. Challenges in collecting and integrating spatial data from different beamlines include unifying sample preparation (thickness) or sample environment (in vacuo; ambient), dealing with different X-ray beam spot sizes and penetration depths, and obtaining spatial alignment (registry) of the regions of interest (ROIs) analyzed. Different beam spot sizes and penetration depths translate to different pixel and voxel sizes. Integration of such nonuniform data into a single data matrix, e.g., for spatial statistical analyses requires extra processing such as binning of smaller pixels or interpolation of larger pixels. But averaging would potentially lose specificity and detail, and interpolation can introduce uncertainties. Nevertheless, once congruent data matrices are generated, these matrices can then be combined into a single, larger-dimensional matrix, for statistical analyses and inference.

In essence, the above discussion conveys the challenges of integrating synchrotron X-ray techniques and other spatially resolved techniques to fully characterize reactive microsites within geochemical matrices. Regardless of these challenges, spatial statistical techniques are useful for gaining insights on element colocalization and potential reactivity. A simple, pairwise (Pearson) correlation analysis of μ-XRF images is a common statistical approach applied to infer (but does not definitively show) bonding mechanisms via colocalization of trace elements with geochemical-matrix elements [3, 36, 37, 173]. However, analyses limited to such simple correlations do not decouple the effects attributed to the association of colocalized matrix elements with themselves, for example, when Fe- and Al- bearing solids are physically associated into submicron-sized assemblages or when Al substitutes for Fe in Al-substituted Fe (hydr)oxides. Simple correlations also do not account for spatial autocorrelation [183]. Spatial autocorrelation in a chemical image refers to the correlation of an element to itself as a function of distance from a given point, where the correlation is high at closer spatial locations than when farther apart. Partial correlation analysis is an appropriate statistical approach that accounts for the effects of colocalized matrix elements and evaluates statistically significant geochemical-matrix elements that are spatially associated with trace elements [3]. Similarly, spatial statistics and Bayesian analysis have been used to account for the spatial interdependence of single and multiple elements in images of soil matrices [184–186]. Sharma et al. [3] recently demonstrated that spatial regression modeling, which accounts for the spatial component, can be used in variable selection for predicting As accumulation in mineral coatings on a soil sand grain. Manceau et al. [12] have described how a cross-correlation function could be used for quantitatively analyzing μ-XRF images to provide suggestive evidence for an association between elements. The authors have also emphasized that such a technique could be misleading if the geochemical matrix consists of two or more minerals with distinct behavior. Similarly, the application of correlation-distance hierarchical clustering and cosine-distance hierarchical clustering has been made in XANES data analyses for speciation modeling [83]. The former analysis groups XANES spectra by pattern similarity and is database independent, whereas the latter analysis is database dependent as it relates to colocated species groups derived from linear combination fitting. Nicholas et al. [83] have used these statistical approaches to assess the similarities in spectral features of As and Fe μ-XANES in samples analyzed from aquitard, aquifer, and aquitard-aquifer interface sediments that had low solid-phase As concentrations. They showed that As speciation was spatially correlated, but also varied with Fe speciation and relative proportions of As, Fe, and S. Furthermore, statistical techniques can also be used to improve the connectivity of data across different spatial scales. For example, Sharma [187] analyzed soil clay particles using scanning transmission electron microscopy (STEM) to assess the nanoscale spatial distribution of reacted arsenic in relation to spatially heterogeneous soil matrix elements (e.g., Fe, Al, Ca, Ti, Si, etc.) and determined how these multiple elements affected As accumulation at different spatial scales. She mathematically created micron-scale pixels from the nanometer-scale STEM imaging datasets. For this, first the data matrix was binned by summing the neighboring pixel values that are adjacent in the matrix to diminish noise. Then, the binned data matrix was divided into two regions based on the highest and lowest As signals, excluding background pixels of zero signal. Finally, the binned data matrix from each region, which consists of nanometer size pixels, were divided into 30 groups based on spatial location, and the group means were calculated. These new data matrices ($n=30$) created from group means had micrometer size pixels. Partial correlation analysis was used on these datasets to make an inference, which suggested that multielement effects on arsenic accumulation on complex soil particles are more substantial when measured with nanoscale spatial resolution than microscale resolution. This nonstationary effect across spatial scale emphasized the shortcomings of the common practice of implying molecular-scale chemical speciation from statistical analysis of microscale imaging datasets.

Limitations of synchrotron methods

Biogeochemical complexity and loss of specificity

A conventional approach to understanding the retention, mobilization, and chemical speciation of elements in geochemical systems is to determine their behavior on pure, model analogs that are synthesized in isolation from the soil. Whereas spectroscopic analyses of these systems provide a means for analyzing chemical bonding mechanisms and the structure of species on mineral surfaces with a high degree of specificity, the multicomponent complexity of soil mineral assemblages is not captured [188]. Spectroscopic analyses produce a signal that is related to average, local molecular bonding environment of atoms of any element and thus create challenges in the spectral interpretation of complex systems [20]. This is because complex, natural systems can potentially host a nutrient or contaminant in multiple chemical bonding environments and determining which types of functional groups are involved in nutrient or contaminant binding cannot be easily deconvolved from an average spectroscopic signal [188]. Also, the analytical limitations of synchrotron techniques to resolve structural details of chemical species limit us from achieving the molecular-level specificity of speciation in increasingly complex geochemical systems. As the degree of specificity of chemical speciation analysis decreases, the uncertainty in the analysis increases [188]. Therefore, integrating different analytical techniques to obtain multiple perspectives of a chemical component would be one approach to increase the specificity of chemical speciation in environmental matrices.

Practical field management

Complexities of the biogeochemical system create challenges in upscaling information from fine-scale analyses for practical land management. A central premise in soil chemistry is that chemical reactions at the molecular scale directly influence the field-scale behavior of chemical elements, however precise management of complex natural systems might become difficult when decisions are made based on knowledge derived from relatively simple, homogeneous systems [8]. On the other hand, although the measurement of field samples helps to regulate the negative impacts of trace elements, the spatial heterogeneity of soil at all spatial scales can limit our confidence in transferring knowledge from one site to the other [20]. Therefore, a better understanding of the biogeochemistry of any contaminants for precise land management requires approaches that address the multi-scale complexity and heterogeneity of natural systems and can potentially unify multidisciplinary information of varying complexity collected at different spatial and temporal scales.

Uncertainties in speciation analysis by synchrotron X-ray analyses

Linear combination fitting (LCF) is a common approach to analyzing synchrotron (μ-) XAS data to model chemical speciation. However, the accuracy of LCF results is contingent upon whether the chosen set of standards truly represents the heterogeneous environmental samples and the complexity of the matrix [78, 188, 189]. Moreover, some standards might lack unique spectral features, and consequently different combinations of standards result in fits of similar quality. This lack of sensitivity increases the uncertainty in LCF analysis. Thus, the choice of fitting standards must also consider other relevant properties of the sample, such as the mass balance on species stoichiometry based on total elemental composition, and mineralogical analysis of the biogeochemical matrix.

Given these sources of uncertainty, speciation models derived from LCF analyses of biogeochemical systems should be interpreted with caution. A fitting result should be considered a model for chemical speciation rather than a quantitative reflection of the absolute species in a sample. For example, Sharma et al. [3] found As(V) adsorbed on boehmite standard in their fits when modeling As speciation in a sand grain reacted with applied As(V). However, a fit with As(V) bound on boehmite is more likely a representation of Al-bonded As(V) rather than this adsorbed species, particularly since boehmite, which was used for preparing the adsorbed standard, is likely not a dominant mineral in the soil sample [190, 191]. One approach to dealing with the uncertainty in absolute species analysis is to design synchrotron experiments that focus on analyzing a trend in spectral and speciation changes across series of well-controlled treatments (e.g., Hesterberg et al. [192]). In summary, uncertainty in the speciation analyses increases with uncertainty in the appropriateness of fitting standards, the lack of distinguishing spectral features [188], and the complexity of the geochemical matrix, which could be diminished by the increased spatial resolution of the analytical techniques [6].

Beam-induced artifacts

As synchrotron light sources deliver greater photon flux on the samples, observation of beam damage in XAS experiments has become more common [193]. X-ray beam-induced reduction or oxidation of redox-sensitive elements during data collection is an example of beam damage [194–196]. Ross et al. [194] studied the effects of photon density, exposure time, and the sample on the reduction of Mn(IV,III) to Mn(II) in soils. The samples irradiated using two orders of magnitude more intense beam showed a greater drop in the energy of the main absorption edge for the same duration of exposure. In general, they found that the shift in energy was contingent on all three factors—the photon density, the exposure time, and the sample. Zachara et al. [196] found X-ray beam-induced reduction of Cr(VI) to Cr(III) during the XAFS study of Cr speciation. Bulk XAS analyses of Se were found to produce rapid photoreduction of Se(VI) to Se(IV) [195]. Various ways can be undertaken to diminish the effects of beam damage. One way to avoid beam induced reduction is to conduct experiments under cryogenic conditions [193, 197]. Beam damage effects can be assessed by collecting repeated scans on the same sample when using bulk XAS, and damage effects might be avoided in μ-XAS by analyzing adjacent voxels in repeated scans aimed at increasing spectral quality [198–200]. The use of fast detector systems also reduces the potential for beam damage by reducing the sample exposure to the X-ray beam [13, 197, 201, 202]. Similarly, Sharma et al. [3] proposed merging μ-XANES spectra from voxels of similar chemistry to improve the signal, which allows for reduced exposure times by not repeating scans in the same location.

Another challenge of using synchrotron microprobe especially on small particles is beam instability. The use of XAS techniques on heterogeneous environmental samples that are of small particle size could sometimes result in artifacts in the data because of the beam motion. For example, for synchrotrons that are not operated in top-up mode, a focused beam might move between synchrotron fills, or changes in the incident power on the optics may also cause the beam position to move [12]. Beam motion is particularly problematic when the beam spot size is nearly the same as the size of unique particles in the sample. In this case, positional instability will move the beam off of the particle. When a focused X-ray beam is larger than a unique particle in a sample, particularly one that has a dominant signal, then minor beam movement can be tolerated if the particle remains fully irradiated. On the contrary, if the beam is smaller than a characteristic particle imaged, then beam movement should be tolerable as long as the beam remains on the particle and the particle is uniform. Various ways to align images affected by beam motion are discussed below.

Besides the development of the state-of-the-art optics, other strategies can be followed to diminish effects of beam motion. One of them includes the acquisition of 2D XRF maps from XANES imaging (also referred to as XANES hyper-spectral mapping) [88, 203]. In this case, even if maps are shifted from one energy to another the beam motion can be corrected afterward, and the images aligned [204]. Because μ-XRF maps are not affected by the varying energy of the incident beam, hyperspectral mapping can be used to monitor the beam drift. For example, Cotte et al. [203] used the Ba map (which showed a contrast distribution) to align images of Cr that were collected at two different energies—5.993 and 6.086 keV, corresponding with unique features of Cr(VI) and Cr(III). Jacobson et al. [88] showed that images collected across different energies could be aligned using cross correlations and quadratic peak interpolation (Fig. 9.10).

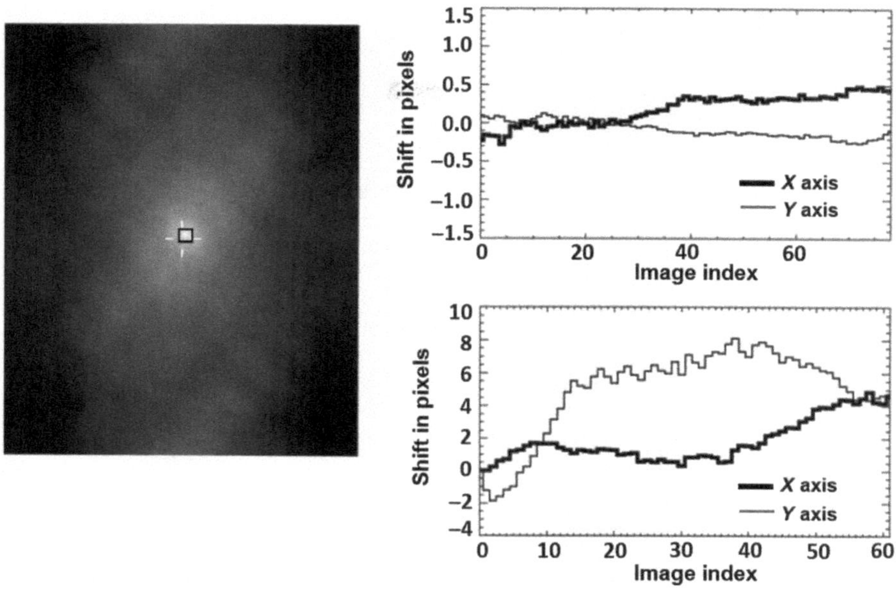

Fig. 9.10 Alignment of images using cross correlation. The logarithm of the cross-correlation image |c| is shown on the left. The center of the array is indicated with four short white lines, whereas the position of the peak of the cross-correlation image |c| is indicated with a small dark box (the box is placed over the nearest integer pixel position). Two examples of the shift in the center position of images in stacks aligned using cross correlations and quadratic interpolation are shown on the right. The stack at upper right consisted of 79 images, each with 250×30 pixels of 150 nm size, acquired over 2 h. The stack at lower right consisted of 62 images, each with 180×220 pixels of 47 nm size, acquired over 6 h. The gradual drift of the image position center in the example at lower right is probably due to some combination of misalignment of the focusing stage relative to the beam direction and thermal drifts in the specimen position. Abrupt changes between image center positions are most likely caused by imperfections in the cross roller bearings of the translation stage used for focusing the zone plate.
Reprinted from C. Jacobsen, S. Wirick, G. Flynn, C. Zimba, Soft X-ray spectroscopy from image sequences with sub-100-nm spatial resolution, J. Microsc. 197 (2000) 173, with permission from John Wiley & Sons. © 2000 The Royal Microscopical Society.

Another way to determine beam drift during data acquisition is to statistically analyze the data using the least sum-of-squares approach. We applied this approach on μ-XRF images of Fe collected from an As(V) reacted soil sand grain. A series of twelve $10 \times 10\,\mu m^2$ images of Fe were created at evenly spaced energies across the absorption edge of As (Fig. 9.11). Sum of squared residuals was calculated from the regression of one map on the other in each pair, i.e., between an element image collected at given energy and the corresponding element image collected at 11,759 eV [the lowest energy of the As μ-XANES (data not shown)]. The sum of squared residuals was calculated for a set of 49 shifts (including no shift) of up to ±3 pixels in the X and Y directions for the overlaid image. The sums of squared

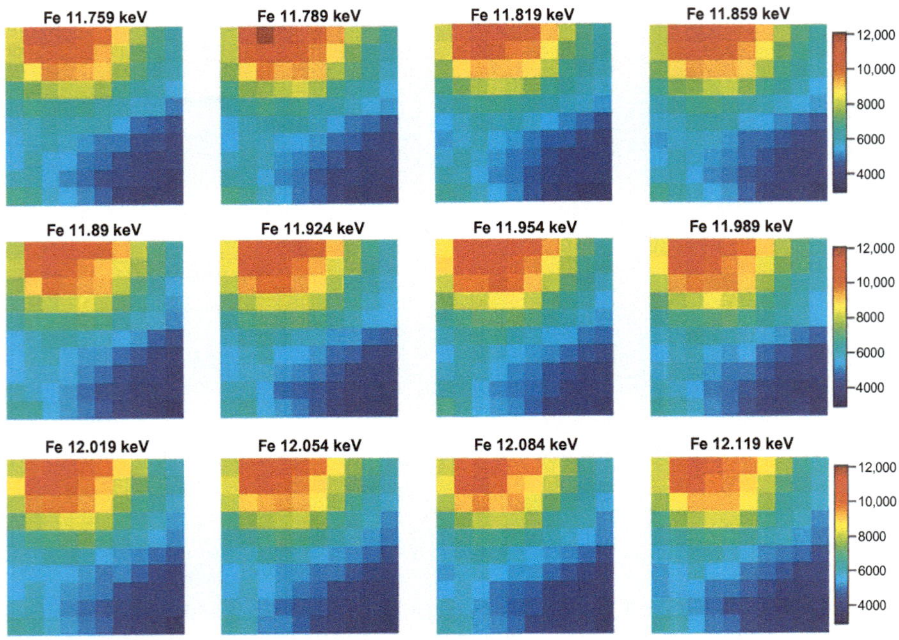

Fig. 9.11 A series of twelve $10 \times 10\,\mu m^2$ Fe images created at evenly spaced energies across the absorption edge of As (i.e., between 11.759 and 12.119 keV), plotted on the same color-intensity scale.

residuals were minimum when no shift was applied (Fig. 9.12), suggesting that the beam shift was always less than one pixel. In summary, these various statistical approaches for aligning spatial data at different energies (and possibly data from different beamlines) would ideally be integrated into software packages at synchrotron beamlines for real-time analysis of the reliability of datasets.

Future directions

With the progressive improvement in synchrotron machines and individual beamlines, the advantages offered by these techniques have continued to grow. Fourth-generation synchrotron sources, such as Max IV, produce a greater proportion of coherent X-rays [205], which should enhance the use of CDI techniques. With the use of improved detector systems such as the MAIA detector [206], which has a high data acquisition rate, and faster electronics, XRF maps that were collected using 0.5–1 s dwell time per pixel can now be collected in milliseconds. However, improvements are needed in handling large multidimensional datasets offered by such capabilities. Standardized approaches for collecting data across multiple beamlines to cover a range of elements important for biogeochemical systems are also needed. Additionally, sample alignment approaches for datasets from various beamlines are critical for data integration and statistical analyses. The issue of data analysis and interpretation of larger datasets

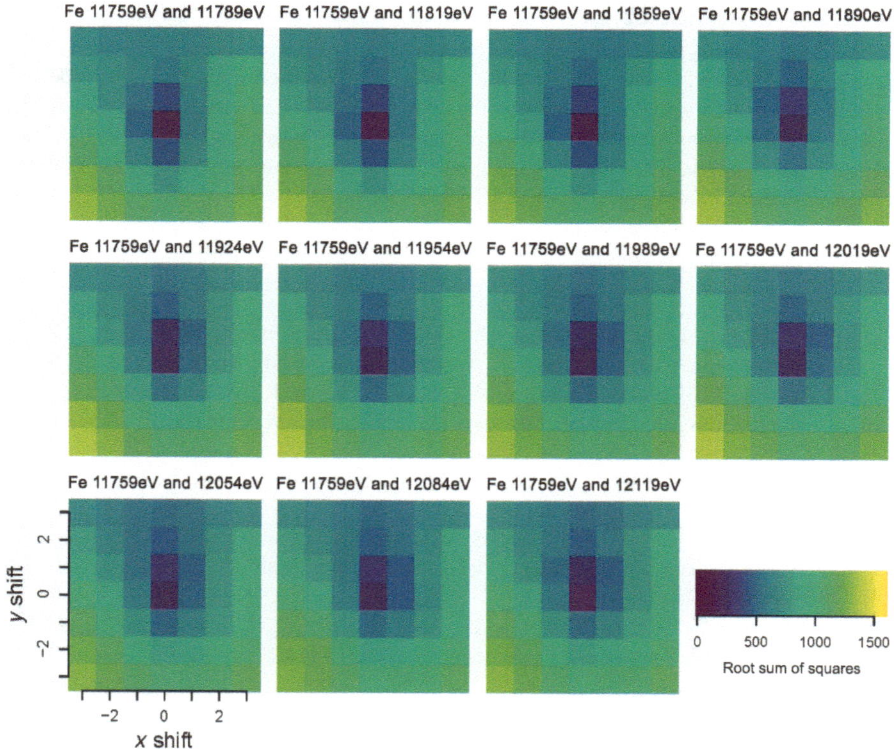

Fig. 9.12 A series of images showing the sum-of-squared residuals for all possible combinations of *X-Y* shifts of up to ±3 pixels (49 shift calculations per image pair) for Fe images shown in Fig. 9.11. Sum-of-squared residuals were minimum when no shift was applied (i.e., at *X* shift = 0 and *Y* shift = 0).

has been a significant bottleneck, especially due to the lack of established data analysis methods, user-friendly software, and the limited computational capacity [15]. A standardized method for data analyses and reporting for large datasets would help biogeochemists to realize the full potential of SR techniques. Development of efficient statistical methods, like the one developed by Muyskens [186], should allow full exploration of large, information-rich multidimensional datasets, which should further advance our current understanding of the biogeochemical processes. Finally, research is needed to develop better approaches for quantitatively upscaling our extensive knowledge of molecular-to-macro-scale mechanisms to more effectively manage and control landscape-to-global scale processes in natural ecosystems. We believe that transdisciplinary collaboration between biogeochemists, physicists, chemists, biologists, and statisticians who conduct research at different scales should help in upscaling chemical information. Developing more powerful ways to upscale information from fine-scale analyses should guide effective policy making and appropriate management practices for alleviating environmental impacts of biogeochemical processes.

Acknowledgments

We are grateful to Dr. Carlos Sato from the Brazilian Synchrotron Light Laboratory (LNLS, CNPEM) for generating plotting data in Fig. 9.9 from the NIST database, to Dr. Joseph Guinness for developing a least-sum-of-squares approach for image alignment, and to Dr. James LeBeau for assisting in STEM analyses. Support for this work was provided by the National Science Foundation through Grant No. EAR-1349374.

References

[1] P.S.P. Arachchige, G.M. Hettiarachchi, C.W. Rice, J.J. Dynes, L. Maurmann, J. Wang, C. Karunakaran, A.D. Kilcoyne, C.P. Attanayake, T.J. Amado, Sci. Rep. 8 (2018) 16810.
[2] T. Rennert, K.U. Totsche, K. Heister, M. Kersten, J. Thieme, J. Soils Sediments 12 (2012) 3.
[3] A. Sharma, A. Muyskens, J. Guinness, M. Polizzotto, M. Fuentes, R. Tappero, Y.-C. Chen-Wiegart, J. Thieme, G. Williams, A. Acerbo, J. Synchrotron Radiat. 26 (2019).
[4] D. Solomon, J. Lehmann, J. Wang, J. Kinyangi, K. Heymann, Y. Lu, S. Wirick, C. Jacobsen, Sci. Total Environ. 438 (2012) 372.
[5] D.L. Sparks, Advances in the Use of Synchrotron Radiation to Elucidate Environmental Interfacial Reaction Processes and Mechanisms in the Earth's Critical Zone, Springer, 2013, p. 93.
[6] A. Manceau, N. Tamura, M.A. Marcus, A.A. MacDowell, R.S. Celestre, R.E. Sublett, G. Sposito, H.A. Padmore, Am. Mineral. 87 (2002) 1494.
[7] T. Sowers, D. Adhikari, J. Wang, Y. Yang, D.L. Sparks, Environ. Sci. Technol. (2018).
[8] G.E. Brown Jr., N.C. Sturchio, Rev. Mineral. Geochem. 49 (2002) 1.
[9] A.L. Foster, C.S. Kim, Rev. Mineral. Geochem. 79 (2014) 257.
[10] G.M. Hettiarachchi, E. Donner, E. Doelsch, J. Environ. Qual. 46 (2017) 1139.
[11] A. Hitchcock, M. Obst, J. Wang, Y. Lu, T. Tyliszczak, Environ. Sci. Technol. 46 (2012) 2821.
[12] A. Manceau, M.A. Marcus, N. Tamura, Rev. Mineral. Geochem. 49 (2002) 341.
[13] P.M. Kopittke, P. Wang, E. Lombi, E. Donner, J. Environ. Qual. 46 (2017) 1175.
[14] L.R. Baker, G.M. Pierzynski, G.M. Hettiarachchi, K.G. Scheckel, M. Newville, J. Environ. Qual. 43 (2014) 488.
[15] E. Lombi, R. Sekine, E. Donner, Environ. Chem. 11 (2014) 1.
[16] D. Hesterberg, I. McNulty, J. Thieme, J. Environ. Qual. 46 (2017) 1190.
[17] R.K. Taggart, N.A. Rivera, C. Levard, J.-P. Ambrosi, D. Borschneck, J.C. Hower, H. Hsu-Kim, Environ Sci Process Impacts 20 (2018) 1390.
[18] B.M. Toner, S.L. Nicholas, J.K.C. Wasik, Environ. Chem. 11 (2014) 4.
[19] D.G. Schulze, Soil Mineralogy With Environmental Applications, 2002 p. 1.
[20] D. Hesterberg, M.C. Duff, J.B. Dixon, M.J. Vepraskas, J. Environ. Qual. 40 (2011) 667.
[21] Y.H. Lo, C.-T. Liao, J. Zhou, A. Rana, C.S. Bevis, G. Gui, B. Enders, K.M. Cannon, Y.-S. Yu, R. Celestre, Sci. Adv. 5 (2019) eaax3009.
[22] A.P. Hitchcock, J. Electron Spectrosc. Relat. Phenom. 200 (2015) 49.
[23] P.I. Guntoro, Y. Ghorbani, P.-H. Koch, J. Rosenkranz, Minerals 9 (2019) 183.
[24] E. Lombi, M.D. de Jonge, E. Donner, P.M. Kopittke, D.L. Howard, R. Kirkham, C. G. Ryan, D. Paterson, PLoS One 6 (2011) e20626.
[25] J. Deng, D.J. Vine, S. Chen, Q. Jin, Y.S. Nashed, T. Peterka, S. Vogt, C. Jacobsen, Sci. Rep. 7 (2017) 445.
[26] J. Miao, T. Ishikawa, I.K. Robinson, M.M. Murnane, Science 348 (2015) 530.

[27] M.A. Pfeifer, G.J. Williams, I.A. Vartanyants, R. Harder, I.K. Robinson, Nature 442 (2006) 63.
[28] F. Pfeiffer, Nat. Photonics 12 (2018) 9.
[29] H.R. Field, A.H. Whitaker, J.A. Henson, O.W. Duckworth, Sci. Total Environ. 697 (2019) 134111.
[30] T.D. Sowers, J.M. Harrington, M.L. Polizzotto, O.W. Duckworth, Geochim. Cosmochim. Acta 198 (2017) 194.
[31] E.C. Gillispie, A.R. Matteson, O.W. Duckworth, R.B. Neumann, N. Phen, M. L. Polizzotto, Geochim. Cosmochim. Acta 245 (2019) 441.
[32] S. Fendorf, M.J. Eick, P. Grossl, D.L. Sparks, Environ. Sci. Technol. 31 (1997) 315.
[33] G.A. Waychunas, J.A. Davis, C.C. Fuller, Geochim. Cosmochim. Acta 59 (1995) 3655.
[34] I. Persson, M. Trublet, W. Klysubun, J. Phys. Chem. A 122 (2018) 7413.
[35] G. Ona-Nguema, G. Morin, F. Juillot, G. Calas, G.E. Brown, Environ. Sci. Technol. 39 (2005) 9147.
[36] D. Strawn, H. Doner, M. Zavarin, S. McHugo, Geoderma 108 (2002) 237.
[37] R.R. Karna, G.M. Hettiarachchi, M. Newville, C. Sun, Q. Ma, J. Environ. Qual. 45 (2016) 1883.
[38] G. Landrot, R. Tappero, S.M. Webb, D.L. Sparks, Chemosphere 88 (2012) 1196.
[39] N. Rivera, N. Kaur, D. Hesterberg, C.R. Ward, R.E. Austin, O.W. Duckworth, Energy Fuel 29 (2015) 954.
[40] M. Gräfe, M. Landers, R. Tappero, P. Austin, B. Gan, A. Grabsch, C. Klauber, J. Environ. Qual. 40 (2011) 767.
[41] L.R. Baker, G.M. Pierzynski, G.M. Hettiarachchi, K.G. Scheckel, M. Newville, J. Environ. Qual. 41 (2012) 1865.
[42] A. Manceau, M.A. Marcus, N. Tamura, O. Proux, N. Geoffroy, B. Lanson, Geochim. Cosmochim. Acta 68 (2004) 2467.
[43] J. Prietzel, S. Müller, I. Kögel-Knabner, J. Thieme, C. Jaye, D. Fischer, Sci. Total Environ. 628 (2018) 906.
[44] M. Obst, G. Schmid, 3D Chemical Mapping: Application of Scanning Transmission (Soft) X-Ray Microscopy (STXM) in Combination With Angle-Scan Tomography in Bio-, Geo-, and Environmental Sciences, Springer, 2014, p. 757.
[45] M.J. Bedzyk, L. Cheng, Rev. Mineral. Geochem. 49 (2002) 221.
[46] J. Zegenhagen, A. Kazimirov, X-Ray Standing Waves in a Nutshell, World Scientific, 2013, p. 3.
[47] T.P. Trainor, A.S. Templeton, G.E. Brown, G.A. Parks, Langmuir 18 (2002) 5782.
[48] M.D. de Jonge, S. Vogt, Curr. Opin. Struct. Biol. 20 (2010) 606.
[49] S.H. Anderson, J.W. Hopmans, Soil-Water-Root Processes: Advances in Tomography and Imaging, Soil Science Society of America, Madison, WI, 2013.
[50] S. Choudhury, D. Agyeman-Budu, A. Woll, T. Swanston, T. Varney, D. Cooper, E. Hallin, G. George, I. Pickering, I. Coulthard, J. Anal. At. Spectrom. 32 (2017) 527.
[51] B.M. Patterson, J. Campbell, G.J. Havrilla, X-Ray Spectrom. Int. J. 39 (2010) 184.
[52] U.E.A. Fittschen, G. Falkenberg, Anal. Bioanal. Chem. 400 (2011) 1743.
[53] T.U. Schülli, S.J. Leake, Curr. Opinion Solid State Mater. Sci. 22 (2018) 188.
[54] K. Carrado, L. Xu, D. Gregory, K. Song, S. Seifert, R. Botto, Chem. Mater. 12 (2000) 3052.
[55] Y. Hu, Q. Li, B. Lee, Y.-S. Jun, Environ. Sci. Technol. 48 (2014) 299.
[56] J.M. Cordoba, D.F. Coral-Coral, DYNA 86 (2019) 142.
[57] R. Garcia-Diez, A. Sikora, C. Gollwitzer, C. Minelli, M. Krumrey, Eur. Polym. J. 81 (2016) 641.

[58] J. Balko, R.H. Lohwasser, M. Sommer, M. Thelakkat, T. Thurn-Albrecht, Macromolecules 46 (2013) 9642.
[59] K. Faber, F. Badaczewski, M. Oschatz, G. Mondin, W. Nickel, S. Kaskel, B.M. Smarsly, J. Phys. Chem. C 118 (2014) 15705.
[60] L.J. Michot, I. Bihannic, F. Thomas, B.S. Lartiges, Y. Waldvogel, C.L. Caillet, J. Thieme, S.R.S. Funari, P. Levitz, Langmuir 29 (2013) 3500.
[61] S.J.S. Qazi, A.R. Rennie, J.K. Cockcroft, M. Vickers, J. Colloid Interface Sci. 338 (2009) 105.
[62] J. Kruse, M. Abraham, W. Amelung, C. Baum, R. Bol, O. Kühn, H. Lewandowski, J. Niederberger, Y. Oelmann, C. Rüger, J. Plant Nutr. Soil Sci. 178 (2015) 43.
[63] J. Baltrusaitis, D.M. Cwiertny, V.H. Grassian, Phys. Chem. Chem. Phys. 9 (2007) 5542.
[64] M. Mallet, K. Barthélémy, C. Ruby, A. Renard, S. Naille, J. Colloid Interface Sci. 407 (2013) 95.
[65] S. Yamamoto, T. Kendelewicz, J.T. Newberg, G. Ketteler, D.E. Starr, E.R. Mysak, K. J. Andersson, H. Ogasawara, H. Bluhm, M. Salmeron, J. Phys. Chem. C 114 (2010) 2256.
[66] T. Kendelewicz, S. Kaya, J. Newberg, H. Bluhm, N. Mulakaluri, W. Moritz, M. Scheffler, A. Nilsson, R. Pentcheva, G. Brown Jr., J. Phys. Chem. C 117 (2013) 2719.
[67] S.J. Parikh, K.W. Goyne, A.J. Margenot, F.N. Mukome, F.J. Calderón, Soil Chemical Insights Provided Through Vibrational Spectroscopy, Elsevier, 2014, p. 1.
[68] H.-Y.N. Holman, Synchrotron Infrared Spectromicroscopy for Studying Chemistry of Microbial Activity in Geologic Materials, Elsevier, 2010, p. 103.
[69] F.N. Mukome, A.L. Kilcoyne, S.J. Parikh, Soil Sci. Soc. Am. J. 78 (2014) 1632.
[70] M.C. Martin, U. Schade, P. Lerch, P. Dumas, TrAC Trends Anal. Chem. 29 (2010) 453.
[71] S. Beauchemin, D. Hesterberg, J. Chou, M. Beauchemin, R.R. Simard, D.E. Sayers, J. Environ. Qual. 32 (2003) 1809.
[72] E. Castorina, E.D. Ingall, P.L. Morton, D.A. Tavakoli, B. Lai, J. Synchrotron Radiat. 26 (2019).
[73] D. Fandeur, F. Juillot, G. Morin, L. Olivi, A. Cognigni, S.M. Webb, J.-P. Ambrosi, E. Fritsch, F.O. Guyot, J. Brown, E. Gordon, Environ. Sci. Technol. 43 (2009) 7384.
[74] S. Fendorf, P.S. Nico, B.D. Kocar, Y. Masue, K.J. Tufano, Arsenic Chemistry in Soils and Sediments, Elsevier, 2010, p. 357.
[75] M. Gräfe, E. Donner, R.N. Collins, E. Lombi, Anal. Chim. Acta 822 (2014) 1.
[76] J. Thieme, I. McNult, S. Vogt, D. Paterson, X-Ray Spectromicroscopy—A Tool for Environmental Sciences, ACS Publications, 2007.
[77] D. Koningsberger, R. Prins, X-ray absorption: principles, applications, techniques of EXAFS, SEXAFS and XANES, in: Chemical Analysis, vol. 92, Wiley-Interscience, Chichester 1988, 673 pp
[78] S. Kelly, D. Hesterberg, B. Ravel, Methods Soil Anal. (Part 5) (2008) 387.
[79] N. Khare, D. Hesterberg, J.D. Martin, Environ. Sci. Technol. 39 (2005) 2152.
[80] I. Arcon, B. Mirtic, A. Kodre, J. Am. Ceram. Soc. 81 (1998) 222.
[81] Y. Chu, Iraqi J. Appl. Phys. 11 (2015) 13.
[82] G. Martinez-Criado, J. Villanova, R. Tucoulou, D. Salomon, J.-P. Suuronen, S. Labouré, C. Guilloud, V. Valls, R. Barrett, E. Gagliardini, J. Synchrotron Radiat. 23 (2016) 344.
[83] S.L. Nicholas, M.L. Erickson, L.G. Woodruff, A.R. Knaeble, M.A. Marcus, J.K. Lynch, B.M. Toner, Geochim. Cosmochim. Acta 211 (2017) 228.
[84] D. Schulze, Minerals in Soil Environments, Soil Science Society of America, Madison, WI, (1989).
[85] E. Lombi, J. Susini, Plant Soil 320 (2009) 1.
[86] N. Tamura, R. Celestre, A. MacDowell, H. Padmore, R. Spolenak, B. Valek, N. Meier Chang, A. Manceau, J. Patel, Rev. Sci. Instrum. 73 (2002) 1369.

[87] P. Langner, C. Mikutta, E. Suess, M.A. Marcus, R. Kretzschmar, Environ. Sci. Technol. 47 (2013) 9706.
[88] C. Jacobsen, S. Wirick, G. Flynn, C. Zimba, J. Microsc. 197 (2000) 173.
[89] K. Diller, R.J. Maurer, M. Müller, K. Reuter, J. Chem. Phys. 146 (2017) 214701.
[90] J. Stöhr, NEXAFS Spectroscopy, Springer Science & Business Media, 2013.
[91] J. Kinyangi, D. Solomon, B. Liang, M. Lerotic, S. Wirick, J. Lehmann, Soil Sci. Soc. Am. J. 70 (2006) 1708.
[92] C. Chen, J.J. Dynes, J. Wang, C. Karunakaran, D.L. Sparks, Environ. Sci. Technol. 48 (2014) 6678.
[93] C. Chen, D.L. Sparks, Environ. Chem. 12 (2015) 64.
[94] J. Lehmann, J. Kinyangi, D. Solomon, Biogeochemistry 85 (2007) 45.
[95] D. Solomon, J. Lehmann, J. Harden, J. Wang, J. Kinyangi, K. Heymann, C. Karunakaran, Y. Lu, S. Wirick, C. Jacobsen, Chem. Geol. 329 (2012) 53.
[96] J. Wan, T. Tyliszczak, T.K. Tokunaga, Geochim. Cosmochim. Acta 71 (2007) 5439.
[97] J.J. Dynes, T. Tyliszczak, T. Araki, J.R. Lawrence, G.D. Swerhone, G.G. Leppard, A. P. Hitchcock, Environ. Sci. Technol. 40 (2006) 1556.
[98] A.P. Hitchcock, H.D. Stöver, L.M. Croll, R.F. Childs, Aust. J. Chem. 58 (2005) 423.
[99] M.J. Bedzyk, Encyclopedia of Condensed Matter Physics, vol. 6, 2005, p. 330.
[100] P. Fenter, L. Cheng, S. Rihs, M. Machesky, M.J. Bedzyk, N. Sturchio, J. Colloid Interface Sci. 225 (2000) 154.
[101] L. Cheng, P. Fenter, N.C. Sturchio, Z. Zhong, M.J. Bedzyk, Geochim. Cosmochim. Acta 63 (1999) 3153.
[102] L. Cheng, P.F. Lyman, N.C. Sturchio, M.J. Bedzyk, Surf. Sci. 382 (1997) L690.
[103] N.C. Sturchio, R.P. Chiarello, L. Cheng, P.F. Lyman, M.J. Bedzyk, Y. Qian, H. You, D. Yee, P. Geissbuhler, L.B. Sorensen, Geochim. Cosmochim. Acta 61 (1997) 251.
[104] S.D. Keyes, K.R. Daly, N.J. Gostling, D.L. Jones, P. Talboys, B.R. Pinzer, R. Boardman, I. Sinclair, A. Marchant, T. Roose, New Phytol. 198 (2013) 1023.
[105] M.J. Pushie, I.J. Pickering, M. Korbas, M.J. Hackett, G.N. George, Chem. Rev. 114 (2014) 8499.
[106] R. Terzano, Z. Al Chami, B. Vekemans, K. Janssens, T. Miano, P. Ruggiero, J. Agric. Food Chem. 56 (2008) 3222.
[107] T.R. Ferreira, L.F. Pires, D. Wildenschild, R.J. Heck, A.C. Antonino, Geoderma 324 (2018) 119.
[108] S.A. Kim, T. Punshon, A. Lanzirotti, L. Li, J.M. Alonso, J.R. Ecker, J. Kaplan, M. L. Guerinot, Science 314 (2006) 1295.
[109] D.H. McNear, E. Peltier, J. Everhart, R.L. Chaney, S. Sutton, M. Newville, M. Rivers, D. L. Sparks, Environ. Sci. Technol. 39 (2005) 2210.
[110] K.G. Scheckel, R. Hamon, L. Jassogne, M. Rivers, E. Lombi, Plant Soil 290 (2007) 51.
[111] J.E. Aravena, M. Berli, M. Menon, T.A. Ghezzehei, A.K. Mandava, E.E. Regentova, N. S. Pillai, J. Steude, M.H. Young, P.S. Nico, Soil–Water–Root Processes: Advances in Tomography and Imaging, The Soil Science Society of America, Inc, 2013, p. 39.
[112] S.D. Keyes, K.C. Zygalakis, T. Roose, Bull. Math. Biol. 79 (2017) 2785.
[113] S.J. Mooney, T.P. Pridmore, J. Helliwell, M.J. Bennett, Plant Soil 352 (2012) 1.
[114] G.C. Iltis, R.T. Armstrong, D.P. Jansik, B.D. Wood, D. Wildenschild, Water Resour. Res. 47 (2011).
[115] S. Rolland du Roscoat, J.M. Martins, P. Séchet, E. Vince, P. Latil, C. Geindreau, Biotechnol. Bioeng. 111 (2014) 1265.
[116] C.M. Klibert, Method for Synchrotron X-Ray Computed Tomographic Imaging of Biofilms in Porous Media, Louisiana State University, 2014, p. 83.

[117] C. Alwmark, B. Schmitz, S. Holm, F. Marone, M. Stampanoni, Meteorit. Planet. Sci. 46 (2011) 1071.
[118] N.L. Cordes, S. Seshadri, G.J. Havrilla, X. Yuan, M. Feser, B.M. Patterson, Spectrochim. Acta B At. Spectrosc. 103 (2015) 144.
[119] L. Vincze, B. Vekemans, F.E. Brenker, G. Falkenberg, K. Rickers, A. Somogyi, M. Kersten, F. Adams, Anal. Chem. 76 (2004) 6786.
[120] E. Bulska, I.A. Wysocka, M.H. Wierzbicka, K. Proost, K. Janssens, G. Falkenberg, Anal. Chem. 78 (2006) 7616.
[121] B. Kanngießer, W. Malzer, M. Pagels, L. Lühl, G. Weseloh, Anal. Bioanal. Chem. 389 (2007) 1171.
[122] R. Terzano, M. Alfeld, K. Janssens, B. Vekemans, T. Schoonjans, L. Vincze, N. Tomasi, R. Pinton, S. Cesco, Anal. Bioanal. Chem. 405 (2013) 3341.
[123] B. Vekemans, L. Vincze, F. Brenker, F. Adams, J. Anal. At. Spectrom. 19 (2004) 1302.
[124] L. Vincze, Microsc. Microanal. 11 (2005) 682.
[125] J. Badger, J. Lal, R. Harder, H. Inouye, S.C. Gleber, S. Vogt, I. Robinson, L. Makowski, Sci. Rep. 3 (2013) 2843.
[126] H. Schnablegger, Y. Singh, The SAXS Guide: Getting Acquainted with the Principles, Anton Paar GmbH, Austria, 2011.
[127] S. Shaw, C. Henderson, B. Komanschek, Chem. Geol. 167 (2000) 141.
[128] A. Masion, J. Rose, J.-Y. Bottero, D. Tchoubar, P. Elmerich, Langmuir 13 (1997) 3882.
[129] A. Masion, E. Doelsch, J. Rose, S. Moustier, J.Y. Bottero, P.M. Bertsch, Langmuir 17 (2001) 4753.
[130] J. Wagner, W. Härtl, R. Hempelmann, Langmuir 16 (2000) 4080.
[131] K. Nakamura, T. Kawabata, Y. Mori, Powder Technol. 131 (2003) 120.
[132] V. Goertz, N. Dingenouts, H. Nirschl, Part. Part. Syst. Charact. 26 (2009) 17.
[133] G. Beaucage, H.K. Kammler, S.E. Pratsinis, J. Appl. Crystallogr. 37 (2004) 523.
[134] K. Frost, D. Kaminski, G. Kirwan, E. Lascaris, R. Shanks, Carbohydr. Polym. 78 (2009) 543.
[135] G. Waychunas, C. Fuller, B. Rea, J. Davis, Geochim. Cosmochim. Acta 60 (1996) 1765.
[136] H. Zhang, B. Chen, Y. Ren, G.A. Waychunas, J.F. Banfield, Phys. Rev. B 81 (2010) 125444.
[137] W. Fan, M. Ogura, G. Sankar, T. Okubo, Chem. Mater. 19 (2007) 1906.
[138] D. Pontoni, J. Bolze, N. Dingenouts, T. Narayanan, M. Ballauff, J. Phys. Chem. B 107 (2003) 5123.
[139] A. Zach-Maor, R. Semiat, H. Shemer, J. Colloid Interface Sci. 363 (2011) 608.
[140] T. Kendelewicz, P. Liu, C. Doyle, G. Brown Jr., E.J. Nelson, S.A. Chambers, Surf. Sci. 453 (2000) 32.
[141] S.D. Senanayake, G. Waterhouse, A. Chan, T. Madey, D.R. Mullins, H. Idriss, Catal. Today 120 (2007) 151.
[142] S. Yamamoto, K. Andersson, H. Bluhm, G. Ketteler, D.E. Starr, T. Schiros, H. Ogasawara, L.G. Pettersson, M. Salmeron, A. Nilsson, J. Phys. Chem. C 111 (2007) 7848.
[143] H. Nesbitt, M. Scaini, H. Hochst, G. Bancroft, A. Schaufuss, R. Szargan, Am. Mineral. 85 (2000) 850.
[144] A. Pratt, Surf. Interface Anal. 36 (2004) 654.
[145] E. Todd, D. Sherman, J. Purton, Geochim. Cosmochim. Acta 67 (2003) 881.
[146] B.C. Smith, Fundamentals of Fourier Transform Infrared Spectroscopy, CRC Press, 2011.
[147] C.J. Hirschmugl, Rev. Mineral. Geochem. 49 (2002) 317.

[148] P. Heraud, B.R. Wood, M.J. Tobin, J. Beardall, D. McNaughton, FEMS Microbiol. Lett. 249 (2005) 219.
[149] H.-Y.N. Holman, K.A. Bjornstad, M.P. McNamara, M.C. Martin, W.R. McKinney, E. A. Blakely, J. Biomed. Opt. 7 (2001).
[150] L.M. Miller, P. Dumas, Curr. Opin. Struct. Biol. 20 (2010) 649.
[151] L. Quaroni, T. Zlateva, Analyst 136 (2011) 3219.
[152] J. Lehmann, D. Solomon, Organic Carbon Chemistry in Soils Observed by Synchrotron-Based Spectroscopy, Elsevier, 2010, p. 289.
[153] D. Solomon, J. Lehmann, J. Kinyangi, B. Liang, T. Schäfer, Soil Sci. Soc. Am. J. 69 (2005) 107.
[154] M.C. Hernandez-Soriano, R.C. Dalal, F.J. Warren, P. Wang, K. Green, M.J. Tobin, N. W. Menzies, P.M. Kopittke, Environ. Sci. Technol. 52 (2018) 12275.
[155] F.-S. Sun, G.-H. Yu, M.L. Polizzotto, W. Ran, Q.-R. Shen, Geoderma 337 (2019) 238.
[156] C. Vogel, C. Adam, R. Sekine, T. Schiller, E. Lipiec, D. McNaughton, Appl. Spectrosc. 67 (2013) 1165.
[157] C. Vogel, C. Rivard, I. Tanabe, C. Adam, Commun. Soil Sci. Plant Anal. 47 (2016) 2088.
[158] U. Ghosh, J.S. Gillette, R.G. Luthy, R.N. Zare, Environ. Sci. Technol. 34 (2000) 1729.
[159] L. Luo, J. Lv, C. Xu, S. Zhang, Anal. Chem. 86 (2014) 11002.
[160] Z. Song, E. Chouparova, K. Jones, H. Feng, N. Marinkovic, NSLS Activity Rep. 2 (2001) 112.
[161] Z. Song, K.W. Jones, N. Marinkovic, X.M. Xiao, H. Feng, E. Tchouparova, CLEAN–Soil Air Water 39 (2011) 1041.
[162] K. Dokken, L. Davis, J. Environ. Qual. 40 (2011) 719.
[163] K. Dokken, L. Davis, L. Erickson, S. Castro-Diaz, N. Marinkovic, Microchem. J. 81 (2005) 86.
[164] K. Dokken, L. Davis, N. Marinkovic, Synchrotron radiation Fourier infrared micro-spectroscopy (SR-IMS) as a tool to study the fate and transport of organic contaminants in plants, Spectroscopy 20 (2005).
[165] H. Holman, Tracking Microbial Interactions With Oxyanions by Synchrotron Infrared Spectromicroscopy, American Chemical Society, Washington, DC, 2004, p. U619.
[166] H.-Y.N. Holman, D.L. Perry, M.C. Martin, G.M. Lamble, W.R. McKinney, J.C. Hunter-Cevera, Geomicrobiol J. 16 (1999) 307.
[167] H.Y. Holman, M.C. Martin, Adv. Agron. 90 (2006) 79.
[168] K.L. Haus, R.L. Hooper, L.A. Strumness, J.B. Mahoney, Appl. Geochem. 23 (2008) 692.
[169] V.S. Ciminelli, D.C. Antônio, C.L. Caldeira, E.T. Freitas, I.D. Delbem, M.M. Fernandes, M. Gasparon, J.C. Ng, J. Hazard. Mater. 353 (2018) 261.
[170] E. Di Iorio, H.G. Cho, Y. Liu, Z. Cheng, R. Angelico, C. Colombo, Geochim. Cosmochim. Acta 237 (2018) 155.
[171] F. Ono, R. Tappero, D. Sparks, L. Guilherme, Environ. Sci. Pollut. Res. 23 (2016) 638.
[172] E.C. Gillispie, E. Andujar, M.L. Polizzotto, Environ Sci Process Impacts 18 (2016) 1090.
[173] D.R. Schwer III, D.H. McNear, J. Environ. Qual. 40 (2011) 1172.
[174] W.P. Adderley, I.L. Alberts, I.A. Simpson, T.J. Wess, J. Archaeol. Sci. 31 (2004) 1215.
[175] J. Frommer, A. Voegelin, J. Dittmar, M.A. Marcus, R. Kretzschmar, Eur. J. Soil Sci. 62 (2011) 305.
[176] L. Jassogne, G. Hettiarachchi, A. McNeill, D. Chittleborough, Soil Res. 50 (2012) 424.
[177] A. Manceau, C. Tommaseo, S. Rihs, N. Geoffroy, D. Chateigner, M. Schlegel, D. Tisserand, M.A. Marcus, N. Tamura, Z.-S. Chen, Geochim. Cosmochim. Acta 69 (2005) 4007.

[178] G. Sarret, J. Balesdent, L. Bouziri, J.-M. Garnier, M.A. Marcus, N. Geoffroy, F. Panfili, A. Manceau, Environ. Sci. Technol. 38 (2004) 2792.
[179] F. Van Oort, A. Jongmans, L. Citeau, I. Lamy, P. Chevallier, Eur. J. Soil Sci. 57 (2006) 154.
[180] J. Lehmann, B. Liang, D. Solomon, M. Lerotic, F. Luizão, J. Kinyangi, T. Schäfer, S. Wirick, C. Jacobsen, Glob. Biogeochem. Cycles 19 (2005).
[181] T.R. Neu, B. Manz, F. Volke, J.J. Dynes, A.P. Hitchcock, J.R. Lawrence, FEMS Microbiol. Ecol. 72 (2010) 1.
[182] A. Dazzi, C.B. Prater, Q. Hu, D.B. Chase, J.F. Rabolt, C. Marcott, Appl. Spectrosc. 66 (2012) 1365.
[183] A. Sharma, A. Muyskens, J. Guinness, M.L. Polizzotto, M. Fuentes, R.V. Tappero, Y.-C. K. Chen-Wiegart, J. Thieme, G.J. Williams, A.S. Acerbo, D. Hesterberg, J. Synchrotron Radiat. 26 (2019).
[184] J. Guinness, M. Fuentes, D. Hesterberg, M. Polizzotto, Spat. Stat. 9 (2014) 93.
[185] M.A. Terres, M. Fuentes, D. Hesterberg, M. Polizzotto, Bayesian Anal. 13 (2018) 1.
[186] A.L. Muyskens, Computationally Efficient Estimation of Non-stationary Gaussian Process Models for Large Spatial Data, NC State University, 2019, p. 100.
[187] A. Sharma, Arsenic Reactivity in Soil Microsites in Relation to Matrix Composition, NC State University, 2019, p. 237.
[188] F. Kizewski, Y.-T. Liu, A. Morris, D. Hesterberg, J. Environ. Qual. 40 (2011) 751.
[189] D. Hesterberg, Macroscale Chemical Properties and X-Ray Absorption Spectroscopy of Soil Phosphorus, Elsevier, 2010, p. 313.
[190] D.L. Sparks, Environmental Soil Chemistry, Elsevier, 2003.
[191] P. Hsu, in: J. Dixon, S. Weed (Eds.), Aluminum Hydroxides and Oxyhydroxides, 2018, p. 331.
[192] D. Hesterberg, J.W. Chou, K.J. Hutchison, D.E. Sayers, Environ. Sci. Technol. 35 (2001) 2741.
[193] G.N. George, I.J. Pickering, M.J. Pushie, K. Nienaber, M.J. Hackett, I. Ascone, B. Hedman, K.O. Hodgson, J.B. Aitken, A. Levina, J. Synchrotron Radiat. 19 (2012) 875.
[194] D.S. Ross, H.C. Hales, G.C. Shea-McCarthy, A. Lanzirotti, Soil Sci. Soc. Am. J. 65 (2001) 736.
[195] P. Wang, N.W. Menzies, E. Lombi, B.A. McKenna, M.D. de Jonge, D.J. Paterson, D.L. Howard, C.J. Glover, S. James, P. Kappen, Plant Physiol. 163 (2013) 407.
[196] J.M. Zachara, C.C. Ainsworth, G.E. Brown Jr., J.G. Catalano, J.P. McKinley, O. Qafoku, S. C. Smith, J.E. Szecsody, S.J. Traina, J.A. Warner, Geochim. Cosmochim. Acta 68 (2004) 13.
[197] A. van der Ent, W.J. Przybyłowicz, M.D. de Jonge, H.H. Harris, C.G. Ryan, G. Tylko, D.J. Paterson, A.D. Barnabas, P.M. Kopittke, J. Mesjasz-Przybyłowicz, New Phytol. 218 (2018) 432.
[198] D.S. Ross, H.C. Hales, A. Lanzirotti, Soil Sci. Soc. Am. J. 77 (2013) 1996.
[199] J. Liu, J. Yang, B.J. Cade-Menun, Y. Hu, J. Li, C. Peng, Y. Ma, Sci. Rep. 7 (2017) 15354.
[200] J. Liu, B.J. Cade-Menun, J. Yang, Y. Hu, C.W. Liu, J. Tremblay, K. LaForge, M. Schellenberg, C. Hamel, L.D. Bainard, Front. Microbiol. 9 (2018) 1643.
[201] F.-J. Zhao, K.L. Moore, E. Lombi, Y.-G. Zhu, Trends Plant Sci. 19 (2014) 183.
[202] E. Lombi, M. De Jonge, E. Donner, C. Ryan, D. Paterson, Anal. Bioanal. Chem. 400 (2011) 1637.
[203] M. Cotte, E. Pouyet, M. Salomé, C. Rivard, W. De Nolf, H. Castillo-Michel, T. Fabris, L. Monico, K. Janssens, T. Wang, J. Anal. At. Spectrom. 32 (2017) 477.
[204] M. Cotte, A. Genty-Vincent, K. Janssens, J. Susini, CR Phys. 19 (2018) 575.
[205] N. Martensson, M. Eriksson, Nucl. Instrum. Methods Phys. Res. 907 (2018) 97.
[206] C. Ryan, R. Kirkham, R. Hough, G. Moorhead, D. Siddons, M. De Jonge, D. Paterson, G. De Geronimo, D. Howard, J. Cleverley, Nucl. Instrum. Methods Phys. Res. 619 (2010) 37.

Aakriti Sharma is a postdoctoral research scholar in the Department of Crop and Soil Sciences at NC State University. Dr. Sharma received a BS degree in Agriculture (2011) from the Institute of Agriculture and Animal Science, Tribhuvan University, Nepal, an MS degree in Plant and Soil Science (2015) from Texas Tech University, Lubbock, Texas, and a PhD in Soil Science (2019) from NC State University, Raleigh, NC, USA. Her research interest centers around understanding the underlying reaction mechanism of contaminants in complex geochemical systems at multiple spatial scales. Part of her PhD research was conducted at the National Synchrotron Light Source II (NSLS-II), Brookhaven National Laboratory in Upton, NY.

Dean Hesterberg is a Distinguished Professor of Soil Chemistry in the Department of Soil Science at NC State University. Dr. Hesterberg received a BS degree in Plant and Soil Science from Southern Illinois University—Carbondale, an MS degree in Agronomy from Purdue University, and a PhD in Soil and Environmental Sciences from the University of California—Riverside. Before joining NC State University in 1993, he worked for Chevron Oil Field Research Company between 1988 and 1990 and the Institute for Soil Fertility Research in the Netherlands between 1990 and 1992. He worked at the National Laboratory for Synchrotron Light (LNLS) at the National Center for Research on Energy and Materials (CNPEM) in Campinas, Brazil between 2019 and 2020. His current research applies synchrotron X-ray absorption spectroscopy and spectromicroscopy to the molecular chemistry of phosphorus and heavy metals in soils and other complex environmental systems.

Author Index

Note: Page numbers followed by *f* indicate figures and *t* indicate tables.

C

E

F

G

I

J

K

M

P

Q

R

T

U

V

Subject Index

Note: Page numbers followed by *f* indicate figures and *t* indicate tables.

E

F

T

U

V

W

X

Z

MIX
Papier | Fördert gute Waldnutzung
FSC® C083411

Kolibri 360 GmbH
Ferdinand-Jühlke-Straße 7
99095 Erfurt, Deutschland
produktsicherheit@kolibri360.de